Brain Inspired Cognitive Systems 2008

For other volumes:
http://www.springer.com/series/5584

Amir Hussain • Igor Aleksander • Leslie S. Smith
Allan Kardec Barros • Ron Chrisley
Vassilis Cutsuridis
Editors

Brain Inspired Cognitive Systems 2008

Springer

Editors
Dr. Amir Hussain
Department of Computing Science
and Mathematics
University of Stirling
Stirling
United Kingdom FK9 4LA
a.hussain@cs.stir.ac.uk

Prof. Dr. Igor Aleksander
Imperial College of Science, Technology
& Medicine
Dept. Electrical & Electronic Engineering
Exhibition Rd.
London
Building Room 1009
United Kingdom SW7 2BT
i.aleksander@imperial.ac.uk

Dr. Leslie S. Smith
Department of Computing Science
and Mathematics
University of Stirling
Stirling
United Kingdom FK9 4LA
lss@cs.stir.ac.uk

Dr. Allan Kardec Barros
Universidade Federal do Maranhão
Centro Tecnológico
Curso de Engenharia Elétrica
Av. dos Portugueses s/n
São Luís-MA
Bacanga
Brazil
allan@ufma.br

Dr. Ron Chrisley
University of Sussex
Dept. Informatics
Falmer
Brighton
United Kingdom BN1 9QJ
R.L.Chrisley@sussex.ac.uk

Dr. Vassilis Cutsuridis
Department of Computing Science
and Mathematics
University of Stirling
Stirling
United Kingdom FK9 4LA
vcu@cs.stir.ac.uk

ISSN 0065-2598
ISBN 978-1-4614-2540-3 e-ISBN 978-0-387-79100-5
DOI 10.1007/978-0-387-79100-5
Springer New York Dordrecht Heidelberg London

Printed on acid-free paper

Springer is part of Springer Science+Business Media (www.springer.com)

Biologically Inspired Cognitive Systems (BICS'2008) - General Preface

This book is based on selected, expanded and significantly revised versions of papers presented at the third International Conference: Brain Inspired Cognitive Systems (BICS 2008) held in Sao Luis, Brazil, 24-27 June 2008, under the auspices of the University of Marahnão and the General Chairmanship of Allan Kardec Barros of that University.

BICS is an established biennial international Conference series (with BICS 2010 to be held in Madrid, Spain from 14-16 July 2010), which grew out of a set of earlier meetings, namely the International ICSC Symposia on Neural Computation (NC 1998) held in Vienna in 1998, and NC 2000 held in Berlin in 2000, which were followed by the first BICS 2004 held in Stirling, Scotland in 2004, and the second BICS 2006 in Lesvos, Greece. BICS aims to become a major point of contact for research scientists, engineers and practitioners throughout the world in the fields of cognitive and computational systems inspired by the brain and biology. As the brain is the most competent information processing system on earth, BICS brings together those who work on trying to understand how the brain achieves its competence and apply this knowledge to the design of ever more intelligent computers. Neurologists, Psychologists, Engineers and Computer Scientists regularly attend BICS meetings which specialise on models of consciousness, computational studies of the brain, and the design of brain-inspired machines and algorithms.

The aim of the third BICS 2008, like its predecessors, was to bring together leading computational scientists and engineers, in order to understand the prodigious processing properties of biological systems and, specifically, of the brain, and to exploit such knowledge to advance computational methods towards ever higher levels of cognitive competence. Four major international symposia were organized as part of BICS 2008, under the headings: Cognitive Neuroscience (CNS 2008), Biologically Inspired Systems (BIS 2008), Neural Computation (NC 2008 and Models of Consciousness (MoC 2008). The symposia were organized in patterns to encourage cross-fertilization across the symposia topics.

This book comprises a selection of 19 chapters written by BICS 2008 authors who were invited to contribute on the basis of originality, technical quality and relevance of the works presented at the BICS 2008 Conference. All invited chapters have been subjected to rigorous peer review by anonymous referees. The Editors are convinced that this selection of works provides the reader with an up to date account of the latest research, development and ideas in the wide arena of disciplines encompassed under the heading of BICS 2008.

The book chapters have been grouped into four Parts, corresponding to the four BICS 2008 symposia: (1) CNS 2008, covering computational models of the brain and brain inspired algorithms and artefacts; (2) BIS 2008, covering broader issues in biological inspiration and neuromorphic systems; (3) NC 2008, covering progress in neural systems; and (4) MoC 2008 covering models of consciousness. A preface for each Part has been written by the respective Symposium Chair that introduces the constituent chapters.

Finally, the Editors would like to express their gratitude to all the BICS 2008 authors who submitted high quality works and the anonymous reviewers who helped ensure the quality of the chapters included in the book. Special thanks go to Jeanny Pontin of the ICSC Interdisciplinary Research (Canada) for instigating and organizing the exciting biennial BICS meetings, and to Ann Avouris, the Springer Publishing Editor, who gave us the opportunity to edit this book.

BICS 2008 Editors:

Amir Hussain, BICS 2008 Publications Chair and Chair NC 2008
(a.hussain@cs.stir.ac.uk)
Igor Aleksander, Chair CNS 2008 (i.aleksander@imperial.ac.uk)
Leslie Smith, Chair BIS 2008 (lss@cs.stir.ac.uk)
Ron Chrisley, Chair MoC 2008 (ronc@sussex.ac.uk)
Allan Barros, General Chair BICS 2008 (allan@elo.com.br)
Vassilis Cutsuridis, BICS 2008 Book co-Editor (vcu@cs.stir.ac.uk)

Contents

Contributors

Rodrigo Varejão Andreão Instituto Federal do Espírito Santo, Av. Vitória, 1729, 29040-780, Vitória, ES, Brazil

Alexandre Balbinot Biomedical Engineering Research Group – NPEngBio, Departamento de Engenharia Elétrica, Universidade de Caxias do Sul - UCS, Alameda João Dal Sasso, 800 – Zip Code: 95700-000 – Bento Gonçalves, RS, Brazil, abalbinot@gmail.com
and
Universidade Federal do Rio Grande do Sul – UFRGS, Escola de Engenharia, Departamento de Engenharia Elétrica – DELET, Laboratório de Instrumentação Eletro-Eletrônica – IEE, Av Osvaldo Aranha, 103, Bom Fim, Porto Alegre, RS, CEP: 90035-190, Brazil

Allan K.D. Barros Department of Electrical Engineering, Federal University of Maranhao, Sao Luis, Maranhao, Brazil, allan@ufma.br

Thiago Bassani Instituto Nacional de Inovação em Diagnósticos para a Saúde Pública – fisica.ufpr.br/INIDSP CPGEI–Programa de Pós-Graduação em Engenharia Elétrica e Informática Industrial, www.cpgei.cefetpr.br
and
UTFPR – Universidade Tecnológica Federal do Paraná, www.utfpr.edu.br, www.ppgia.pucpr.br/~tbassani, thiago.bassani@gmail.com

Henrique E. Borges CEFET-MG, Av. Amazonas 7675, Belo Horizonte, MG, CEP 30510-000, Brazil, henrique@lsi.cefetmg.br

Antônio Pádua Braga UFMG, Av. Antônio Carlos 6627, Belo Horizonte, MG, CEP 31270-010, Brazil, apbraga@cpdee.ufmg.br

Giovani Carra Biomedical Engineering Research Group – NPEngBio, Departamento de Engenharia Elétrica, Universidade de Caxias do Sul – UCS, Alameda João Dal Sasso, 800 – Zip Code: 95700-000 – Bento Gonçalves, RS, Brazil, gcarra4@ucs.br

Maria T. Carrillo-de-la-Peña Department of Clinical Psychology and Psychobiology, University of Santiago de Compostela, 15705 Santiago de Compostela, Spain

Antonio Chella Dipartimento di Ingegneria Informatica Universitá di Palermo, Viale delle Scienze, I-90128 Palermo, Italy, chella@unipa.it

Sandro Tomassoni Coelho Instituto Nacional de Telecomunicações, Av. João de Camargo, 510, Santa Rita do Sapucaí, MG, Brazil, sandrot@inatel.br

L. Andrew Coward Department of Computer Science, Australian National University, Canberra, ACT 0200, Australia

Vassilis Cutsuridis Department of Computing Science and Mathematics, University of Stirling, Stirling FK9 4LA, UK, vcu@cs.stir.ac.uk

Draulio B. de Araujo Department of Physics and Mathematics, FFCLRP, University of Sao Paulo, Ribeirao Preto, SP, Brazil, draulio@usp.br

Leandro de Campos Teixeira Gomes Telecommunications Research Center (CPqD), Rod. Campinas–Mogi-Mirim (SP 340), km 118.5, Campinas, SP 13086-902, Brazil, tgomes@cpqd.com.br

Julio de Lima do Rego Monteiro University of Sao Paulo, Escola Politecnica, Av. Prof. Luciano Gualberto, 158, Tr. 3, Sao Paulo, SP 05586-090, Brazil

Lorena Sophia Campos de Oliveira Graduate Program on Electrical Engineering, Federal University of Espírito Santo, Av. Fernando Ferrari, 514, 29075-910, Vitória, ES, Brazil

Carlos A. Estombelo-Montesco Department of Physics and Mathematics, FFCLRP, University of Sao Paulo, Ribeirao Preto, SP, Brazil, estombelo@gmail.com
and
DCOMP/UFS Depto. de Computaçao da Universidade Federal de Sergipe, Cidade Universitaria Prof., Jose Aloisio de Campos, Jardim Rosa Elze, CEP 49100-000, São Cristóvão, SE, Brazil

Carlos A. Franco Associate Professor, Computer Science Department, Mathematical Institute, Mental Health Area, Psychiatric Institute, UFRJ, Brazil, carlosfranco@sensesac.org

Sulamita Frohlich Psychologist, Member of the Research Group in Cognitive Neuroscience, Neuropsychology, Mental Health Area, Psychiatric Institute, UFRJ, Brazil, sulafroh@arka2.com.br; arka2@arka2.com.br

João Carlos Gluz Pós-Graduação em Computação Aplicada (PIPCA), Universidade do Vale do Rio dos Sinos (UNISINOS), São Leopoldo, RS, Brazil, jcgluz@unisinos.br

Rogério Martins Gomes CEFET-MG, Av. Amazonas 7675, Belo Horizonte, MG, CEP 30510-000, Brazil, rogerio@lsi.cefetmg.br

Jaime Gómez Autonomous Systems Laboratory (ASLab-UPM), Universidad Politecnica de Madrid, Jose Gutierrez Abascal 2, 28045 Madrid, Spain, jd.gomez@upm.es

Ricardo Gudwin Department of Computer Engineering and Industrial Automation, FEEC, State University of Campinas, Brazil, gudwin@dca.fee.unicamp.br

Carlos Hernández Autonomous Systems Laboratory (ASLab-UPM), Universidad Politecnica de Madrid, Jose Gutierrez Abascal 2, 28045 Madrid, Spain, carlos.hernandez@upm.es

Adolfo Hernando Autonomous Systems Laboratory (ASLab-UPM), Universidad Politecnica de Madrid, Jose Gutierrez Abascal 2, 28045 Madrid, Spain, adolfo.hernando@upm.es

Josef Kittler Centre for Vision, Speech and Signal Processing, Faculty of Engineering & Physical Sciences, University of Surrey, Guildford, UK, kittler@surrey.ac.uk

Joao Eduardo Kogler University of Sao Paulo, Escola Politecnica, Av. Prof. Luciano Gualberto, 158, Tr. 3, Sao Paulo, SP 05586-090, Brazil, kogler@lsi.usp.br

Angelo Loula Department of Exact Sciences, State University of Feira de Santana, Brazil, angelocl@ecomp.uefs.br
and
Department of Computer Engineering and Industrial Automation, FEEC, State University of Campinas, Brazil

Jeremias Barbosa Machado School of Electrical and Computer Engineering, FEEC - UNICAMP, Av. Albert Einstein - 400, Cidade Universitária Zeferino Vaz, Distrito Barão Geraldo, 13083-852 Campinas, SP, Brazil, jeremias@dca.fee.unicamp.br

Gaëlle Spielmann Moura Department of Clinical Psychology and Psychobiology, University of Santiago de Compostela, 15705 Santiago de Compostela, Spain, gaellemoura@gmail.com

Edson José Nagle Telecommunications Research Center (CPqD), Rod. Campinas–Mogi-Mirim (SP 340), km 118.5, Campinas, SP 13086-902, Brazil, nagle@cpqd.com.br

Mário Uliani Neto Telecommunications Research Center (CPqD), Rod. Campinas–Mogi-Mirim (SP 340), km 118.5, Campinas, SP 13086-902, Brazil, uliani@cpqd.com.br

Marcio Lobo Netto University of Sao Paulo, Escola Politecnica, Av. Prof. Luciano Gualberto, 158, Tr. 3, Sao Paulo, SP 05586-090, Brazil

Julio Cesar Nievola Knowledge Discovery and Machine Learning Research Group, Programa de Pós-Graduação em Informática, Pontifícia Universidade Católica do Panama, Panama, nievola@ppgia.pucpr.br

Stefan Oliveira Pós-Graduação em Computação Aplicada (PIPCA), Universidade do Vale do Rio dos Sinos, (UNISINOS), São Leopoldo, RS, Brazil, stefanoliver@gmail.com

João Queiroz Graduate Studies Program on History, Philosophy, and Science Teaching, Federal University of Bahia/State University of Feira de Santana, Brazil, queirozj@semiotics.pro.br

Joao Henrique Ranhel Ribeiro University of Sao Paulo, Escola Politecnica, Av. Prof. Luciano Gualberto, 158, Tr. 3, Sao Paulo, SP 05586-090, Brazil

Sidarta Ribeiro Edmond and Lily Safra International Institute of Neuroscience of Natal (ELS-IINN), Brazil, ribeiro@natalneuro.org.br
and
Department of Neuroscience, Federal University of Rio Grande do Norte, Brazil

Fernando Oscar Runstein Telecommunications Research Center (CPqD), Rod. Campinas–Mogi-Mirim (SP 340), km 118.5, Campinas, SP 13086-902, Brazil, runstein@cpqd.com.br

Ricardo Sanz Autonomous Systems Laboratory (ASLab-UPM), Universidad Politecnica de Madrid, Jose Gutierrez Abascal 2, 28045 Madrid, Spain, ricardo.sanz@upm.es

Mario Sarcinelli-Filho Graduate Program on Electrical Engineering, Federal University of Espírito Santo, Av. Fernando Ferrari, 514, 29075-910, Vitória, ES, Brazil

Flávio Olmos Simões Telecommunications Research Center (CPqD), Rod. Campinas–Mogi-Mirim (SP 340), km 118.5, Campinas, SP 13086-902, Brazil, simoes@cpqd.com.br

Leslie S. Smith Department of Computing Science and Mathematics, University of Stirling, Stirling FK9 4LA, UK, l.s.smith@cs.stir.ac.uk

Marilda Spindola Biomedical Engineering Research Group – NPEngBio, Departamento de Engenharia Elétrica, Universidade de Caxias do Sul – UCS, Alameda João Dal Sasso, 800 – Zip Code: 95700-000 – Bento Gonçalves, RS, Brazil, mschiara@ucs.br

Marcio Jr. Sturzbecher Department of Physics and Mathematics, FFCLRP, University of Sao Paulo, Ribeirao Preto, SP, Brazil, marcio@biomag.usp.br

Yolanda Triñanes-Pego Department of Clinical Psychology and Psychobiology, University of Santiago de Compostela, 15705 Santiago de Compostela, Spain, yolandatp80@hotmail.com

David Windridge Centre for Vision, Speech and Signal Processing, Faculty of Engineering & Physical Sciences, University of Surrey, Guildford, UK, d.windridge@surrey.ac.uk

Carlos Alberto Ynoguti Instituto Nacional de Telecomunicações, Av. João de Camargo, 510, Santa Rita do Sapucaí, MG, Brazil, ynoguti@inatel.br

Milton A. Zaro Universidade Federal do Rio Grande do Sul – UFRGS, Programa de Pós-graduação em Informática na Educação, Av. Paulo Gama, 110, prédio 12105, 3° andar sala 332, 90040-060 Porto Alegre, RS, Brazil, zaro@ufrgs.br

Part I
Cognitive Neuroscience

Preface

Among the topics that make up the Brain Inspired Cognitive Systems group of pardigms, Cognitive Neuroscience (CNS) is distinguished through the use of computational methods in one specific way. Where in the rest of BICS an aim is to use constructive methods to either understand or design systems with advancing brain-like properties, CNS aims to use computational knowledge to analyse, model and consequently understand the *actual* working of the neuro-chemical brain. To be more specific, workers in CNS attempt to look at the properties of the brain that have been identified through cognitive psychology (classically, attention, memory, language, volition and emotion) and provide an analysis of the neural substrates thought to be responsible for these properties. So rather than provide equivalent models as one does in neuromorphic modelling, CNS attempts to provide a computational science of the brain.

In this volume four examples of this type of work are published. The first, by Moura, Pego and Carillo-de-la-Pena of the University of Santiago De Compostela in Spain, relates to the brain's ability to adjust its sensitivity according to the novelty of stimuli – reduction of sensitivity for predictable input and an increase for novel patterns. The paper addresses the auditory modality. Neurophysiological measurements are modelled and used to show that important gating mechanisms exist, which encourages further research in the vein of the paper. The second contribution by Do Rego Monteiro, Kogler, Ribeiro and Netto of the University of Sao Paulo in Brazil has more of a methodological flavour and looks for useful techniques for a broad application to bridge the gap between the operation of neurons and the resulting ability of an agent to gather analyse and transform knowledge. They concentrate on a Memory Evolutive System and indicate that category theory forms a good basis for general analysis. They go on to consider useful techniques for implementation of an explanatory theory and show that the Izhikevich model of the neuron with spike-dependent plasticity has considerable advantages.

The third paper by Oliveira and Gluz, addresses the question of human competencies as need to be known in job-selection is an exception in the sense that agent rather than neural models are applied to cognitive ideas. This type of work still needs verification through comparisons made with human evaluators of competency. The final paper in this group, by Custuridis of the University of Stirling uses the neural domain for attempting to model the cognitive ability of humans to make decisions. This is distilled to decisions made in the making of eye movements.

The above papers demonstrate two things. First there is no clear boundary between it and other cognitive sciences or, even, some areas of artificial intelligence. Second, the computational methodologies that are used in explanations of cognitive abilities of living organisms are continually being expanded. Neither of these observations is a criticism of the field, merely a statement that a true understanding of the cognitive behaviour of the brain need not and should not be curtailed by the definitions or analytic techniques that are fashionable at any particular time.

Igor Aleksander
Allan Barros

Effects of Stimuli Intensity and Frequency on Auditory P50 and N100 Sensory Gating

Gaëlle Spielmann Moura, Yolanda Triñanes-Pego, and Maria T. Carrillo-de-la-Peña[1]

Abstract Sensory gating is the brain's ability to adjust its sensitivity to incoming stimuli, i.e., to diminish its response to irrelevant or repetitive stimuli (*gating out*) and to increase it when a novel stimulus is presented (*gating in*). Most of the existing studies have investigated the *gating out* mechanism, giving little attention to the *gating in* function. Although both the P50 and N100 components of the auditory ERPs (event related potentials) show amplitude reductions to stimuli repetition, it is not clear if both components are part of a common gating system or if their sensory modulation is uncorrelated. In order to respond to these questions and to further characterize the sensory gating functions, we examined to what extent P50 and N100 are influenced by changes in the stimuli parameters and whether the sensory modulation of both components are interrelated. To this end, we obtained ERPs from 23 healthy volunteers using pairs of auditory stimuli which could be identical (S1 = S2), different in frequency (S1 = 1000 Hz; S2 = 2000 Hz) or different in intensity (S1 = 80 dB SPL; S2 = 100 dB SPL). As expected, the amplitudes of P50 and N100 decreased in response to the second stimuli of the identical pairs. With non-identical pairs, amplitude increases of P50 and N100 were observed only in pairs with different intensity, but not frequency. Thus, the results showed that both P50 and N100 are sensory modulated, showing that amplitude decreased to stimuli repetition (*gating out*) and increased when the two stimuli of a pair differed in intensity (*gating in*). A correlational analysis of the sensory gating indices (S2/S1 ratio and S1–S2 difference) obtained for P50 and N100 suggested that the sensory gating function of both components may be of a different nature. The reliability of the ratio and the difference indices of sensory gating is also discussed.

Keywords P50 · N100 · Sensory gating · Reliability · Intensity · Frequency

---

G.S. Moura, Y. Triñanes-Pego, and M.T. Carrillo-de-la-Peña (✉)
Department of Clinical Psychology and Psychobiology, University of Santiago de Compostela, 15705 Santiago de Compostela, Spain
e-mail: gaellemoura@gmail.com; yolandatp80@hotmail.com

[1] Corresponding author. Tel.: +34-981-563-100, ext. 13798; fax: +34-981-521581.
e-mail: pcpbmtc@usc.es (M.T. Carrillo-de-la-Peña).

A. Hussain et al. (eds.), *Brain Inspired Cognitive Systems 2008*, Advances in Experimental Medicine and Biology 657, DOI 10.1007/978-0-387-79100-5_1,

1 Introduction

The sensory gating mechanism is the capacity of the nervous system to modulate or to adjust its sensitivity to incoming stimuli, reducing its response to repeated or irrelevant information – "*gating out*" – or increasing it in response to changing stimulation – "*gating in*" (Freedman et al. 1987; Braff and Geyer 1990; Freedman et al. 1991; Freedman et al. 1996). Both mechanisms are equally important for normal brain functioning because, on the one hand, they protect it against information overload, avoiding the transmission of irrelevant signals to the consciousness (Venables 1964) and, on the other, they allow the detection of significant changes in the environment.

The sensory gating mechanism has been studied in the auditory modality through the P50, a component of the event related potentials (ERPs) that peaks around 50 ms. The P50 reflects mainly preattentional processing and shows an amplitude reduction to the second of a pair of identical stimuli (Freedman et al. 1983; Guterman et al. 1992; Clementz et al. 1998a; Clementz et al. 1998b; Boutros et al. 1999; Boutros and Belger 1999; Grunwald et al. 2003; Boutros et al. 2004; Hirano et al., 2008). Very few studies have investigated the *gating in* function of P50, and so far they have produced inconsistent results (Boutros et al. 1999). It would be interesting to observe how P50 amplitude is modulated by changes in intensity and frequency, using non-identical pairs of stimuli.

It is likely that the sensory gating occurs at different phases of the information process (Smith et al. 1994). In this vein, in addition to the P50, the sensory modulation of other later ERPs components such as the N100 has been also studied (Boutros et al. 1999; Clementz and Blumenfeld 2001; Boutros et al. 2004; Bramon et al. 2004; Hsieh et al. 2004). Although the N100 also displays amplitude reductions to repetitive stimuli, the majority of studies have found an absence of association between P50 and N100 sensory gating. Although the two components differ by nature (preattentional vs. attentional), one would expect that deficits at one stage of the information processing would have an effect in a later phase.

The sensory gating function of P50 has been formulated as the ratio (Qr) of P50 amplitude to the second (S2) and first (S1) stimulus of an identical pair (S2/S1). With reasonable consistency, pathological groups such as schizophrenics have shown larger S2/S1 ratios (smaller amplitude reductions to repetitive stimuli) than healthy subjects. Nevertheless, the existing studies have shown a great variability, to the extent that Qrs considered normal in one investigation are pathological in other (Adler et al. 1982; Freedman et al. 1987; Nagamoto et al. 1989; Kathmann and Engel 1990; Clementz et al. 1998b; Clementz and Blumenfeld 2001; Wilde et al. 2007; Patterson et al. 2008). Alternatively, S1 minus S2 difference (Qd) has also been calculated to evaluate the sensory gating mechanism. However, this index is not free of low consistency problems. Thus more research work is necessary to analyze the reliability of Qr and Qd, and to provide more normative data in healthy people that could help to interpret the data in pathological groups.

In light of the above reflections, the present work has three main objectives. First, to evaluate whether P50 and N100 amplitude changes in response to stimuli parameters reflect both *gating in* and *gating out* processes. Second, to evaluate to what extent the sensory modulation at one stage (around 50 ms) is related to the modulation at a later stage (around 100 ms) of the information process. Finally, to provide normative data of Qr and Qd for P50 and N100 components in young healthy subjects and to analyze the consistency of these indexes. To this end, the standard auditory paradigm used to obtain P50 was modified and the ERPs were recorded while the subjects listened passively to a series of pairs of stimuli which could be of three types: identical (S1 = S2), different in frequency (S1 = 1000 Hz; S2 = 2000 Hz) or different in intensity (S1 = 80 dB SPL; S2 = 100 dB SPL).

2 Materials and Methods

2.1 Subjects

Twenty three healthy psychology students (12 males and 11 females; mean age 21.52 with a range from 20 to 27 years) participated in the experiment. None had a history of psychiatric or neurological disease, nor were undergoing pharmacological treatment. The normal audiological status was verified by an evaluation of the auditory threshold that revealed a mean hearing level around 30 dB SPL. Each student was asked not to smoke and not to drink coffee two hours before the recording session.

2.2 Auditory Evoked Potential Procedure

Pure tone stimuli (Ss) with duration of 50 ms were presented in a series that had three different pairs of stimuli. The first type of pair had two identical Ss (both 80 dB SPL and 1000 Hz). The second pair had stimuli with identical intensity (80 dB SPL) but different frequency (S1 = 1000 Hz; S2 = 2000 Hz). The third pair had stimuli with identical frequency (1000 Hz) and different intensity (S1 = 80 dB SPL; S2 = 100 dB SPL). To distinguish the S2 of the pairs, we call the second stimulus with different frequency, S3, and that with different intensity, S4.

Forty identical pairs, forty pairs with different frequencies and forty pairs with different intensities were presented pseudo randomly in the sequence. The interstimulus interval (ISI) was 500 ms, and the inter-trial interval (ITI) was 8 seconds.

Participants were seated in a comfortable armchair in an electrically-isolated, sound-light attenuated room. They were instructed to fix their eyes on a point in front of them, to avoid movements during the test, and to listen attentively to the series of stimuli.

For this report, EEG activity was recorded with a 20 electrode electrocap (Bionic Electrics Inc.) following the 10–20 International System and referred to both ear lobes. An electrode placed on the forehead served as ground. The vertical and horizontal EOG were also registered for a posterior elimination of its influence over the EEG recording.

For EEG data acquisition, amplification and filtering, a SynAmps amplifier connected to NeuroScan version 4.1 (Neuroscan Labs.) was used. The acquisition rate was 500 Hz. The signal was amplified 10.000 times and a 0.1–100 Hz bandpass and a 50 Hz notch filters were used. Impedances were kept under 10 KΩ.

Data treatment to obtain ERPs was performed with Vision Analyzer program (Brain Vision). First, the EEG was segmented in epochs from −100 to 500 ms. Segments over +/− 100μV were automatically rejected. Other contaminated segments were then eliminated after visual inspection. Ocular artifacts, baseline and linear trends were corrected. Digital filters of 10–50 Hz and of 0.1–100 Hz were applied off-line to obtain P50 and N100, respectively.

P50 was identified as the major positive component between 35 and 80 ms. When two positive deflections were observed in this range, we opted for the latest. Although P50 amplitude was referred to in the literature as preceded by a negative peak (Nb), due to the difficulty in identifying this component in some subjects, we decided to calculate P50 amplitude (in μV) in reference to the baseline. This methodology seems to produce similar results as the peak to peak method. N100 amplitude was measured at the most negative peak between 70 and 120 ms.

The sensory gating indices, Qr (S2/S1) and Qd (S1–S2), were calculated for both P50 and N100 amplitudes.

2.3 *Data Analysis*

To verify the effect of stimuli repetition (*gating out*) on P50 and N100 amplitudes, t-tests were applied for the difference between amplitudes obtained from the S1 and S2 of identical pairs.

In order to check whether P50 and N100 present *gating in* as a result of the 2nd stimuli of a pair being different from the 1st, we used a repetitive measures ANOVA with Stimuli Change as within-subjects factor (3 levels: S2 – or amplitudes for the 2nd stimuli of the identical pairs, S3 – amplitudes for the 2nd stimuli of the pairs that differed in frequency, and S4 – amplitudes for the 2nd stimuli of pairs with different intensity). The ANOVAs were performed separately for the following variables: P50 amplitude and P50 Qr and Qd, N100 amplitude and N100 Qr and Qd. We used t-tests for *a posteriori* comparisons.

The possible influence of gender of the participants was evaluated in a preliminary analysis. Since this variable was not significant in any case, the corresponding results were not presented.

For the second objective, Pearson correlations between sensory gating indices obtained for P50 and N100 were calculated.

To assess the consistency of the sensory gating indices (3rd objective), we also analyzed the association between Qr and Qd using Pearson coefficients.

3 Results

3.1 P50 and N100 Amplitudes in Response to Stimuli Repetition and Change

The mean P50 and N100 amplitude and latency values for each pair of stimuli are presented in Table 1. The grand averages of ERPs filtered 10–50 Hz to obtain P50 are presented in Figure 1 and those corresponding to N100 (ERPs filtered 0.1–100 Hz) are presented in Figure 2 (figure "a" is for identical pairs, "b" for pairs differing in frequency, and "c" for pairs of different intensity).

Figure 1a suggests that P50 amplitude is smaller in response to the second stimulus when it is identical to the first (*gating out* function). The t-test confirmed a significant difference between P50 amplitude to S1 and to S2 ($t = 6.26$; $p < .001$).

The trend in the N100 data followed a similar pattern in response to stimuli repetition, with amplitude reductions to the 2nd stimuli of the identical pairs (see Figure 2a). The t-test for the difference between N100 amplitudes for the 1st and 2nd stimuli were also significant ($t = -4.93$; $p < .001$). As can be seen in Figure 2a, the component P200, which was not analyzed in this study, also showed an amplitude reduction in the traces corresponding to S2.

In response to non-identical pairs (*gating in* function), the repeated measures ANOVA showed that the Change factor (identical pairs, pairs with different frequencies and pairs with different intensities) was significant for both P50 and N100 amplitudes, as well as for the Qr and Qd indices obtained from those components (with the exception of N100 Qr; see Table 2).

Post-hoc comparisons performed for P50 amplitudes and Q indices revealed that P50 amplitude of the 2nd stimulus of non-identical pairs is significantly larger than that of the 2nd stimulus of identical pairs, but only when the pairs are different in intensity and not in frequency (S4 > S2 but S3 is not significantly different from S2; see Table 3). Thus, the change in tone frequency (1000 to 2000 Hz) did not cause the *gating in* phenomenon as expected. On the contrary, this amplification function was observed for intensity changes. The same results were obtained when P50 Qr were

Table 1 Mean P50 and N100 latencies (ms) and amplitudes (μV) to three pairs of stimuli: S1–S2 (identical stimuli), S1–S3 (stimuli with different frequency) and S1–S4 (stimuli with different intensity). Standard deviations are in parentheses

	P50		N100	
Pairs	Latency	Amplitude	Latency	Amplitude
S1	56.82 (9.63)	2.65 (1.12)	101.52 (12.15)	−9.74 (4.72)
S2	52.27 (11.34)	1.30 (0.77)	92.00 (14.71)	−5.35 (2.83)
S1	60.00 (6.02)	2.65 (1.31)	101.33 (11.14)	−9.83 (4.65)
S3	56.55 (7.98)	1.35 (0.87)	96.10 (13.21)	−6.39 (3.30)
S1	59.55 (7.40)	2.77 (1.60)	103.14 (10.82)	−10.33 (4.80)
S4	58.36 (8.50)	2.84 (1.69)	101.33 (12.59)	−11.76 (6.32)

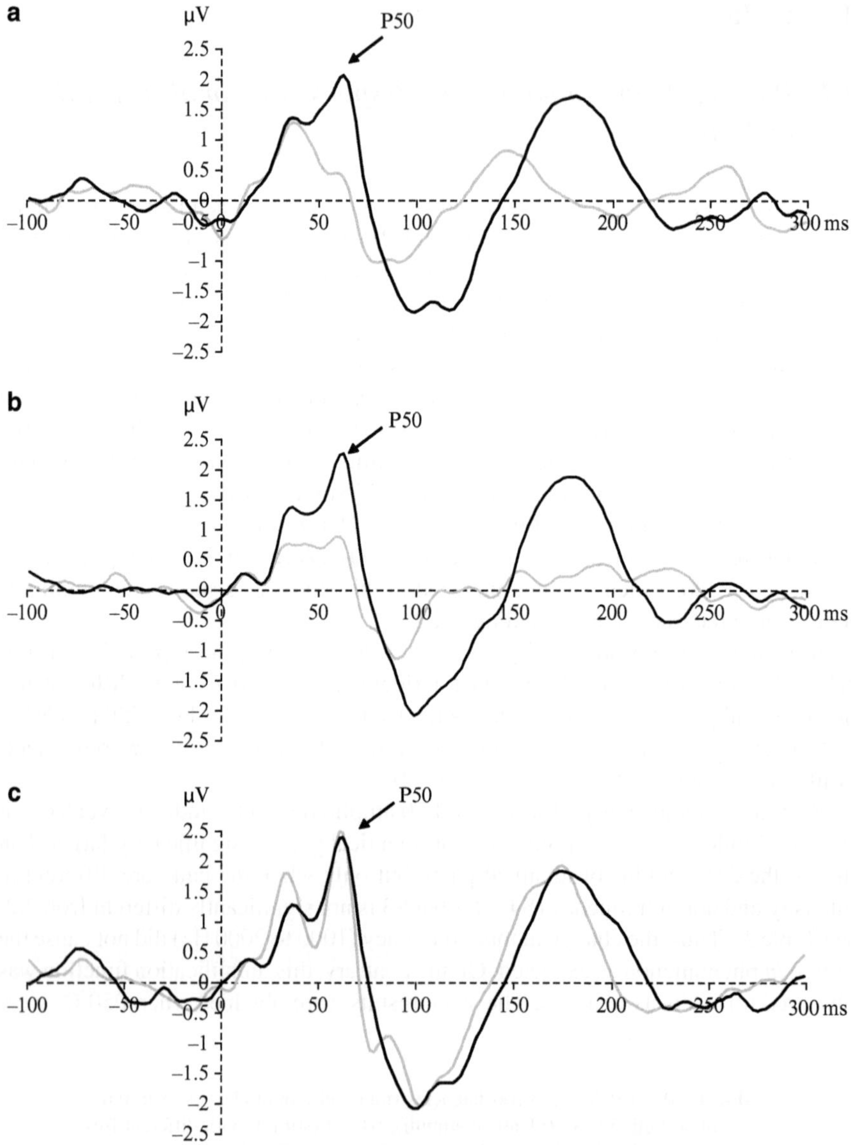

Fig. 1 P50 to identical and different pairs of stimuli. The black line corresponds to the 1st stimulus of the pair (S1) and the gray line to the 2nd stimulus of each pair (S2 in figure 1a; S3 in figure 1b; S4 in figure 1c)

used as the dependent variable. The difference in response to the 2nd stimuli which differ in frequency or intensity can be observed in Figures 1b and 1c, respectively.

The above pattern of results was replicated for N100 amplitude, which also increased in response to the second stimuli of the pair only when they had different

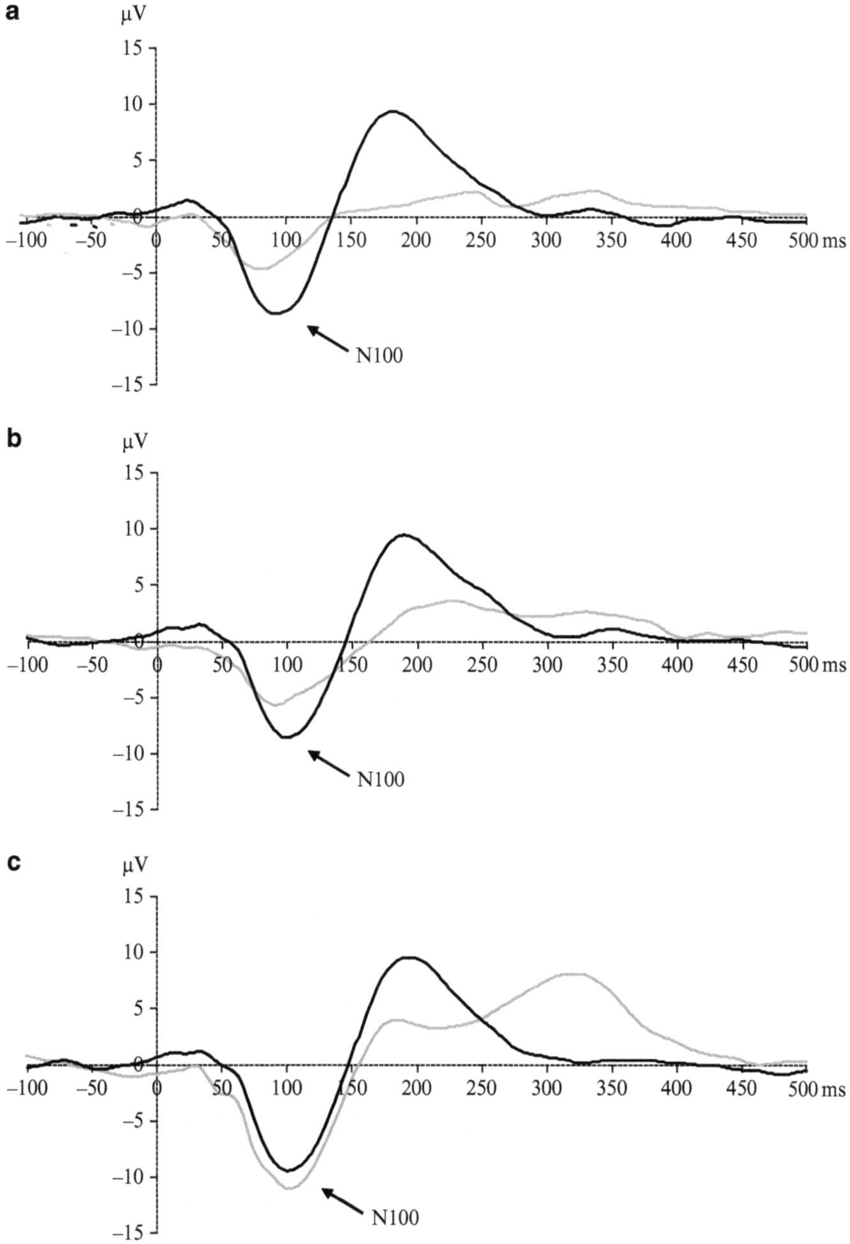

Fig. 2 N100 to identical and different pairs of stimuli. The black line corresponds to the first stimulus of the pair (S1) while the gray line corresponds to the second stimulus of each pair (S2 in figure 2a; S3 in figure 2b; S4 in figure 2c)

Table 2 Intra-subjects effects tests for the Change factor for each dependent variable considered

	F	D.F.	P VALUES
P50 AMPLITUDE	17.54	2;42	.000
P50 Qr	9.46	1.15;42	.004
P50 Qd	12.36	1.47;42	.000
N100 AMPLITUDE	22.62	2;40	.000
N100 Qr	5.36	1.03;40	.094
N100 Qd	13.96	1.35;40	.000

Table 3 Post-hoc contrasts using paired samples t-tests for P50 and N100 amplitudes and Qrs, with different types of second stimuli

	T	D.F.	P VALUES		T	D.F.	P VALUES
P50 Amplitude				N100 Amplitude			
S2–S3	−0.23	22	.823	S2–S3	1.71	23	.103
S2–S4	−5.33	22	.000	S2–S4	5.05	23	.000
S3–S4	−4.30	22	.000	S3–S4	4.98	23	.000
P50 Qr				N100 Qr			
Qr12–Qr13	−0.37	22	.710	Qr12–Qr13	−0.30	23	.764
Qr12–Qr14	−3.65	22	.002	Qr12–Qr14	−1.69	23	.103
Qr13–Qr14	−2.84	22	.001	Qr13–Qr14	−1.85	23	.079
P50 Qd				N100 Qd			
Qd12–Qd13	0.14	22	.885	Qd12–Qd13	−1.39	23	.178
Qd12–Qd14	4.81	22	.000	Qd12–Qd14	−3.92	23	.001
Qd13–Qd14	3.64	22	.001	Qd13–Qd14	−4.25	23	.000

intensity but not when they differed in frequency (see Table 3 and Fig. 2b and 2c). As shown in Fig. 2, generally the second stimuli caused a reduction in P200 amplitude that was more accentuated when the pairs of stimuli were identical (Fig. 2a) than when they were different (Fig. 2b and 2c). Moreover, an increase in the intensity of the 2nd stimulus (S4) was associated with a positive component in the latency around 300 ms that did not appear under the other conditions.

3.2 Sensory Gating of P50 and N100. Are They Related?

The analysis of Pearson coefficients between Q indices obtained from P50 and N100 revealed scarce significant correlations (see Tables 4 and 5, for Qr and Qd, respectively).

As can be seen in Tables 4 and 5, the only significant correlations between P50 and N100 indices were for the pair S1–S4 (stimuli with different intensity). For the rest of the pairs of stimuli, no significant correlation between the indices of sensorial modulation calculated for P50 and N100 were found.

Table 4 Pearson correlation coefficients between P50 Qr and N100 Qr for each pair of stimuli

	P50 Qr – N100 Qr
S1–S2	−0.15 n.s
S1–S3	−0.19 n.s.
S1–S4	−0.85***

(***significant to $p < .001$; n.s. non significant)

Table 5 Pearson correlation coefficients between P50 Qd and N100 Qd for each pair of stimuli

	P50 Qd – N100 Qd
S1–S2	−0.23 n.s.
S1–S3	−0.29 n.s.
S1–S4	−0.74 ***

(***significant to $p < .001$; n.s. non significant)

Table 6 Mean Ratio (Qr) and difference (Qd) indices obtained from P50 and N100 amplitudes. Standard deviations are in parentheses

	P50			N100		
	S1–S2	S1–S3	S1–S4	S1–S2	S1–S3	S1–S4
Qr	0.52 (0.36)	0.56 (0.40)	1.54 (1.59)	0.65 (0.40)	0.68 (0.31)	1.54 (2.24)
Qd	1.35 (1.01)	1.30 (1.17)	−0.07 (1.89)	−4.39 (4.08)	−3.44 (3.15)	1.44 (5.44)

3.3 *Normative Data and Reliability of Sensory Gating Indices*

The ratio (Qr) and difference (Qd) indices obtained from P50 and N100 amplitude data are presented in Table 6. As can be seen, the P50 Qr ratio was 0.52, a value lower than a unity (indexing "*gating out*"). The value is still slightly superior to those found in other investigations that use the standard paradigm in normal groups (in a range of 0.14 to 0.51). The ratio increased when the second Ss parameters changed, although it was only higher than a unity (indexing "*gating in*") when the 2nd stimuli were different in intensity (0.56 for frequency change and 1.54 for intensity change).

The Qrs calculated for N100 followed a similar pattern, although the values for S1–S2 and S1–S3 pairs were slightly superior to those calculated for P50.

The P50 sensory gating index calculated as the amplitude difference between the 1st and 2nd Ss (Qd) followed a similar but inverse pattern than Qr, i.e., values decreased when the parameters of the second stimuli changed, especially with changes in intensity. As expected (given the negative amplitudes of N100), an inverse pattern was found for N100 Qd.

The correlations between P50 Qr and P50 Qd were moderate but significant for all of the pairs (Table 7). The correlations between N100 Qr and N100 Qd were moderate but only significant for identical pairs and for the pairs with different frequencies (Table 8).

Table 7 Pearson correlation coefficients of P50 Qr and P50 Qd to each pair of stimuli

Pairs	P50 Qr-P50 Qd
S1–S2	−0.73***
S1–S3	−0.75***
S1–S4	−0.69***

(*** significant to $p < .001$)

Table 8 Pearson correlation coefficients of N100 Qr and N100 Qd to each pair of stimuli

Pairs	N100 Qr-N100 Qd
S1–S2	0.80 ***
S1–S3	0.82 ***
S1–S4	0.08 n.s.

(*** significant to $p < .001$; n.s. non significant)

4 Discussion and Conclusions

The results showed that the presentation of pairs of identical auditory stimuli provoked a reduction in the amplitude of P50 and N100 in response to the second stimuli of the pairs. This suppression supports the existence of a sensory gating mechanism to repetitive and irrelevant information (*gating out*) that, as found by other authors, is manifested at different indices of the ERPs (Adler et al. 1982; Boutros et al. 1999; Clementz and Blumenfeld 2001; Boutros et al. 2004). The waveforms showed that, besides P50 and N100, other components such as P200 also presented a *gating out* function (also found by Fuerst, Gallinat and Boutros, 2007). Thus, it seems that the sensory gating mechanism operates at different phases of the information process.

Although the sensory modulation process occurring at 50 ms is possibly of a different nature than that occurring at 100 ms, it is logical to expect that an earlier process would influence a later stage of information processing. Contrary to this expectation, the correlation analysis did not support any relationship between sensory gating indices (Qr, Qd) obtained from P50 and N100 amplitudes using identical pairs. This result supports the argument that despite the fact that these components are modulated by the characteristics of the sensory information, the nature of both modulation functions is different. It is in partial accordance with a recent study by Brockhaus-Dumke et al. (2008), who did not find any significant correlation between P50 and N100 when considering the ratio index and only a weak to moderate correlation for the difference index. Nevertheless, Fuerst et al. (2007) found significant correlations between P50 and N100 sensory gating for both indices, although the coefficients were of higher magnitude when considering the Qds. The root of this discrepancy may lie on the fact that Fuerst et al. used peak-to-peak, instead of peak-to-baseline measurements of P50 and N100. In summary, there is an open debate in the literature about whether the sensory modulation of P50 and N100 components are part of a common gating system or, on the contrary, they are independent functions.

In response to non-identical pairs, the results partially confirm that P50 and N100 present a *gating in* function. When the stimuli of the pair had different frequencies, no significant increase in P50 or in N100 amplitude was observed in response to the second stimulus. And, although Qr from pairs of 1000–2000 Hz was higher (0.56) than that obtained from identical pairs (0.52), the difference was not significant. These results contrast with those of Boutros et al. (1999) and Boutros and Belger (1999) that showed that P50 increases when the second stimulus has a different frequency compared to the first one. A possible reason for this discrepancy could be that, in the present study, the pairs composed by grave and acute tones appeared pseudorandomly mixed with pairs of different intensity. Possibly, in the comparison between both kinds of pair, the frequency change could have lost its novelty value, which is indispensable for observing a *gating in* effect.

This may be explained because for the human race and from an adaptive point of view, it is more important to discriminate the intensity than the frequency of the sounds. This explanation seems to be supported by the data obtained with pairs of different intensity, which had a significant increase of P50 and N100 amplitudes to the 2nd stimuli (with indices $Qr > 1$). High intensity stimuli also produced an effect in components of long latency, like the positive deflection observed around 300 ms (P300). The appearance of this ERP component suggests that intense stimuli involuntarily attracted the subject's attention, involving some kind of *bottom-up* mechanism. The P300 presence also suggests that the high difference in the intensity between S4 and the rest of the stimuli could have determined that it became a target stimulus, and thus that the series became an oddball-like paradigm (S4, of 100 dB, presented 16.66% of the times).

The pattern founded agrees with the investigation of Ninomiya et al. (2000) who observed that P50 amplitude only modifies with an intensity change but not with a frequency change. Nevertheless, at odds with that study, we found an increase, but not a P50 amplitude reduction, in response to the 100 dB stimuli compared with the lower intensities. This contradiction can be understood by considering that the brain's response to really intense sounds is subject to inter-individual differences, a phenomenon which is the object of an area of research (Carrillo-de-la-Peña 1992). Other studies have also shown inconsistent P50 changes in response to intensity increases (for example, Griffith et al. 1995 found P50 amplitude increases in response to the first stimulus but, in response to the second stimulus, normal subjects showed amplitude increases related to loud intensity while schizophrenics showed the opposite). The lack of studies applying paradigms that include pairs of stimuli with different intensities does not allow a comparison of the results of this research with similar investigations.

The ratio index (Qr) obtained for P50 in response to identical pairs of stimuli was 0.52, a value slightly superior to those considered characteristic of healthy people, that were in a range between 0.14 to 0.51 in other studies (Freedman et al. 1987; Waldo et al. 1992; Hetrick et al. 1996, Clementz et al. 1998a, Clementz et al. 1998b; Boutros et al. 1999; Light et al. 1999; Cadenhead et al. 2000). Maybe this difference is due to the fact that the paradigm employed in the present study deviates from the standard, because we presented different pairs of stimuli and an inferior number of repetitions of each kind of pair (40) in comparison with previous papers. As found in the review made by Patterson et al. (2008), the heterogeneity among the studies explains the existence of a wide range of variability for both the schizophrenic groups (with ratios from 56 to 158 %) and for the controls (from 9 to 73.4 %).

The results obtained for the P50 Qr and for the P50 Qd indices followed the same pattern. This suggests that the S1–S2 difference index may be a valid alternative to the S2/S1 ratio. In fact, higher test-retest reliability has been found for the difference than for ratio index (Fuerst et al., 2008; Rentzsch et al., 2008). Nevertheless, the moderate correlations found in the present study indicate that more research about the reliability and consistency of sensory gating indices is needed. In addition, the relationship between the sensorial modulation of the components P50 and N100 should be further analyzed.

References

Adler L, Pachtman E, Franks R, Pecevich M, Waldo M, Freedman R (1982) Neurophysiological evidence for a defect in neuronal mechanisms involved in sensory gating in schizophrenia. Biol Psychiatry 17:639–654

Boutros N, Belger A (1999) Midlatency evoked potentials attenuation and augmentation reflect different aspects of sensory gating. Biol Psychiatry 45:917–922

Boutros N, Belger A, Campbell D, Souza C, Krystal J (1999) Comparison of four components of sensory gating in schizophrenia and normal subjects: a preliminary report. Psychiatry Res 88:119–130

Boutros N, Korzyukov O, Jansen B, Feingold A, Bell M (2004) Sensory gating deficits during the mid-latency phase of information processing in medicated schizophrenia patients. Psychiatry Res 126:203–215

Braff D, Geyer M (1990) Sensorimotor gating and schizophrenia. Arch Gen Psychiatry 47:181–188

Bramon E, Rabe-Hesketh S, Sham P, Murras R, Frangou S (2004) Meta-analysis of the P300 and P50 waveforms in schizophrenia. Schizophr Res 70:315–329

Brockhaus-Dumke A, Schultze-Lutter F, Mueller R, Tendolkar I, Bechdolf A, Pukrop R, Klosterkoetter J, Ruhrmann S (2008) Sensory gating in schizophrenia: P50 and N100 gating in antipsychotic-free subjects at risk, first-episode, and chronic patients. Biol Psychiatry 64:376–384

Cadenhead KS, Light GA, Geyer MA, Braff DL (2000) Sensory gating deficits assessed by the P50 event-related potential in subjects with schizotypal personality disorder. Am J Psychiatry 157:55–59

Carrillo-de-la-Peña MT (1992) ERP augmenting/reducing and sensation seeking: a critical review. Int J Psychophysiology 12:211–220

Clementz B, Blumenfeld L (2001) Multichannel electroencephalographic assessment of auditory evoked response suppression in schizophrenia. Exp Brain Res 139:377–390

Clementz B, Geyer M, Braff D (1998) Multiple site evaluation of P50 suppression among schizophrenia and normal comparison subjects. Schizophr Res 30:71–80

Clementz B, Geyer M, Braff D (1998b) Poor P50 suppression among schizophrenia patients and their first-degree biological relatives. Am J Psychiatry 155:1691–1694

Freedman R, Adler L, Gerhardt G, Waldo M, Baker N, Rose G, Drebing C, Nagamoto H, Bickford-Wimer P, Franks R (1987) Neurobiological studies of sensory gating in schizophrenia. Schizophr Bull 13:669–678

Freedman R, Adler L, Myles-Worsley M, Nagamoto H, Miller C, Kisley M, McRae K, Cawthra E, Waldo M (1996) Inhibitory gating of an evoked response to repeated auditory stimuli in schizophrenic and normal subjects – Human recordings, computer simulation, and an animal model. Arch Gen Psychiatry 53:1114–1121

Freedman R, Adler L, Waldo M, Pachtman E, Franks R (1983) Neurophysiological evidence for a defect in inhibitory pathways in schizophrenia: comparison of medicated and drug-free patients. Biol Psychiatry 18:537–551

Freedman R, Waldo M, Bickford-Winner P, Nagamoto H (1991) Elementary neuronal dysfunction in schizophrenia. Schizophr Research 4:233–243

Fuerst D, Gallinat J, Boutros N (2007) Range of sensory gating values and test-retest reliability in normal subjects. Psychophysiol 44: 620–626

Griffith J, Hoffer L, Adler L, Zerbe G, Freedman R (1995) Effects of sound intensity on a mid-latency evoked response to repeated auditory stimuli in schizophrenic and normal subjects. Psychophysiology 32:460–466

Grunwald T, Boutros N, Pezer N, Oertzen J, Fernández G, Schaller C, Elger C (2003) Neuronal substrates of sensory gating within the human brain. Biol Psychiatry 53:511–519

Guterman Y, Josiassen R, Bashore T (1992) Attentional influence on the P50 component of the auditory event-related brain potential. Int J Psychophysiol 12:197–209

Hetrick W, Sandman C, Bunney W, Jin Y, Potkin S, White M (1996) Gender differences in gating of the auditory evoked potential in normal subjects. Biol Psychiatry 39:51–58

Hsieh M, Liu K, Liu S, Chiu M, Hwu H, Chen A (2004) Memory impairment and auditory evoked potential gating deficit in schizophrenia. Psychiatry Res: Neuroimaging 130:161–169

Light GA, Malaspina D, Geyer MA, Luber BM, Coleman EA, Sackeim HA, Braff DL (1999) Amphetamine disrupts P50 suppression in normal subjects. Biol Psychiatry 46:990–996

Kathmann N, Engel R (1990) Sensory gating in normals and schizophrenics: A failure to find strong P50 suppression in normals. Biol Psychiatry 27:1216–1226

Nagamoto H, Adler L, Waldo M, Freedman R (1989) Sensory gating in schizophrenics and normal controls: effects of changing stimulation interval. Biol Psychiatry 25:549–561

Ninomiya H, Sato E, Onitsuka T, Hayashida T, Tashiro N (2000) Auditory P50 obtained with a repetitive stimulus paradigm shows suppression to high-intensity tones. Psychiatry Clin Neurosci 54:493–497

Patterson J, Hetrick W, Boutros N, Jin Y, Sandman C, Stern H, Potkin S, Bunney WE (2008) P50 sensory gating ratios in schizophrenics and controls: a review and data analysis. Psychiatry Res 158:226–247

Rentzsch J, Jockers-Scherübl, MC, Boutros NN, Gallinat J (2008) Test-retest reliability of P50, N100 and P200 auditory sensory gating in healthy subjects. Int J Psychophysiol 67: 81–90

Smith A, Boutros N, Schwarzkopf B (1994) Reliability of P50 auditory event-related potential indices of sensory gating. Psychophysiology 31:495–502

Venables P (1964) Input dysfunction in schizophrenia. In: Maher BA (ed.) Progress in Experimental Personality Research, Academic Press, Orlando

Waldo M, Gerhardt G, Baker N, Drebing C, Adler L, Freedman R (1992) Auditory sensory gating and catecholamine metabolism in schizophrenic and normal subjects. Psychiatry Res 44:21–32

Wilde O, Bour L, Dingemans P, Koelman J, Linszen D (2007) A meta-analysis of P50 studies in patients with schizophrenia and relatives: differences in methodology between research groups. Schizophrenia Res 97:137–151

On Building a Memory Evolutive System for Application to Learning and Cognition Modeling

Julio de Lima do Rego Monteiro, Joao Eduardo Kogler, Joao Henrique Ranhel Ribeiro, and Marcio Lobo Netto

Abstract We address here aspects of the implementation of a memory evolutive system (MES), based on the model proposed by A. Ehresmann and J. Vanbremeersch (2007), by means of a simulated network of spiking neurons with time dependent plasticity. We point out the advantages and challenges of applying category theory for the representation of cognition, by using the MES architecture. Then we discuss the issues concerning the minimum requirements that an artificial neural network (ANN) should fulfill in order that it would be capable of expressing the categories and mappings between them, underlying the MES. We conclude that a pulsed ANN based on Izhikevich's formal neuron with STDP (spike time-dependent plasticity) has sufficient dynamical properties to achieve these requirements, provided it can cope with the topological requirements. Finally, we present some perspectives of future research concerning the proposed ANN topology.

Keywords Cognitive architectures · Category theory · Memory evolutive systems · Pulsed neural networks with STDP

1 Introduction

Cognition refers to how knowledge can be gathered, analyzed, transformed and organized by a (living) agent. The use of cognition is observed in the behavior of some animal species; it is a quality associated with their nervous systems activity. So, how can one explain the neural basis of these cognitive processes? Or, put in other words, how can one correlate the activity of networks of neurons with the knowledge manipulation tasks that we have above mentioned? This imply in understanding how the neural activity is able to encode semantic knowledge, not only in terms of memory, but also in terms of knowledge processing.

By doing experiments that search for these correlations, one has to make hypotheses founded in some representational model that serves as a guide for the

J. de Lima do Rego Monteiro, J.E. Kogler (✉), J.H.R. Ribeiro, and M.L. Netto
University of Sao Paulo, Escola Politecnica, Av. Prof. Luciano Gualberto, 158, Tr. 3 Sao Paulo SP 05586-090, Brazil
e-mail: kogler@lsi.usp.br

A. Hussain et al. (eds.), *Brain Inspired Cognitive Systems 2008*, Advances in Experimental Medicine and Biology 657, DOI 10.1007/978-0-387-79100-5_2,

rationale of the experiment. That is, for the decisions about what to look for and measure, about which procedures will define the course of action during the experiment and to set guidelines for the interpretations of the experimental results. Thus, for the investigation of the neural basis of cognition, it would be desirable to have at hand some kind of theory of cognition that considers its relation with the network of neurons. Such theory should give hints about how and where to look for the encoding of semantic aspects of the knowledge, about what kind of structure and/or dynamics could express the processes that transforms this knowledge, interpretable at the semantic point of view.

Representations of the activity of the nervous system alternately reflect the patterns of activity registered on neural units, on networks of cells or on regions in the brain, depending on the scale considered. These patterns are then interpreted in terms of the cognitive tasks under investigation. These interpretations can show, in principle, the structural relationship among cells in a group, among groups in a region or among regions, depending on the scale of the experiment. It depends on the method used to model and process the data. The choice of this method depends on what one is looking for. So, the tough part seems to be to decide on what specifically to search, in order to answer the question about the neural basis of cognition.

One way to tackle this challenge is to do simulations with simplified models that can be used to test the principles that will guide the design of our experimental investigation with actual cells and brains. The degree of simplification will depend in what is essential and what is feasible to simulate. Our aim in this work is to discuss these topics and to lead to the implementation of a model proposed by A. Ehresmann and J. Vanbremeersch (2007)[7], by means of a simulated network of spiking neurons with time dependent plasticity. The model that they proposed is called Memory Evolutive System, abbreviated as MES. It is founded in the mathematical theory of categories, which is a suitable tool for both functional and semantic modeling. The MES can provide useful insight for the problem of understanding the mechanisms of neurocognition, however the authors did not make clear how to implement it. We are currently investigating possible solutions on this way, and here we point to some results, after discussing the rationale that lead to this implementation. In the following we first make some considerations about the level of detail on simulations that can portrait meaningful conclusions referring to actual nervous systems. Then, we present an overview of category theory, concentrating on the concepts that are related to this work. After this, we review the related works that employ category theory in the domain of neurocognitive models. Finally, we shall go through the MES, discussing the requirements and constraints that it imposes for building a simulation model and we present our solution, based of the use of a pulsed artificial neural network (ANN) based on Izhikevich's formal neuron [18] with STDP (spike time-dependent plasticity) [33]. Our final considerations will concern about the topological requirements of such a model, which point to our current and future research.

1.1 Models for Neurocognitive Processes

The modeling of cognitive aspects of animal behavior as a result of their nervous systems activity (even considered as an emergent property) is strongly dependent on simulations that can, at the same time, display neural and cognitive features, enabling one to find the correlations between neural and cognitive phenomena. The computer modeling of neurobiological processes can be a very hard task. Even a model of a "simple" animal like *C. elegans*, which has only 302 neurons, and is considered completely mapped since 1986 [2, 34], is a complex entrepreneurship. The *C. elegans* genome encodes at least 80 different types of potassium-selective ion channels, 90 ligand-gated receptors, and approximately 1000 G proteins-linked receptors [1]. It means that neurons in such system may assume several configurations and behave, individually, like a complex dynamical system. As pointed in [3] the combinatorial possibilities are astounding, even for a so small nervous system. The topology of the neural system of *C. elegans* is known and if we choose a neuron model that reflects a realistic biological-like dynamic behavior for each neuron, the computational costs and long-term numerical accuracy are difficult challenges to deal with. However, even if one could run a simulation that would be very accurate from the neurobiological point of view, the interpretation of the results in terms of information and knowledge processing would still be difficult, without an adequate framework for treating the cognitive processes at neural level.

Now we shall consider how to make a representation of semantic knowledge, first into a more abstract way. Then, we shall use it to build the Memory Evolutive System, which will lead to a particular case that can be interpreted as a neurocognitive model. We start by looking to an animal as an agent that uses sensors and effectors to interact with the environment. With the sensors it gathers information that can be used to make decisions about the actions that it will take upon the environment via its effectors. In the process of decision making, it can use prior knowledge that is stored in its memory as a result of its past experience. All these stages and processes involving them uses some knowledge representation. The agent receive messages from the environment via its sensors and send messages to the environment via its effectors. Here, when we say information, we mean something that can influence the state of knowledge of an agent. Information is conveyed by the messages. The knowledge is also made of information. By knowledge we mean an internal representation of the invariant aspects that the agent got from messages that it received during his existence. One can wonder what kind of language does this agent use to make these representations. This idea of language is suggested when we say that the information is conveyed by messages, so one would expect that behind these messages there would be some alphabet of symbols used to decode or compose the messages. This certainly happens in higher levels of communication. But first, we shall consider a very early stage of interaction with the environment, one in which the agent still does not posses an elaborate alphabet of symbols. The messages it receives are streams of primitive symbols that arise from its sensors. In nature, these are so primitive that they are the variation of some physical entity, like a voltage, a current an so on. By analyzing messages, the agent eventually finds invariants and patterns of correlations among them, enabling the construction of more complex symbols from these patterns. So this is the cognitive process taking place. More elaborate and general pieces of knowledge are then constructed from simpler, more primitive ones. Complex objects and relations among them arises from simpler ones. The relations among complex objects are also more sophisticated, because they can involve more properties and classes of relations. This is a scenario that category theory can describe very well.

1.2 Category Theory and Cognitive Processes

Let's start with an example to illustrate the abstract concepts that are going to be handled here. Consider the situation depicted on figure 1. It shows two invertebrate animals involved in some kind of relationship. They have sensors in their legs and heads, that enable them to identify situations of danger or the approach of a prey, for instance. They also have effectors, like legs, wings, tweezers and stings, that can be used for defense or attack, respectively.

Consider one of these animals, the scorpion, for instance. The sensors in its legs inform the approach of a victim, by sending to its nervous system a pattern of stimuli that it can interpret and use as a guide for approaching the target and to decide when to proceed to the attack. All this sequence of events arises from a succession of states, each one describing relationships among the objects and their parts involved. We could call them "knowledge elementary states". So, a knowledge elementary

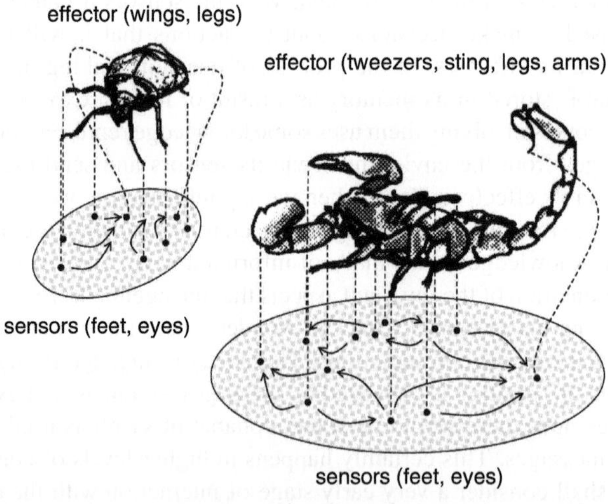

Fig. 1 Functional associations presented as categories. The collection of functional relationships between pairs of organs (sensors and effectors) creates a network of associations, that determines how the inputs gathered by the sensors result in the animal behavior. Each node represents an organ, and each arrow represents a functional relation between two nodes

state is then an instantaneous portrait of the relationships between objects and parts involved in describing a certain scenario. In figure 1, we show under each animal a pictorial representation of one of these knowledge elementary states. They are collections of small circles and arrows between them, forming some kind of graph. For each sensor or effector there is a corresponding small circle, or a node of the graphs. The arrows indicates that there are some kind of relationship between each pair of nodes. This model, although simple, is illustrative to help understanding the idea of a graph representing the relationships among cognitive components. It was inspired on a model of how an actual scorpion can determine the spatial location of a prey, by analyzing with a small group of neurons, the stimuli produced by vibrations detected with sensors in its legs [32].

Each of the so-called "knowledge elementary states" represents one instant in a chain of events. There will be mappings from each of these states to the next, forming a chain of mappings that constitutes the temporal evolution of the knowledge state. This is an evolutionary process that describes the progressive changes of the knowledge acquired by the animal. It translates into an interpretation of the incoming patterns of stimulus, but it emerges from the relationships between each pair of interacting elements in a network of sensing and processing elements. These elements maps sources of information, which we shall call *objects*. The objects are related by *links* which in fact are mappings that describe their interactions or functional dependencies. This model shall include some mechanism that enables generalization and induction. With such mechanism, it will provide means for producing complex concepts and symbols derived from simpler ones. These requirements are fulfilled by the mathematical concept of *category*.

A category is a mathematical construct composed by objects and the relations among them. The category theory was introduced by Eilenberg and Mac Lane in the early 1940s for studying problems of algebraic topology [7, 24]. Formally , a category is a collection of *objects* and *morphisms* between each pair of objects. If we have two objects A and B from a certain category K, a morphism f between them is denoted by $f : A \rightarrow B$. This definition reminds a function mapping a set A to a set B. However, a function is just a particular case of morphism, when the objects that compose the category are sets. In fact a category can consider collections of arbitrary types of objects, provided they are related by morphisms that satisfies the following rules:

1. Composition: if f and g are morphisms that relates objects A to B and B to C, respectively, then the composition $f \circ g$ relates A to C. For simplicity, usually $f \circ g$ is simply denoted by fg.
2. Associativity: If (f,g,h) are morphisms composable as $f(gh)$, then this composition is equal to $(fg)h$ and fgh.
3. Identity: If A is an object, then the morphism $i : A \rightarrow A$ must be included in the category.

The figure 2 illustrates the first two rules. This figure is called a diagram. The identity morphisms are usually omitted in this kind of diagrams. Each object is called a *node* of the diagram and each morphism is an *arrow* or *link* of the diagram. In this

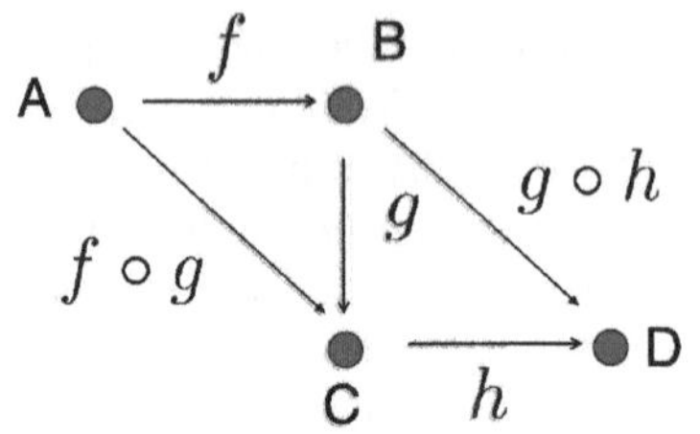

Fig. 2 Objects and morphisms. The morphisms, indicated by the arrows, are identified by the letters f,g and h. In a category, the compositions $f \circ g$ and $g \circ h$ are also included, and they obey the associative law, that is the same to say that the diagram commutes

figure we have the objects A,B,C and D, and the morphisms f, g and h, and their compositions. Again, the identity morphisms are implicit. The alternative paths formed by the compositions shown in the diagram displays graphically the associative property. We then say that the diagram commutes.

Given that categories are used to model the knowledge that some agent obtained from the world, it is expected that they reflect the outside organization up to some extent, accessible to the agent. Given that the world can be described by events involving its objects, temporally and causally organized, taking place in the space, a certain order, at least partial, must be reflected in these categories. Technically, we say that the categories are *indexed* by a set of indexes, which ordering reflects the category organization. By the terminology used in [7], the indexed categories are said *coherent*. Besides the indexing, there is another complementary way of characterizing the organization of the category. This can be done by using a directed graph, that represents the objects as its nodes and the morphisms, as its links. A graph differs from a category in the sense that it does not include links describing the compositions comprehended by concatenating adjacent links. Also, the identity morphisms are not usually included in the graph. However, there are some technical expedients that allows the use of graphs for a correct representation of categories. we shall not go into these details here, and the reader should refer to the bibliography; for more information see [7, 24]. The graph equipped with all these necessary technicalities is called the *underlying graph* of a category and here we shall call it the graph of a category, for simplicity.

Categories have a high representational power. The objects of a category can be quite abstract entities, as sets, spaces, algebraic structures and even complete categories also. A category made of categories is called a *large category*. In this case, the morphisms between objects will be maps between categories. They have a special structure and are called *functors*. A functor F between two categories K and L is a map $F : K \to L$ that maps objects of K in objects of L and also maps morphisms of K in morphisms of L. Further we shall discuss about the time evolution of a category, to describe the evolution of the knowledge state of an agent. At each time instant the agent's knowledge state is described by a category. From one time instant to the other, there will be a functor mapping the corresponding categories. It seems simple to figure out this scenario. However, the power of this model is that from

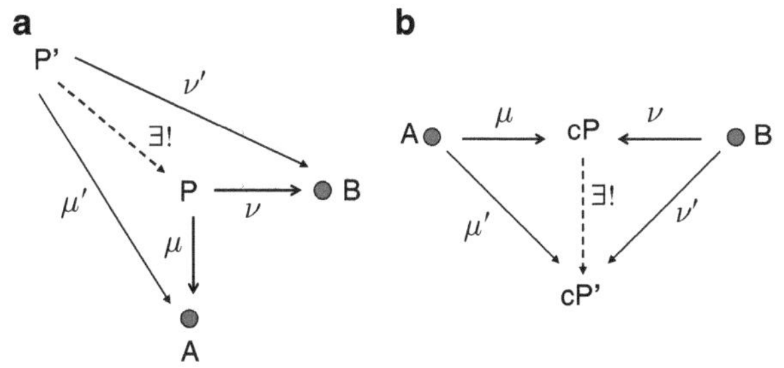

Fig. 3 Product and coproduct. (**a**) The product P is defined by the mappings μ and ν. If another product P' would be defined by other mappings, then the two products would be the same, up to an isomorphism between them. (**b**) The coproduct cP is obtained in a similar way, by just inverting the arrows

its apparent simplicity, one can derive the functionalities that models the cognitive process. The mechanism underlying it is called the *complexification*, which will be treated on section 2. Now we shall set the basic concepts behind this scenario.

The graph of a category displays the relations among its objects by means of its links. In the traditional sense given by set theory, a relation between sets A a B is defined as a sub-set of a Cartesian product $A \times B$, the set of all ordered pairs $\{(a, b), a \in A\ ,\ b \in B\}$. As pointed before, the objects of a category can be very general, can even be other categories. So, in this case the concept of relation as a sub-set of a Cartesian product defined in this way is not adequate, because it assumes that the objects in question are sets. We need then a new definition of Cartesian product. In fact, in category theory we have a wider definition of product between objects, that does not require to look them as sets. The product is defined by the diagram depicted on figure 3a. The advantage of this definition is that many products can be defined between distinct objects. However, given two particular objects, any product defined between them will be equivalent to the other, up to an isomorphism. This is called an *universal property*. One can also define a dual concept of the product, called the *coproduct* (and also called *sum*). It can be obtained from de definition of a product by just inverting the arrows. The product is a concept that enables to create association rules between objects, as the relations, the coproduct enables to create selection rules, or projections. With a coproduct one can select an object from an association of objects.

Assume now that a set of indexes $\mathcal{I} = \{i\}$ can index a category $\mathcal{K}$ with objects A_i, reflecting the internal order in $\mathcal{K}$. In other words, there is a suitable functor $F : \mathcal{I} \rightarrow \mathcal{K}$ that provides this indexing. We define a product A of A_i and A_j and call it an object limit of $\mathcal{K}$ along with morphisms $f_i : A \rightarrow A_i$ if the diagram of figure 4a commutes, given that $m : i \rightarrow j$ is a morphism in $\mathcal{I}$. Analogously, one can define its dual concept, the colimit cA, by just inverting the arrows, as shown by the diagram in figure 4b.

Despite the easiness of definition of colimit by simply flipping the arrows, the colimits have a peculiar behavior that brings important properties. The importance

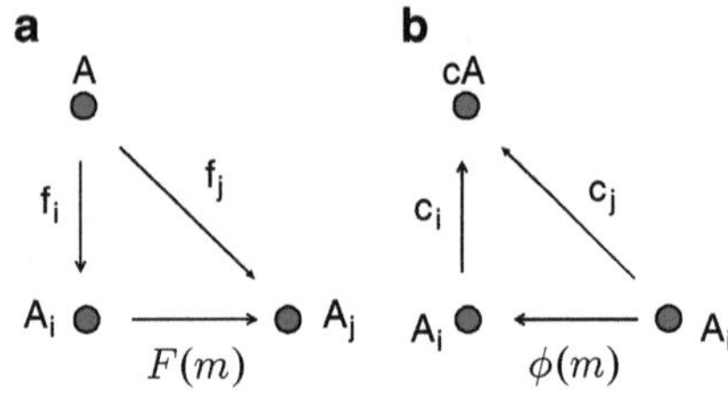

Fig. 4 Limit and colimit. (**a**) The product A defined by the mappings f_i and f_j. (**b**) The colimit cA obtained by inverting the arrows

of limits and colimits derives from their nature as products and coproducts respectively, jointly with the properties inherited from the indexing sets through their respective associated indexing functors. The colimit enables one to select a representative from the class of objects that points to it. The limit, on its own way, enables to state the equivalence among the objects that share the same limit. Together, they form the basis of generalization in our model for knowledge representation. The limit, if exists, enables to group concepts under a certain criterion of similarity, equivalence or generality, that is related to the structure of relationship among the corresponding objects. The colimit, if exists, enables one to select one of these related concepts as a representative of the whole class, as a symbol of generalization. Now we shall use all these basic concepts to show how a cognitive architecture can be built and further, how it can be mapped on a particular case, a neural architecture.

2 The Memory Evolutive System

When using categories to describe complex evolutive systems one can identify the parts of the system as objects and the relations between them as morphisms. However to represent a system that evolves in time, these categories can only represent states of the system (called *state categories*) at a specified time. The evolution of the system is achieved by using partial or total functors that transforms each discrete state category into the next, adding, removing or altering nodes and links.

The system can also be represented hierarchically, with nodes divided onto levels that we shall index by $l \in \mathcal{N}$, so that the complex objects of a given level $l = k + 1$ (the so-called colimits) are composed of pattern of objects belonging to levels $l \leq k$. In a hierarchical system, if two or more patterns are allowed to have the same colimit, this is called multifold object, enabling the emergence of complex links that are lost on a reductionist break-up of the system (see figure 5). In this case, the transformations can also bind patterns to colimits or limits.

According to the multiplicity principle [6, 7], if a system supports multifold objects, then all future transformations on its state category will also support multifold objects. In fact by successively binding patterns to colimits and limits the category representing the system increases in complexity, by gaining more hierarchical levels. This sequential transformation process is called *complexification*.

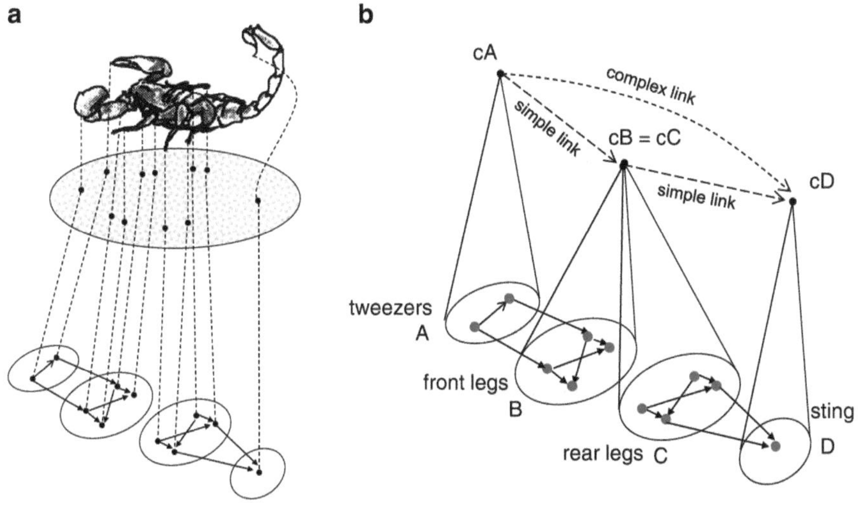

Fig. 5 Complex link formation. In (**a**) it is shown a sample mapping of the relation between groups of active neurons by clusters of links. In (**b**), this same groups, identified by capital letters are associated to higher level colimits, and the groups B and C have the same colimit. The composition rule for categories implies the formation of a complex link from cA to cD, representing a complex, emergent relation between the tweezers and the sting

The MES [7] can be defined as a hierarchical evolutive system over a continuous time scale, with a hierarchical evolutive sub-system named Memory, with the same time scale, that develops over time by the process of complexification. This Memory is subdivided in three:

- Empirical - where nodes represent Records of previous experiences;
- Procedural - with nodes representing Procedures bound to known actions options or strategies;
- Semantic - a higher level memory containing nodes named Concepts, representing classes of equivalences between records or procedures;

The MES is connected to receptors and actuators that translates information from the environment in form of nodes and links. The patterns of active nodes and links in the receptors are bound to Record colimits in the empirical part of the Memory and active patterns in the actuators are bound to Procedures limits in the procedural memory. Concepts are created later by the high level regulative system of the MES.

The MES has no central regulating mechanism, instead it has a hierarchical network of partial regulatory organs, called co-regulators (CRs) that are sub-systems with specific functions and their own time scales, serving to collect information, select and implement responses, and evaluate the result of these actions, at least locally. They have differential access to the Memory as they are also organized hierarchically. The CRs operate in cycles performing successive actions:

(a). Observation - formation of an internal representation of the observable environment, called landscape, with partial information received at that date;
(b). Regulation - selection of objectives and admissible procedures to implement them, sending commands to the effectors and forming an anticipated landscape;
(c). Control - at the end of each step, it evaluates the final results and takes part to their storage in the memory at the beginning of the next step;

The procedures chosen by each CR are summed to create one unified global procedure. However, the various CRs can have conflicting procedures, so the generated global procedure may cause fractures in their previsions. A CR in a fracture state can be inactive until a higher order CR, with longer cycle, have a priority when resolving conflicting procedures. If no high level CRs are able to resolve the conflicts in low level CRs the system can lose functionality as certain CRs will be disabled. If this happens too often, the system can be lead to effectively die, allowing the representation of the aging process.

The MES model can be used to represent and study any kind of complex evolutive system and can be applied in many disciplines such as biology, sociology, economy and cognitive sciences. This article focuses on the specific implementation of the model for neural cognitive systems.

2.1 A MES for Cognitive Systems

One possible application of the MES is the study of the neural system, using the Memory Evolutive Neural System (MENS), an extension of the MES imbued with special properties to deal with networks of neurons and category neurons (cat-neurons).

The basis for modeling the neural system lies in neurons and the synapses between them, forming an intricate graph that indicates the state of the system by the activity (instantaneous firing) of the neurons and the strengths of the synapses. Some models also include a propagation delay for the synapses.

In order to recognize the various feature patterns in the environment and better adapt to them, neurons firing jointly can form assemblies that are patterns to represent complex mental objects. In some cases there is a coordination neuron which binds the assembly [5] and is activated synchronously with the assembly. Frequently a neuron like this can't be observed in the brain. In such cases, learning takes place only by reinforcing the strengths of the synapses in the assembly, according to Hebb's rule [13] or some variant. Experimental studies [5] indicate that many assemblies can lead to the same outputs, and the same item or process can activate several more or less similar assemblies depending on the context.

In the Memory Evolutive Neural System (MENS), a conceptual object, called category-neuron (cat-neuron), is introduced to model the class of these assemblies activated by the same item. This multifold dynamic object can be viewed as a "higher order" virtual neuron, or a "mental object", but it is activated by a physical

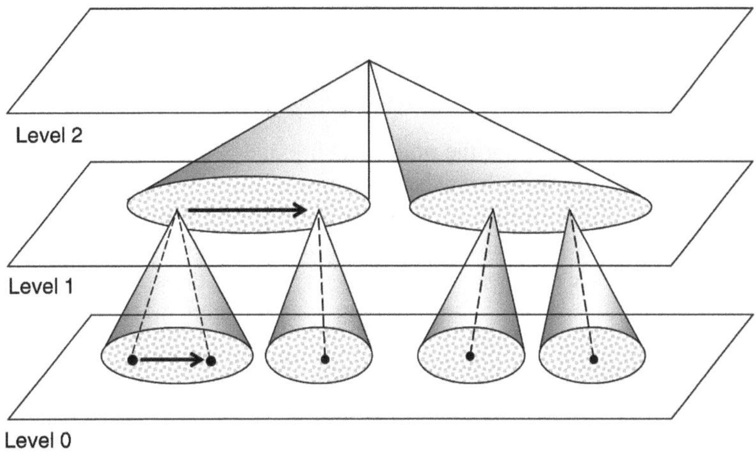

Fig. 6 Cat-neurons. In this picture, level 0 nodes are real neurons, when they form assemblies, they receive a level 1 cat-neuron representative; level 1 cat-neurons can also form assemblies, in this case they receive a level 2 cat-neuron representative, the same process is used for higher levels

event, namely the activation of any of the neural assemblies it represents (possibly non-connected). This process can be iterated obtaining cat-neurons for patterns of cat-neurons, representing synchronous hyper-assemblies.

The **MES** is then defined as a memory evolutive system over the lifetime of the animal, which has for the first hierarchical level the evolutive sub-system **Neur**, generated by the graph of neurons, synapses and synaptic paths, where the synapses are labeled links containing the synaptic strength and the propagation delays. The higher hierarchical levels are obtained by a category of cat-neurons and the links between them (see figure 6).

When dealing with cluster of links between assemblies, the propagation delay of the cluster is defined as the maximum of the propagation delays of the links in the cluster and strength of the cluster as an increasing function of the strengths of its links. The strength of complex links is calculated, like any composite path, by adding propagation delays and multiplying the strengths of simple links composing it.

2.2 Requirements for Implementing the MES

The MES is a mathematical model of an open self-organizing system. So far it has never been fully implemented in a computer. The model does not specify many aspects, that have to be selected according to the context of implementation.

The main aspect that needs to be specified is the network underlying the MES, that remains compatible with category theory and at the same time supports the temporal and stability notions needed for correctly implementing the model. According to [7], this network must satisfy the properties of identity and composition, and the links must have strength and propagation delays. When dealing with the MENS, another important requisite for this network is to have a rule similar to Hebb's to change the link strengths and possibly even the propagation delays according to the temporal order of activation of the nodes, having nodes that fire in a way to contribute to a second node firing have the link between them increased, and otherwise decreased.

Hence it is necessary to choose a network model coherent with the dynamical aspects required for the MES operation. Here we evaluate the use of some different neural network models as candidates to a computational implementation of the MES model.

3 The MES and Artificial Neural Networks

According to [25], artificial neural network research evolved into three generations:

- **first generation** networks studied the McCulloch-Pitts neuron, or perceptrons, using digital inputs and outputs, usually binary. Multi-layer perceptrons with a single hidden layer were able to compute any binary function;
- **second generation** networks had neurons with activation functions, usually some logistic function like the sigmoid, with continuous inputs and outputs, having probabilistic weights in the synapses, and learning algorithms based on gradient descent. This kind of network was able to compute boolean functions (after a threshold) with fewer neurons and was able to approximate any continuous analog function.
- **third generation** networks are based on spiking or pulsating neurons, that incorporate recent neurobiology discoveries, modeling the neuron as a dynamic electrical system with bifurcation in its equilibrium, usually following some variant of the Hodgkin-Huxley model. This model allows the correct representation of time, enabling neurons to pass information by temporal spike coding, also being able to approximate continuous functions, using temporal coded inputs and outputs, requiring fewer neurons in some cases [11, 26], and usually with less network iterations.

Networks composed of spiking neurons can transmit larger ammounts of data through the relative timing of only a few spikes by having groups neurons activating each other sequentially in a short timespan. These groups, or assemblies, represent the information coded from input time patterns, and can lead to the activation of groups representing output patterns.

Therefore, second generation networks are not sufficient as an implementation base for the MES, specifically because they do not support the notions of propagation time for synapses and refractory time for neuron just activated. This two temporal properties are required to express stability times for patterns, necessary to the CR operation and the iterative complexification of the model.

There are many variations for the neuron model in the pulsed networks [11, 17, 25], from the computationally inexpensive "integrate-and-fire" to the costly Hodgkin-Huxley model. One of these models, proposed by Izhikevich [16] seems to be well suited for the task due its balance between computational cost and mathematical accuracy, allowing interesting experiments using thousands of neurons, while enabling more than 20 classes of biologically plausible neuron profiles, by just changing the four parameters in a quadratic differential equation.

3.1 *Izhikevich's Neuron Model*

It is widely accepted that the Hodgkin-Huxley class of equations (or model) that describe the behavior of the giant squid axon, stands as the most successful quantitative computational model in the neural science [1]. But it is also known that such model has a high computational cost, particularly for large networks. Another issue related with Hodgkin-Huxley model is that it cannot describe the functional behavior of all types of neurons found in nervous systems. There are hundreds of morphologically different neuron cells only in mammalians. Along with these morphological features, neurons have physiological specializations. The cellular diversity undoubtedly underlies the capacity of the system of forming complicated networks to mediate sophisticated behaviors [31].

Recently, Izhikevich [16, 18, 20] have presented a model of a single neuron that may represent many biophysically accurate Hodgkin-Huxley-type neural models. By treating neurons as dynamical systems, the model considers that the resting state of neurons corresponds to a stable equilibrium. Neurons are excitable because the equilibrium is near a bifurcation, and despite the existence of many ionic mechanisms of spike generation there are only four generic bifurcations of equilibrium. By analyzing the phase portrait at neuron bifurcations the model can explain why neurons have many different behaviors like well-defined threshold, all-or-none spikes, hysteresis, and frequency preferences among others. As pointed in [18] these features determine the kind of computation a neuron do, not the overall ionic current per se.

A concise explanation of the model may be found in [16] and the full explanation of how the model was achieved may be found in [18]. As pointed in [20], bifurcation methodologies had enabled the reduction of neuron models to a two-dimensional (2-D) system of ordinary differential equations of the form:

$$\mathbf{v}' = 0.04\mathbf{v}^2 + 5\mathbf{v} + 140 - \mathbf{u} + \mathbf{I} \tag{1}$$

$$\mathbf{u}' = \mathbf{a}(\mathbf{bv} - \mathbf{u}) \tag{2}$$

with the additional computational artifact of the after-spike resetting:

$$if\, \mathbf{v} \geq 30mV\; then \begin{cases} \mathbf{v} \leftarrow \mathbf{c} \\ \mathbf{u} \leftarrow \mathbf{u} + \mathbf{d} \end{cases} \tag{3}$$

- **v** represents the membrane potential (scaled to millivolts);
- **u** represents the membrane recovery variable (a negative feedback to **v**);
- **I** is a variable that represents synaptic currents or injected dc-currents.

The parameters **a**, **b**, **c**, and **d** represent:

- **a** describes the time scale of the recovery variable **u**. Smaller values result in slower recovery (typically $\mathbf{a} = 0.02$);
- **b** describes the sensitivity of the recovery variable **u** to the sub-threshold fluctuations of the membrane potential **v** (typically $\mathbf{b} = 0.2$);
- **c** describes the after-spike reset value of the membrane potential **v** (typically $\mathbf{c} = -65mV$);
- **d** describes the after-spike reset of the recovery variable **u** (typically $\mathbf{d} = 2$).

Also in the above equations, $\mathbf{v}'$ and $\mathbf{u}'$ denote the derivatives, where t is the time, given in milliseconds, to correspond to the other parameters.

The individual neuron model reproduces a wide range of neuronal biological behaviors such as spiking, bursting, and mixed mode firing patterns, continuous spiking with frequency adaptation, spike threshold variability, bi-stability of resting and spiking, sub-threshold oscillation and resonance etc. But when trying to model open self-organizing systems by using this neuron model other factors must be taken into account.

More precisely, neurons must be connected one another and, by taken the biology as a model, it requires a synaptic model. It most also be taken into account that in biological world neurons generates action potential (spikes) in time, and also, it takes time to such spikes to propagate from one site to another. Furthermore, when operating together, it is expected from groups of neurons the emergence of synchronous operation, also called coalition.

The mentioned model works with a resolution of one millisecond, so time control is incorporated to the model due the nature of the spiking network to which it is connected to. Regarding to delay propagation, it is suggested in [18] that a possible extension of the model is to treat **u, a,** and **b** as vectors, and use $\Sigma\mathbf{u}$ instead of **u** in the voltage equation (1). Such procedure will account for slow conductance in multiple time scale, although the model's author warns that it would be unnecessary for networks that simulate cortices.

When coupled, spiking neurons may present patterns that resemble collective behavior as well as Poissonian patterns of firing as shown in [20].

3.2 Synaptic Dynamics

There are two types of synapses in the nervous systems: electrical and chemical. In electrical synapses the pre-synaptic and the post-synaptic terminals are not completely separated, so it behaves most like a short-circuit for spikes. By the other side, chemical synapses have a gap junction and a chemical transmitter is responsible for the continuity of propagation of the spike from pre-synaptic to post-synaptic terminals. The former presents virtually no delay on spike propagation, while the last presents typical delays of 1 to 5 ms [22].

Perhaps the most important characteristic of chemical synapses is that they can chance the strength of their connection. Recent researches have shown that the strength of the connection between two chemical synapses can be modified by activity, revealed by a direct dependence on the timing of neuronal firing on either side of the synapse [10, 19, 21, 27, 33].

The so-called STDP (spike timing-dependent plasticity) is a powerful computational characteristic of neurons because, according to the temporal delay between pre and post-synaptic spiking activity, a connection between neurons can be strengthened (when the pre-synaptic spike precedes the post-synaptic one) or weakened (when the post-synaptic spike happens before the pre-synaptic one). Therefore, the temporal order in the precise millisecond-scale is a mechanism that provides biological neural networks with a learning system. This is the equivalent of the Hebb's rule for pulsed networks.

In [19] a STDP model of synaptic plasticity is presented in a neural network implemented with the model described above. According the authors, the dynamics of passive change of the synaptic weight c_{ij} from neuron j to neuron i is described by the second-order linear equation:

$$c''_{ij} = -(c'_{ij} - a)10^4 \tag{4}$$

where a describes slow, activity-independent increase of synaptic weight. Such implementation is a particular one. Others forms of implementation of STDP rules may be implemented. What the example shows is that its implementation may represent another equation that probably must be calculated to each pair of synapses present into the network.

As presented in [20], by combining the Izhikevich neurons, the STDP plasticity rule and synaptic propagation delays, neuronal assemblies emerge in the form of time-locked (or polychronous) groups, neurons that does not file synchronously, but fire in an orderly, repeating time pattern, that changes according to the network topology and synaptic strengths.

A key feature that allows the emergence of neuronal groups in the existence of nonzero propagation times in the synapses, meaning that once a neuron fire, a post-synaptic neuron will only receive the potential change after some specified propagation time. Another important requisite for having polychronous groups is the existence of excitatory and inhibitory neurons, with different firing equations, in a way that once a certain amount of inhibitory neurons are activated, almost every neuron is inhibited, causing the network to have a rhythmic behavior.

The polychronous groups can be understood as assemblies of neurons, that fire in the same temporal pattern repeatedly. Groups are dynamic entities. As soon as the synaptic weights begin to change, some groups disappear and new ones emerge. If the same inputs are presented over and over, it is observed that the number of groups attain an equilibrium [20], as the network aligns to the given inputs. In this sense, groups can be understood as memories in the network, as the network learns to associate this groups to sets of inputs and coordinate them with related outputs.

This kind of neural assemblies shaped as polychronous groups are very useful when modeling the MES, because they form the neuronal building blocks submitted as inputs and observed as outputs, to span a vast array of limits and colimits as higher level cat-neurons.

So far, polychronous network has been simulated to study network dynamics [16], computational capacity [28] and has been used for reservoir computing [30]. However, embedding this kind of network in an agent in a simulated virtual world had not yet been attempted.

4 Towards an Implementation of the MES

When implementing the MES it is important to understand how each of its structural elements works (see section 2). The **Receptors** and **Effectors** are the elements of contact with the environment, they are implemented as single layered networks that transfer information by means of active patterns, directly connected to the **Memory**, a highly structured hierarchically layered network that changes with time by the use of *transition functors* and serves as a medium of communication for all the **Co-regulators** (CRs) immersed in any of its levels. The CRs act as distributed regulatory organs operating in cycles, receiving partial information from the Memory and from lower level CRs, by means of the *difference functor* they build a landscape, a sub-network containing only the nodes each one of them is perceiving as active at each step. Then, based on the constructed landscape, usually associated to a Record in the Memory, they select a Procedure, by looking at the admissible Procedures that the active Record in the Memory connects to. Applying the selected procedure to the current landscape, the CR obtain an anticipated landscape, that will be useful at the beginning of the next cycle, when each CR check to see if the previously selected Procedure was indeed executed, making use of the comparison functor from the anticipated landscape to the newly formed landscape. In case of success, the CR will enforce the selected procedure, by strengthening the link between the selected Record and Procedure. If there was a mistakes executing the Procedure, the CR may create a new Procedure accommodating the unexpected elements and weakening links to expected elements that did not show.

This structural elements and its dynamics is strictly defined by the mathematical model detailed in [7], and the evolution of this model is also predicted by the authors, when coupling to an agent, as leading to cognitive behaviour, such as the emergence of Concepts as high level nodes in the network created by high level CRs when they find classes of equivalence between patterns they obtain from lower level CRs.

4.1 Implementation Model

The first challenge in implementing the MES is converting the detailed mathematical model into a computer algorithm, without losing any significant mechanics. To this intent, it is important to understand how to map the functors used by the MES into network operations, to build and update its state categories. When using a pulsed network, as Izhikevich's, as a base, every propagation in the network is mapped to a functor:

- *transition functor*, corresponding to activating the pushout (sum) of all procedures resulting from the interplay among the procedures chosen by each CR, plus the patterns activated by the inputs provided by the environment;
- *difference functor*, used to build the landscape for each CR, taking the difference between the category of all active patterns and the one that represents what the CR can perceive;
- *comparison functor*, also used internally for each CR, during its evaluation phase, maps the category given by the previous anticipated landscape to the current landscape, identifying what went wrong in the execution of the previous selected procedure;

The implementation of the transition functor can be accomplished just by selecting a good pulsed neural network model, containing a delayed propagation scheme, such as Izhikevich's, and by connecting inputs and outputs to it. For the pulsed network simulation we have chosen to use a simulator called Brian [12], that allows for a fast and reliable implementation of the Izhikevich network with STDP.

The two other functors had to be implemented by hand, using the following algorithm:

1. For each new pattern unmapped by any CR, create new level 1 CR
2. CR Cycle, comprising of 4 substeps:
    a. Build the landscape, by assigning a Record node in the memory to the active pattern, if more than one Record is active, create a level +1 Record, if already at level+1, create a new level+1 CR to handle that new Record;
    b. Choose a Procedure, by following the activator links from the current Record to next Procedures, if none, add one to recreate the current pattern, if more than one create a level +1 Procedure, if already at level+1, create a new level+1 CR to handle that new Procedure;
    c. Evaluation, in case the last cycle had selected a Procedure, if the pattern activated by that Procedure is indeed active, strengthen the link between the previous Record and Procedure, otherwise create a new activator link between the previous Record and the Procedure that originated the current pattern, weakening the activator link that selected the faulty Procedure;

d. Handle Fracture, in case any of the previous steps did not finish before the end of cycle duration, trying to alter the CR's time constants to cope with the new environment stability;

3. Assemble the Operating Procedure from the Procedures selected by every CR and inputs from the environment;

4.2 Preliminary Results

To test the algorithm and the MES implementation we developed a multiagent simulation to allow observing the emergence of interesting behaviour. The agent roams a virtual environment in search of food, receiving stimuli about its own energy level and from the environment (the nearest food direction), and being able to choose the direction to turn. Without any kind of rule to indicate that the agent should feed, the use of Izikevitch network, STDP plasticity and the initial conditions will influence the behavior of the agent and some food searching strategy will emerge [29]. We still could not finish implementing the MES algorithm to compare with the plain Izhikevich network results, but as soon as it is ready this kind of comparison should be very interesting to see.

The multiagent simulation was designed in the Breve 3D simulator [23], that accepts scripting in the Python language, being compatible with the Brian pulsed neural network simulator, used for the Izhikevich network part.

The simulated agent consists of a blue tetrahedron that starts in the center of the field and moves about with constant speed in a flat 20x20 field containing 50 randomly placed small green spheres that represent food. The agent also has an energy level, that starts at 1.0 and gradually decreases towards 0.0. Nothing special happens if the energy reaches zero. As soon as the agent touches some food, its energy level is increased and the food is replaced with another one placed at random on the field. The agent is blocked from leaving the field by invisible walls.

The agent has two sensors, the first, called directional sensor, indicates the angle towards the nearest food object, with a 30 degrees arc and a radius of 10 units of distance. The second sensor is called energy sensor and measures the current energy level of the agent. A challenging aspect of this kind of simulation is converting the inputs from the environment to neural activation patterns, and convert neural patterns into output values to command the effectors. Our approach, not detailed here, was based on the way the desert scorpion finds its prey, using 8 neurons located in the tips of its legs, that measures oscillations from waves in the sand [32]. Using only 8 excitatory and 8 inhibitory neurons, in a way that each 3 inhibit the opposite one, it is possible to convert an incoming wave into a firing pattern. In our case, with a simple adaptation of the same algorithm, we convert numbers representing angle or energy levels to a linear pattern of activation of neurons representing inputs and convert timed activations into numbers in the reverse way.

To act in the world, the agent may choose the angle to turn using its directional actuator, that received a value from 0.0 to 1.0 indicating the turning direction, in a 30 degrees arc. Each of the sensors and actuators comprise a defined set of neurons, usually 5 for each that can be activated in patterns depending on the inputs or the network activity.

While comparing to an agent that moves entirely at random, the agent using the Izhikevich and STDP network with propagation delays and polychronous groups fares much better on the average, specially when a high energy condition is imposed for a long time at the beginning of the simulation allowing the groups to form around, learning that input patterns. This is already significant, but since we still lack results for the MES algorithm to compare, the results are not sufficient to confirm the mechanics of the model embedded inside a virtual agent.

5 Final Remarks

The choice of the Izhikevich's model for implementing the MES is founded in three issues:

- the dynamic properties of the Izhikevichs model are very suitable to match temporal requirements implied in the MES, because they arouse stability spans of correlations among neural assemblies (or sub-networks, in the Healy terms) thanks to the polychronization feature;
- the existence of a rich dynamics with possible choice of diverse firing patterns that can cope with several different mappings, namely functors, between two time instances of a evolving category;
- a satisfactory trading between node complexity and topological connectivity among nodes.

This third issue refers to the number of connections that one neuron should have with the others in an assembly in order to allow sufficient capacity to the network to express a category. We are currently considering formal characterizations of such relations between complexity or expressiveness of a node (neuron) and the number of connections that it should make to form its neighborhood. A difficult aspect found in this formulation is that the dynamical aspects shall be considered when concerning to an evolutive model like the MES.

Although Izhikevich network model alone can display notable capacity for expressing cognitive systems, only by having STDP as synaptic plasticity model is not enough to reorganize the network properly for high level complex systems, perhaps by using the proposed MES algorithm coupled to this kind of network will allow for such complexity increase.

References

1. Bargmann, C.I.: Neurobiology of the Caenorhabditis elegans Genome. Science **282** pp. 2020–2033. (1998)
2. Chalfie, M., Sulston, J.E., White, J.C., Southgate, E. , Thomson , J.N. , Brenner, S.: The neural circuit for touch sensitivity in Caenorhabditis elegans. Journal of Neuroscience, **5**, pp. 959–964. (1985)
3. Koch, C., Laurent, G.: Complexity and the Nervous System. Science Vol 284 - 2 April 1999 pp. 96–98. (1999)
4. Dayan, P., Abbott, L. F.: Theoretical Neuroscience: Computational and Mathematical Modeling of Neural Systems. The MIT Press - Cambridge, MA-USA; London, England. (2005)
5. Edelman, G.M.: The Remembered Present, Basic Books, New York. (1989)
6. Ehresmann, A., Vanbremeersch, J.P.: Multiplicity Principle and emergence in the Memory Evolutive System, Journal of Systems Analysis, Modelling, Simulation **26**, pp.81–117. (1996)
7. Ehresmann, A., Vanbremeersch, J.P.: Memory Evolutive Systems - Hierarchy, Emergence, Cognition, Elsevier, Amsterdam. (2007)
8. Fingelkurts, A.A.: Mapping of Brain Operational Architectonics. In Chen,F.J. (ed.) Focus on Brain Mapping Research, pp. 59–98. Nova Science Publishers, Inc. (2006)
9. Freeman, W.J., Kozma, R., Werbos, P.J.: Biocomplexity: Adaptive behavior in complex stochastic dynamical systems. Biosystems, **59**, 109–123. (2001)
10. Froemke, R.C., Dan, Y.: Spike-timing-dependent synaptic modification induced by natural spike trains. Nature **416**, pp.433–438. (2002)
11. Gerstner,W., Kistler,W.: Spiking neuron models. Cambridge Univ. Press, Cambridge, England. (2002)
12. Goodman, D., Brette, R. 2008. Brian Neural Network Simulator. http://brian.di.ens.fr/ (2008)
13. Hebb, D.O.: The Organization of Behaviour, Wiley, New York. (1949)
14. Healy, M.J.: Colimits in memory: category theory and neural systems. In Proceedings of the International Joint Conference on Neural Networks, IJCNN '99 - Volume 1, pp. 492–496. (1999)
15. Healy, M.J., Caudell, T.P., Yunhai, X.: From categorical semantics to neural network design. In Proceedings of the International Joint Conference on Neural Networks, IJCNN'03 - Volume 3, pp.1981– 1986. (2003)
16. Izhikevich, E. M.: Simple Model of Spiking Neurons. IEEE Transactions on Neural Networks, **14** pp.1569–1572. (2003)
17. Izhikevich, E. M.: Which Model to Use for Cortical Spiking Neurons? IEEE Transactions on Neural Networks, **15**:1063–1070. (2004)
18. Izhikevich, E. M.: Dynamical Systems in Neuroscience: The geometry of Excitability and Bursting. The MIT Press - Cambridge, MA-USA; London, England. (2007)
19. Izhikevich, E. M., Desai, N. S.: Relating STDP to BCM. Neural Computation **15** pp.1511–1523. (2003)
20. Izhikevich, E. M., Gally, J. A., Edelman, G. M.: Spike-timing Dynamics of Neuronal Groups. Cerebral Cortex, **14**(8), pp. 933–944. (2004)
21. Jacob, V., Brasier, D. J. , Erchova, I. , Feldman, D. , Shulz, D. E.: Spike Timing-Dependent Synaptic Depression in the In Vivo Barrel Cortex of the Rat. The Journal of Neuroscience, **27**(6) pp.1271–1284. (2007)
22. Kandel, E. R., Schwartz, J. H., Jessel, T. M.: Principles of Neural Science 4th edition. McGraw-Hill - New York. (2000)
23. Klein, J.: breve: a 3D simulation environment for the simulation of decentralized systems and artificial life. Proceedings of Artificial Life VIII, the 8th International Conference on the Simulation and Synthesis of Living Systems. The MIT Press. (2002)
24. Mac Lane, S.: Categories for the Working Mathematician. Springer, Berlin. (1971)
25. Maass, W.: Networks of Spiking Neurons: The Third Generation of Neural Network Models. Neural Networks, **10**(9):1659–1671. (1997)

26. Maass, W.: Computing with spiking neurons. In W. Maass and C. M. Bishop (eds.), Pulsed Neural Networks. MIT Press, Cambridge, Mass. (1999)
27. Markram, H., Lubke, L., Frotscher, M., Sakmann M.: Regulation of Synaptic Efficacy by Coincidence of Postsynaptic APs and EPSPs. Science **275**, pp.213–215. (1997)
28. Maier, W., Miller, B.: A minimal model for the study of polychronous groups. arXiv:0806.1070v1 [cond-mat.dis-nn]. (2008)
29. Monteiro, J. L. R., Caillou, P., Netto, M. L.: An Agent Model Using Polychronous Networks (Extended Abstract). Proc. of 8th Int. Conf. on Autonomous Agents and Multiagent Systems (AAMAS 2009), Decker, Sichman, Sierra, and Castelfranchi (eds.), To appear in May, 10–15. Budapest, Hungary. (2009)
30. Paugam-Moisy, H., Martinez, R., Bengio, S.: Delay learning and polychronization for reservoir computing. Neurocomputing **71**, pp.1143–1158. (2008)
31. Purves, D. et al : Neuroscience / edited by Dale Purves [et al.] - 3rd ed. Sinauer Associates, Inc. Publishers. Sunderland, Massachusetts, U.S.A. (2003)
32. Sturzl, W., Kempter, R., van Hemmen, J.L.: Theory of arachnoid prey localization. Physical Review Letters **84**, 24, pp.5668–5671. (2000)
33. Turrigiano, G.G., Nelson S.B.: Homeostatic plasticity in the developing nervous system. Nature - Neuroscience, VOL 5 - February 2004 pp. 97–107. (2004)
34. White, J. G., Southgate, E., Thompson, J. N., Brenner, S.: . The structure of the nervous system of the nematode C. Elegans. Phil. Trans. R. Soc. London **314**, pp. 1–340. (1986)

Agent-Based Cognitive Model for Human Resources Competence Management

Stefan Oliveira and João Carlos Gluz

Abstract This chapter presents an agent-based cognitive model aimed to represent human competency concepts and competence management processes of psychological nature. This model is implemented by a multiagent system application intended to help managers of software development projects to select, based on the competence management model, the right professionals to integrate a development team. There are several software engineering methodologies that can be used to design and develop multiagent systems. However, due to the necessity to handle human competency concepts of cognitive nature, like aptitudes, interests, abilities and knowledge, we were driven to choose methodologies that can handle these concepts since the inception of the system. To do so, we integrated the TROPOS methodology, and a set of software engineering methods derived from intelligent tutoring systems research to successfully analyze and design the proposed system. At the end of the paper we present a study case, showing how the proposed system should be applied to the domain of website development.

Keywords Competence-Based Management · Human Resource Management · Agent-Oriented Software Engineering · Competencies Cognitive Model

1 Introduction

In this work we try a new approach to the question of how to find an appropriate software development professional to integrate a development team, based on the *Competence Management* (CM) psychological model of the developer. It is a top-down model-driven approach directed to understand and analyze application domains with strong psychological and anthropomorphic characteristics. The approach is top-down, because it starts from the analysis of high-level cognitive functions of

S. Oliveira (✉) and J.C. Gluz
Pós-Graduação em Computação Aplicada (PIPCA) - Universidade do Vale do Rio dos Sinos (UNISINOS) - São Leopoldo - RS - Brazil
e-mail: stefanoliver@gmail.com; jcgluz@unisinos.br

A. Hussain et al. (eds.), *Brain Inspired Cognitive Systems 2008*, Advances in Experimental Medicine and Biology 657, DOI 10.1007/978-0-387-79100-5_3,

the application domain's actors. It is model-driven, because the analysis of the cognitive functions, and the design of the application is based on cognitive models and software architectures derived from *MultiAgent System* (MAS) research (the BDI - *Belief-Desire-Intention* - Model [9]) and from *Intelligent Tutoring Systems* (ITS) research (the Student Model [4]).

However, is important to point out that the work do not intends to propose any new model, theory or insight in the field of human competencies. The goal of this work is to define a computational model for established models for human competencies, particularly focusing on those models that work with psychological dimensions or aspects of competencies. The main expected result of the work is to show that an agent based computational model, derived from BDI cognitive models, is indeed possible.

There are several *Agent Oriented Software Engineering* (AOSE) methodologies that can be used to design and develop MAS architectures [11]. However, due to the necessity to handle key CM concepts of cognitive nature, like interest, aptitudes, abilities and knowledge, we were driven to choose AOSE methodologies that can handle these concepts since the inception of the system.

Based on previous experience on the development of agent and multiagent based learning environments [6, 7, 12], we do not see any problem in use cognitive models to built the application. We consider that cognitive models can be successfully used in the analysis, design and development of multiagent applications if these models: (a) present viable computational interpretations, otherwise they are not useful to design and implement the application; (b) have clear epistemological and psychological foundations, not being based only on naive intuition or common-sense psychology, but based on scientifically established epistemological and psychological principles; (c) have precise formal specifications, which provide the answer to avoid excessive anthropomorphism (for example, the formal definition of "subjective belief" is a strictly objective definition).

In practical terms, we integrated the TROPOS methodology [2], the I* framework [13], and a set of AOSE methods derived from ITS research [12] to analyze and design a MAS intended to help managers of software development projects to select, through CM model, the right professionals to integrate a development team. The utilization of these AOSE concepts and techniques is justified by the possibility of consider psychological dimensions of CM models in the project of the system, aiming to make the decisions taken by this system more realistic with respect to human traits.

The concept of human competency is the base for the human resources CM process, which goal is to manage the competence gap usually existent in organizations or teams, trying to reduce or eliminate this gap through the identification of what professionals can do (current competencies) and what the organization expects from them (desired competencies) [3]. The human competency is expressed when someone generates a result in his or her work, which is caused by the combined application of knowledge, skill and attitude. This competency should add social and economical value to individuals and organizations and, at the same time, contribute for the realization of organizational goals, and express the social recognition of people's abilities.

As will be seen in the *CM Model* Section, this characterization of human competence proved to be very akin to concepts and abstractions used to understood and develop MAS, at least, when they are formed by the cognitive sort of agents.

2 Theoretical Basis

The general approach of this work is to apply agent-based abstractions and concepts in the analysis, specification and implementation of a computational model for human competencies. To do so, we relied strongly on current AOSE methodologies, focusing on those methodologies that take into account the characteristics of cognitive BDI agent models since the beginning of the analysis and development process. For this purpose, we choose two AOSE methodologies: the TROPOS/i* methodology [2, 13] and the AOSE methods derived from ITS [12]. These methodologies were integrated to solve the problem of how to computationally represent not only the factual aspects of human competencies, but also the psychological subjective dimensions of these competencies. These sections present a brief introduction on these subjects.

2.1 TROPOS Methodology and I Framework*

TROPOS is an AOSE methodology, which supports high-level agent abstractions, like actors, goals, plans and social dependencies applied to all phases of development, since the initial phase of analysis to the implementation phase. Another key feature of TROPOS is the great importance attributed to the initial stage of analysis and specification of application requirements, allowing a more detailed assessment of the impact caused by the system's introduction as well of the possible types of interactions that may occur between the system and its users.

The TROPOS methodology sets five stages for the process of MAS development. The first two stages *Early Requirements* and *Late Requirements* deal basically with elicitation and analysis of requirements and specification of the application's domain. The third stage, *Architectural Design* deals with structural aspects, which involve the choosing of an architectural style and social standards for the system. The fourth stage, *Detailed Design*, brings questions regarding communication and behavior of each system's component. In the fifth stage, *Implementation*, a mapping is built from concepts of TROPOS to the elements of a platform of agent's implementation.

The I* framework [13] has a major importance during the early stages of TROPOS. This framework aims to represent organizational aspects involved in social processes, describing motivations and aspects of intentionality that involve actors in an organizational environment. Among a number of techniques of organizational environment modeling, I* has been highlighting for making possible a better expression of the "whys" related to existing practices and organizational structures. Since we can high-light the organization's general objectives, its actors' intentions in regard with the intended systems, detailing of the reasons associated with actors to reach certain goals and description of non-functional requirements. Thus, it enables developers to investigate strategic dependencies among actors, as well as such actors' strategic reasons.

In order to describe the organizational environment, I* proposes two models: the *Strategic Dependencies* (SD) model and the *Strategic Reasons* (SR) model. The SD model is a graphical model composed of nodes and links. Nodes represent actors in the environment and links represent dependencies among actors. Actor is the name given to an entity that performs actions in order to obtain objectives in the environment context. Actors have relations of dependency among themselves in order to obtain objectives. The actor that somehow depends on another actor is named *Depender* and the actor that meets *Depender's* demands is called *Dependee*. The element of dependency between *Depender* and *Dependee* is named *Dependum*. As a result, we have the following pattern of relationship: *Depender* → *Dependum* → *Dependee*. Unlike the SD in which only the external relationship among actors is represented, SR allows a wider comprehension of the strategic reasons of the environment's actors regarding organization's processes and the way they are represented. Such strategic reasons can be more easily decomposed through the observation of how *Dependees* can satisfy the *Dependums* associated to them and, from that point on, observe and decompose the intentions and the organization's strategic reasons as a whole. The symbology used in SD and SR models can be observed on Fig. 1.

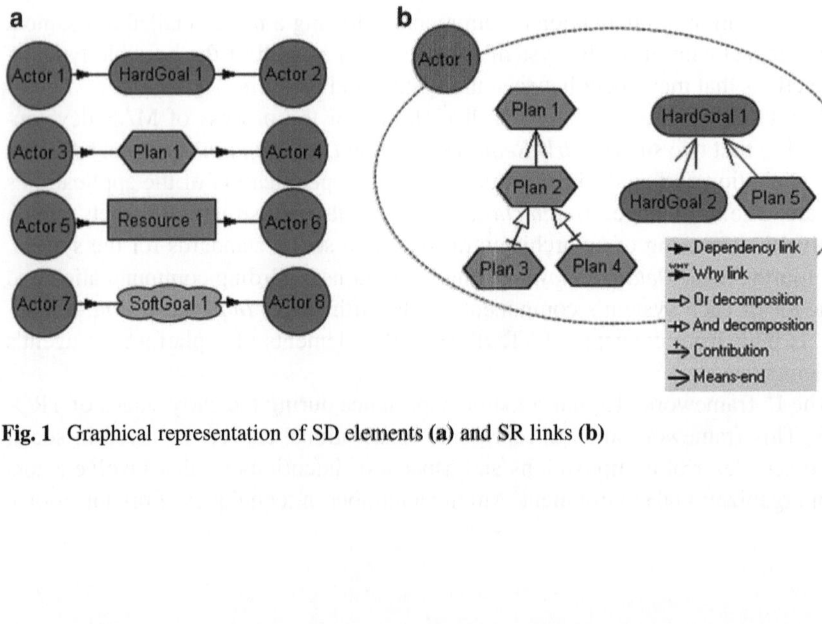

Fig. 1 Graphical representation of SD elements **(a)** and SR links **(b)**

2.2 AOSE Methods Derived from ITS

The set of software engineering methods proposed in [12] provide several applicability criteria, design principles and development guidelines that can be useful to analyze, design and develop a BDI MAS. These methods were abstracted from the empirical observation of how several ITS based on MAS concepts and technologies were built. The main reason for the proposal of these methods is that, according to [12], current AOSE methodologies (possibly with the exception of TROPOS) do not consider agents' cognition abstractions from the beginning of the system's development process, including the stage of requirements engineering. Cognition abstractions such as beliefs, objectives, intentions and social relationships based on cognitive models, should provide the base for the extraction of high-level properties of domains, which not only can be intuitively understood, but will form the basis of the application's requirements specification.

The ITS derived methods are based on a top-down approach for the analysis, project and development of multi-agent applications. The first step is to check if several applicability criteria are satisfied in the requirements elicitation phase of software engineering process. The main idea behind these criteria is to ensure that the design principles and development guidelines proposed for the project of BDI MAS can be applied after the analysis phase. Such criteria force requirements engineers to consider the agent abstractions since the beginning of process of analysis of domain and elicitation of requirements. If the application satisfies the proposed criteria, the specification will naturally incorporate agent concepts and abstractions.

The applicability criteria cover six different aspects that should be considered when the development of a multi-agent system is intended. First, it is necessary to check if the domain of application really includes entities that can be understood as agents working together in an organized system. The steps that follow establish criteria over agents' belief structuring, the use of cognitive models by agents, and how communication interactions should occur, and social relationships be established among agents. They still establish that application requirements attributed to agents should be clearly listed by a specification that determines what knowledge (extracted out of a base of beliefs) is necessary to satisfy those requirements.

Following the successful application of these criteria, then several design principles are proposed, showing how the application domain should be divided, at least, into three distinct sub-domains: the *Users and Agents Modeling* (UAM) sub-domain, which covers knowledge about the users (or external agents) that interact with the application; the *Social Mediated Interactions* (SMI) sub-domain, which covers the knowledge about social interaction mechanisms necessary to mediate the communication between the application and its users (or external agents); and the *Problem Solving* (PS) sub-domain, which covers the knowledge about the specific problems that the application is intended to solve (and are not covered in the other sub-domains).

A specific MAS architecture (which can be called the "triad" architecture) is then proposed to handle these domains (see Fig. 2). This architecture is composed by three types of agents: PS agents that work in the PS sub-domain and solve

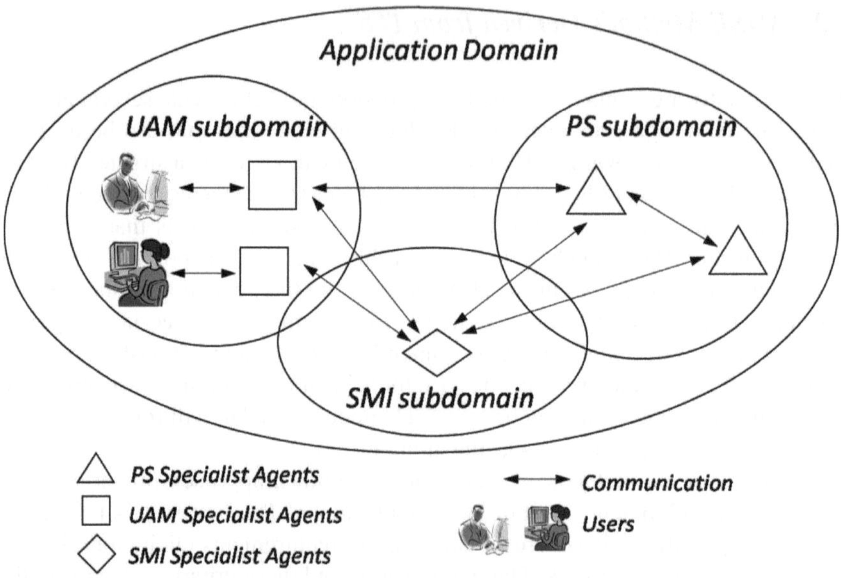

Fig. 2 Sub-domains and agent types of Triad architecture

application problems; UAM agents that provide the interface of the application, and make cognitive models of its users and external agents; and SMI agents that incorporate the social interaction mechanisms. The sub-domains do not need to be completely separated. It is possible (and even necessary) that there exist non-null intersections between sub-domains, containing knowledge that interrelate concepts in both sub-domains.

Development guidelines that must be applied during the implementation process and system's test are also presented in [12]. Those guidelines provide useful ideas on how to transform application architecture designed in accordance with the proposed design principles in effective systems.

2.3 Human Competencies

According to [3], human competencies are synergic combinations of knowledge, skills and attitudes, expressed by the performance of professionals inside an organization, which add value to the people and to the organization. As suggested by [5], this conceptualization emerges from the junction of two great currents of study in human competencies: the school of North-American authors, which is based on three dimensions of competences (knowledge, skills, and attitudes), and the French school which is focused on objective and measurable (observable) competence referentialities.

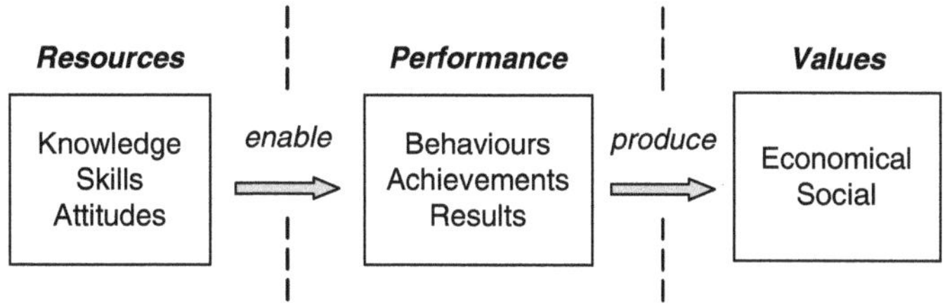

Fig. 3 Human competencies model (adapted from [3])

The human competency is expressed when someone generates a result in his or her work, which is caused by the combined application of knowledge, skills and attitudes (see Fig. 3). This competency should add social and economical value to individuals and organizations and, at the same time, contribute for the realization of organizational goals, and express the social recognition of people's abilities.

Knowledge corresponds to information about some domain or area that, when recognized and understood by an individual, influences his or her decision process in respect to this knowledge area. The knowledge forms a necessary condition to the emergence of competent behaviors of the individual in respect to this area. Skills are related to the "know how" of the individual, in respect to specific tasks in the knowledge domain or area. It is the productive application of knowledge in the definition and execution of actions. Attitudes, otherwise, are related to what the individual wants or desire to do. They are the intentions, inclinations and predispositions of the individual, and they determine his or her general conduct in respect to other people, to the job and to live circumstances [3].

The concept of human competency is the base for the human resources CM process, which goal is to manage the competence gap usually existent in organizations or teams, trying to reduce or eliminate this gap through the identification of what professionals can do (current competencies) and what the organization expects from them (desired competencies) [3].

2.4 Cognitive and BDI Agents

Agents form the basic element of computation in MAS, which can be defined simply as systems formed by several agents working together. In this context, an agent is a computational process situated in an environment, which is designed to achieve a purpose in this environment through an autonomous and flexible behavior [15].

The BDI cognitive model for agents assumes that all agent's purposes can be fully specified by the definition of its beliefs and desires, and that the behavior of the agent is clearly implied by its intentions. The BDI model is one of the possible cognition models of the Mental State approach for agent modeling (see Fig. 4). In this model, the set of *beliefs* represent provisional knowledge of the agent, which can change with the passing of the time. Beliefs define what the agent knows about the

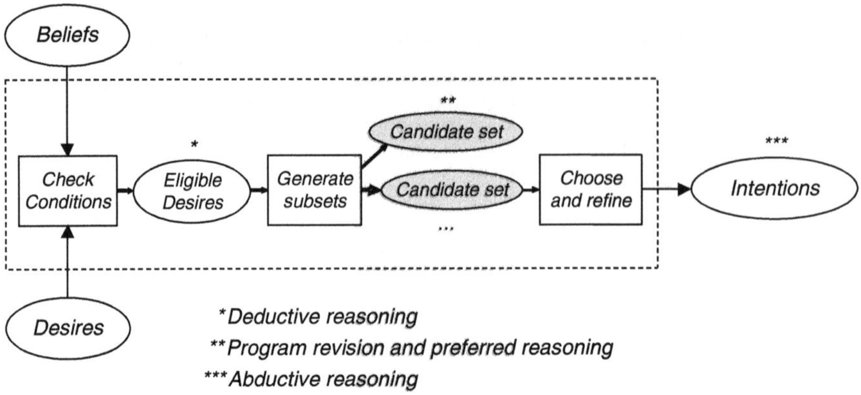

Fig. 4 DI Cognitive Structure

environment, what it knows about the other agents and what it knows about itself. Beliefs are specified by logical properties concerning other agents, the environment, and about the very own agent. Agents should update its beliefs to reflect changes detected (perceived) in other agents, the environment, and itself. They must maintain the consistency of the beliefs after this update.

Desires specify the set of *state of affairs* the agent eventually wants to bring about. One particular state of affairs is specified by a logical property to be held in this future state and by a list of attributes that define the admissibility criteria of the desire. The admissibility criteria attributes specify agent's beliefs about the desire. They define, at least, the priority of the desire, the ability of the agent to achieve the desire and the estimated possibility of the desire to become true. In the cognitive model of agents that we are using, we suppose that the purpose of the agent is explicitly stated as the set of highest-priority desires of the agent.

The fact that an agent has a desire does not mean it will act to satisfy it. Acts are governed by *intentions* that are characterized by a choice of a state of affairs to achieve, and a commitment to do this choice (here we follow the definition of Cohen & Levesque [16]). Intentions are related to desires by admissibility criteria attributes. The agent will choose those desires that are possible, according to these attributes and to its current base of beliefs. It is important to note that intentions are also beliefs of the agent. One particular intention is a compromise the agent has to reach a specific possible future, that is, the agent believes that the state of affair it wants to achieve does not hold now, and that it must work to reach that state. It means that before an agent decides what to do, it will be engaged in a reasoning process, confronting its desires with its possibilities, defining its intentions and then planning its actions is respect to this intention.

In other words, an intention poses a *decision problem* (or a *planning problem*) for the agent. The agent should solve this problem and decide the course of actions, or *plan of actions*, to be followed in order to achieve the intention. A plan of actions is composed by a set of actions structured by sequence, iteration and test/choice order relations (operators). These plans do not need to be fully specified from the beginning, they can be partial and the agent can start to follow the plan and reassess or complete it during execution.

The interaction of the agent with its environment is done by *actions* and *perceptions*. An action is an alteration in the external environment caused directly by the agent. From an intentional point of view, it also represents a way to attain an end (intention). Therefore, internally the agent should know (believe) the basic effects produced by possible actions and what are the relations of these actions to their intentions. Agents detect properties in the environment, or more commonly, changes in these properties through perceptions. These changes can occur independent of the agent, or they can be caused by actions executed by the agent or by other agents, but the only way the agent has to detect them is through its perceptions. Perceptions produce updating in the base of beliefs of the agent, but, the exact update produced by a particular perception depends on the current state of beliefs of the agent.

Agent architectures are also usually divided into several distinct *abstraction layers*. InteRRap [18] and the architecture of Glaser and Morignot [17] are representative examples of layered agent architectures. Glaser and Morignot suggest that the design of agents should be organized in: social, cooperative, cognitive, and reactive layers. Higher social and cooperative layers incorporate knowledge and reasoning processes related to social issues. A medium cognitive layer incorporates knowledge and reasoning processes necessary for decision taking and action planning. A low reactive layer is responsible for reflexive behaviors directly associated to the basic perceptions and actions of the agent. InteRRap architecture is divided in three layers: the *Cooperative Planning Layer* (CPL) responsible for the social model employed by the agent, the *Local Planning Layer* (LPL) responsible for the agent's mental model and the *Behavior Based Layer* (BBL) responsible for the modeling the world and environment.

3 Cognitive Competence Management Model

The main idea behind the cognitive CM model proposed in this work is based on what is, for us, a strong analogy between human competency concepts and the BDI concepts reviewed in the last sections. For instance, if this analogy is viewed from a simulation perspective, then the knowledge resources of some agent can be represented by its high-level conceptual beliefs about the environment (domain) where it will operate. The skills of this agent should be represented by its beliefs about what are the appropriate decision processes and planning methods to be used to solve problems in this domain, and its attitudes are represented or directed by the set of desires (goals) it wants to achieve in the domain. As a consequence, behaviors, when considered as objective evidences of performance, should correspond to the effective behaviors (actions) generated by the agent's reasoning processes in response to environment perceptions. Following the simulation perspective, social and economic values should increase (or decrease) in correlation with the achievements and results obtained by the agent's behaviors. The agent should also have some knowledge about desirable social or economic values, modeled as beliefs and desires residing in its social and/or cooperative layers.

This analogy derived from the simulation perspective is interesting as a general guide of how to build agent-based computational models for CM processes, but it does not need to be followed to its complete extent in all domains of applications. In the case of the application being considered in this work, which aims to help managers to select software professionals to integrate a development team, it is not necessary to emulate the behavior of these professionals, but only try to discover and represent the professional competences and the manager needs. This is presented in the following sub-sections.

3.1 The Organizational Environment

Following TROPOS methodology [2], the organizational environment of the application's domain should be divided in several categories of *actors*, which are related one to each other through several *social dependencies* necessary to achieve common objectives. In the case of our application, we will focus the organizational environment of small and medium sized software companies. The existing actors in the environment are representations of roles usually found in software development environments of this kind of companies. These roles are: the Project Manager (*PManager*), the Team Leaders (*TLeader*), the Software Developers (*SWDeveloper*), the Human Resources Consultant (*HRConsultant*) and the final *Customer*.

The *PManager* is the type of actor who has direct contact with the *Customer* and therefore deals with administrative issues related to the project, in addition to being responsible for receiving new software development requests. The *TLeader* is the actor who controls and determines the characteristics and resources involved in a project. *TLeader* should delegate specific development duties to the *SWDeveloper* actors that take part in the team under its responsibility. To do so, it can ask the *HRConsultant* whenever it turns out to be necessary to get information about human resources. A same *TLeader* can take part in one or more development projects, in which, not necessarily, the professional assumes the role of leader in all the projects they participate in. The *SWDeveloper* actor performs activities delegated by the *TLeader* of the project he/she participates in. In order to become a member of a new project, the *SWDeveloper* must meet some requirements that are necessary for the good course of the development activities. The characteristics requested to those professionals vary according to the domain of application of the software to be developed.

The *HRConsultant* actor provides information on professional and personal characteristics of every collaborator of the company. The information provided is collected with the aid of some techniques that normally are based on documents and forms with answers supplied by the collaborators themselves, where some specific characteristics related to technologies or competences are taken into account.

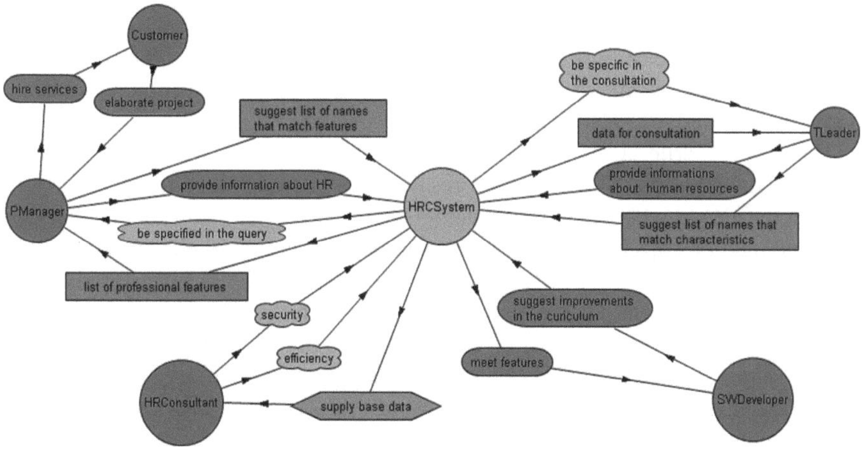

Fig. 5 SD model of HRCSystem domain – Late Requirements

The early and late requirements phases of TROPOS methodology are intended to elicit the several dependencies between the organization's actors and the reasons behind these dependencies. The difference between the two phases is basically the inclusion of the system in the organizational environment, through the modification of the original SR and SD diagrams, and the analysis of what dependencies the system will change. The diagram presented in Fig. 5 shows the resulting SD model of the late requirements analysis of *HRCSystem*. In this model the dependence relationships between *HRCSystem* (**the yellow node**) and the other actors in the domain are already clearly delimited, providing the functional and non-functional requirements of this application.

From the perspective of TROPOS methodology the applicability criteria proposed by the ITS derived AOSE methods [12] work as a checklist that should be verified between the domain requirements elicitation phases and the architectural design phase. These criteria should be considered during the process of analysis and elicitation of application requirements. They must be considered specifically when functional requirements of system's architecture are being analyzed.

The first three criteria set the basic context for the application domain: first (AC.1), this domain should contain entities better understood as agents, second (AC.2), it should be possible to classify the knowledge used by these agents in knowledge about other agents (or actors) and knowledge about non-agent entities, and third (AC.3), the communication between agents is symbolic and occur at the knowledge level.

In our case, (AC.1) is easily satisfied because, if necessary, any of the roles (actors) presented in the problem's domain are better represented computationally by agents other than, for example, objects. That can be justified considering the need of information (knowledge) exchange among the roles (agents) of system in order to reach their goals. The (AC.2) criterion is also satisfied, because it simply states that the agent modeling knowledge, which represents developer's professional and personal characteristics (competences), can be cleared distinguished, for example, from the knowledge of how projects are structured and managed. The (AC.3) criterion is satisfied by all actors on the SR model, because they should exchange symbolic knowledge to achieve their goals.

The next two criteria (AC.4) and (AC.5) are the core criteria of the methods proposed in [12]. The (AC.4) criteria requires that the system contains agents that form cognitive (BDI) models about the system's users, and (AC.5) says that all interactions and relationships between the system and their users should be based on these cognitive models. For the specific case of *HRCSystem* this will imply, for instance, that for any real software developer that interacts with the system, should exist an agent inside the system that will create a cognitive model about the developer by means of its perceptions over the observable behavior of developer (for example, what technologies the developer usually uses, what degree of critically the developer presents toward a project, what sort of professional experience and how long the developer has been in the company). In the SR model of *HRCSystem*, there are several relationships and social interactions between the agents, such as, for instance, the main social relationship between the *HRCSystem* to the PManager actors is related to the proper selection of some developer to join into a development team. The establishment of a successful relationship depends on the cognitive model that the agents inside this system had built about the developer.

The last criterion (AC.6) requests that the application requirements attributed to the agents must be clearly listed in the requirements specification. Starting from SR models it is possible to list all application requirements associated to the actors, and additionally, after a deeper analysis, it is possible to list what knowledge are necessary to satisfy these requirements. In this case, the requirements associated to *HRCSystem* node should be distributed among the agents that will form this system.

With this list of requirements in hand, is necessary to select an appropriate MAS architecture to model the system. TROPOS methodology suggests several architectures for implementation of MAS solutions. However, in the case of HRCSystem, due to the satisfaction of all applicability criteria defined in [12], the most appropriate MAS architecture is the "triad" architecture proposed in this same work. In this architecture it is possible to classify the actors of the system in accordance with sub-domains with well defined features, as shown in Fig. 6.

Based on the model SR of the last requirements phase of TROPOS, the actor who represents the system undergoes a "breakup" being organized as a system composed of several distinct agents. Following DP.1 to DP.3 design principles defined in [12], these agents are classified as follows: (a) the *HRConsultant* mediator agent, which belongs to SMI subdomain, knows the CM model, and is responsible to identify *SWDevelopers* that meet *PManagers* and *TLeaders* requirements; (b) the *ProjMgmt* problem solving agent, which belongs to PS subdomain, knows what a project is, what are the projects of the organization, and is responsible to solve questions related to the importance of tasks and projects; (c) the *SWDeveloper*, *PManager*, and *TLeader* interface agents, which belong to UAM subdomain, and are responsible to interact and make cognitive models of users of *HRCSystem*.

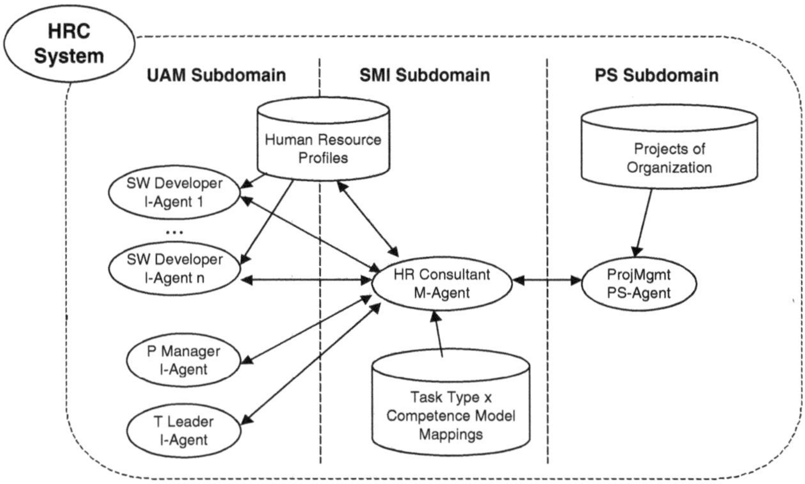

Fig. 6 Triad Architecture for *HRCSystem*

The knowledge bases *Human Resource Profiles*, *Task Type x Competence Model Mappings*, and *Projects of Organization* are being analyzed and structured according DP.4 principle (see [12]). They are organized as belief bases that represent: (1) the significant entities of the sub-domain and their main properties; (2) the basic identification abilities, and possible actions and perceptions of the agent, in respect of these entities; (3) the planning and problem-solving skills necessary to achieve desires (goals) related to these entities.

3.2 Agent Models

The Fig. 7 presents the cognitive CM model proposed to *HRConsultant* agent. This model intends to represent information relative to personal competences, structured in one side according to the organization needs and on the other side according to the professionals available. It provides the reference for which kind of social skills and relationships the *HRConsultant* agent should establish and maintain.

The *Class of Tasks* category specifies the different types of tasks possible in the organization projects. The *Role* attribute defines which kind of professional can assume the task in the organization. A particular task is performed by a particular professional acting in the corresponding role. However, for the task to be successfully accomplished is necessary a minimum level of performance from the professional. This is represented by the *Competence Referential* (CREF) profile, which specifies the set of referentialities for a particular competence.

A CREF profile is a representation of competences that focus on objective and observable behaviors of professionals, teams or organizations directly related to the competence. It is formed by a set of referentialities, each one pointing out an objective and observable behavior that represent the competence.

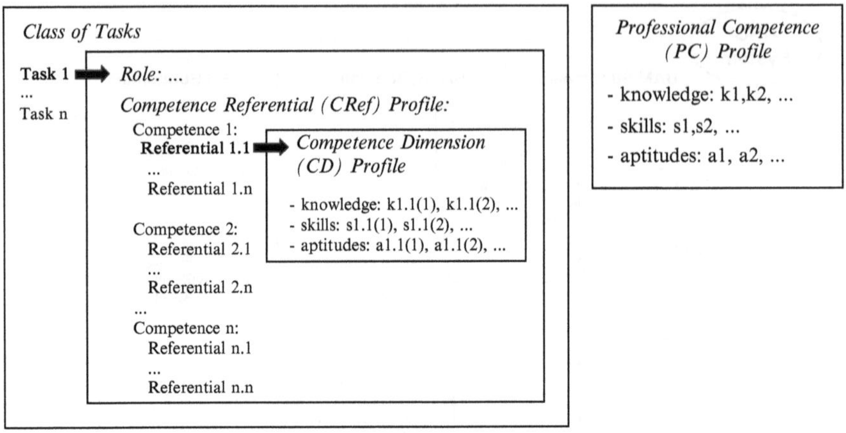

Fig. 7 Cognitive CM Model for HRConsultant Agent

Table 1 Meta-level Cognitive Model of *SWDeveloper* Agent

Beliefs	*Knowledge*	Beliefs about what is the developer's knowledge in programming
	Skills	Beliefs about the developer's professional experience in IT
	Attitudes	Beliefs about the developer's attitudes in the working environment
Goals	To create a faithful belief model about the external corresponding actor (the real developer)	
	To identify what organization roles this model satisfy	
	To answer the requests of the *HRConsultant* agent	
Plans	Methods, algorithms and heuristics that allow the agent to achieve its goals	

Other way to represent competences is by the classification of the competence in three dimensions: knowledge, skills and attitudes, which are necessary for the professional to efficiently perform the competence. This kind of representation is incorporated in the *Competence Dimensions* (CD) profile, and in the *Professional Competence* (PC) profile of the cognitive CM model. The CD profile is responsible to map, for a particular competence, which are the expected knowledge, skills and attitudes required for the efficient performing of the competence. Once the CD profile for a particular competence is identified, then it can be compared with PC profile for each professional of the organization, looking for the best professional profile able to perform the competence.

The cognitive model of *SWDeveloper* interface agent is presented in Table 1. This model intends to represent the real software developer actors inside the system. The information mental states of *SWDeveloper* are represented by beliefs that correspond to the three dimensions of competence. Thus, *SWDeveloper* agent has a belief model composed by the knowledge, skills and attitudes, which should represent the PC profile of the corresponding external actor, the software developer actor. The proactive mental states of this agent are formed by its goals and plans. The main goal of these agents is how to create a faithful belief model of the PC profiles. Secondary purposes of this agent are the identification of what roles this profile should satisfy inside the organization, and to answer the requests of other agents of the system (in particular the requests of the mediator agent, the *HRConsultant* agent). Agent plans define exactly what actions the agent should perform to achieve these goals. Plans are the most implementation and programming oriented tasks in the design of the agent.

3.3 The Website Construction Study Case

The CM cognitive model can be applied to several situations. The abilities of this model through a study case, which shows how the model can be applied to website software construction tasks. This study case is based on empirical data obtained with interviews made with website software development specialists from the *S1 Soluções Inteligentes* software development company [14]. The case was elaborated in conformity with the mission and vision of the company, supporting its strategic planning. Following [3], it is necessary to consider that competence management should be oriented to the strategic planning of the organization. In this work we will suppose that this strategic planning will be based on the fulfilling of the practices and methods proposed by the Personal Software Process (PSP) [8], which specify a set of productivity characteristics that software development professionals should satisfy.

Considering the model presented in Fig. 7, the website construction development process should be decomposed in a series of tasks, each one, being fulfilled by some role of the organization and needing the competences specified by the CREF profile attribute. In this context, the several activities necessary for the website construction development process will form the basic competences required to the professionals involved in this process. These competences are the following: (1) *Requirements analysis*: the identification of client's needs; (2) *Layout creation*: the elaboration of illustrative images, which represent the services contained in the final stage of the website; (3) *XHTML implementation*: conversion of layouts to XHTML; (4) *Data modeling*: development of classes that represent the resources used on the website's implementation; (5) *Data entry*: inclusion of contents made by the client, through an user friendly interface of content management; (6) *Business logic programming*: writing of the source code (Python, PHP) of the website, taking into account the client's business rules, and the template rendering; (7) *Template configuration*: implementation of the business rules in XHTML (interface layer); (8) *Compatibility testing*: verification the compatibility of the website code with several platforms (PCs and mobile), browsers, search softbots, according to a previously defined checklist of tests; (9) *Content publishing*: handling of the publishing process on the client server, or, in a third part server, if the client's site is hosted in a third part company.

Each one of these competences forms also an activity that must be performed by some professional of the organization. The roles of the professionals involved on website construction task are the following: (a) *External relationship professional*:

the professional which establish the communication channel between the client and the development team; (b) *Developer*: it is responsible by coding the information and structure of the website; (c) *Designer*: this professional elaborates the layout, consisting in the visual programming of the website's static pages/screens; (d) *Tester*: this professional is responsible by the verification of functionalities and resources implemented in the website, assuring that all requirements specified to the website were satisfied.

The relation between role and activities (competences) defines which competences will be necessary to the role. The *External relationship* role must be competent in (must have the competencies of) *Requirements analysis* and *Data entry activities*; the *Designer* in *Layout creation*; the *Developer* in *XHTML implementation*, *Data modeling*, *Business logic implementation*, *Template configuration* and*Content publishing*; and the *Tester* in *Compatibility testing*.

To identify the specific resources (knowledge, skills and attitudes) necessary for some competence, the corresponding activity should be analyzed to determinate the information necessary to the realization of the activity. Following the model defined in Fig. 7, this information will form the CD profile of the competence. The identification of these resources is made by the matching of keywords (signs) related to the performing of the activity in the related domain. Table 2 shows the CD Profile for the Requirements analysis competence. This table relates through keywords, the kind of knowledge, set of skills and set of attitudes necessary to efficiently perform the Requirement Analysis of websites.

After the identification of the resources related to the activities/competences it is necessary to define profile of competences of some particular professional. Following the model defined in Fig. 7, this information will form the PC profile of the professional.

The PC profile for some professional is created based on the analysis of different sources, likes: documents (résumé, historical records of previously performed activities), interviews, surveys, and examinations (personal account, evaluation of other professionals, internal tests). This profile should be a reliable internal representation of the professional inside the system. The profile is an active and flexible model implemented by a specific agent of the system, the *SWDeveloper* agent, which has the responsibility to assure that it is a trustworthy model of the corresponding professional. Table 2 shows the PC profile of some fictitious professional.

Table 2 Example of PC Profile

Professional: *John Doe*	
Resources	Keywords
Knowledge	Graduation in Computer Science; Data modeling; Requirements elicitation; Software development methodologies; Process documentation
Skills	UML data modeling; Object oriented programming; Basic knowledge of LATEX; Knowledge of basic electronic; Programming logic; Teaching; APIs documentation
Attitudes	Attentive; Communicative; Pro-active; Observant; Patient; Collaborative

All goals of *SWDeveloper* agent are related to the PC profile: the agent will try to keep this model updated and trustworthy; it also will have methods to identify what roles of the organization match this profile, and will answer requests of other agents about information on this profile and on these matches.

3.4 HRCSystem Development

The *HRCSystem* is being implemented as a MAS in the JASON *Integrated Development Environment* (IDE) [1]. JASON IDE is a complete and fully functional development environment for the AgentSpeak(L) [10] agent programming language. AgentSpeak(L) was selected to implement the proof-of-concept prototype of the *HRCSystem* due to several important features: it is a very high-level logical language that can represent BDI concepts directly; it allows a direct mapping between formalized concepts and operational programs; it has an effective interpreter and programming environment; the interpreter is implemented in Java, what allows an easy action/perception interface to legacy applications, including the ability to integrate *HRCSystem* with other project management tools and applications; it allows interoperation with other agent platforms due to the support to KQML and FIPA-ACL agent communication languages.

Each agent presented in Fig. 6 is implemented by a distinct AgentSpeak(L) source code module. In AgentSpeak(L) the several goals of the agent must be supported by, at least, one plan which defines how the agent can achieve the goal. Plans define the specific algorithms, methods and heuristics used to achieve agent goals. For instance, the plans related to the construction of the PC profiles are based in actions that give access to information extracted from three sources: interviews with the professional, the organization human resources database, and interviews with managers. Fig. 8 presents an excerpt of the belief base that models the example PC profile described in Table 2. The AgentSpeak(L) code presented in Fig. 8 specifies the beliefs that an instance of *SWDeveloper* agent has about it corresponding external actor, the John Doe (fictitious) developer.

```
hasKnowledge (johnDoe, graduationComputerScience). [source(johnDoe) ]
hasKnowledge (johnDoe,dataModelling). [source(johnDoe) ]
    ...
hasKnowledge (johnDoe, processesDocumentation). [source(johnDoe) ]
hasSkill (johnDoe, objectOrientProgram). [source(johnDoe) ]
    ...
hasSkill (johnDoe, apisDocumentation). [source(johnDoe) ]
    ...
hasAttitude(johnDoe, collaborative). [source(johnDoe) ]
```

Fig. 8 Formalization of PC Profile in AgentSpeak

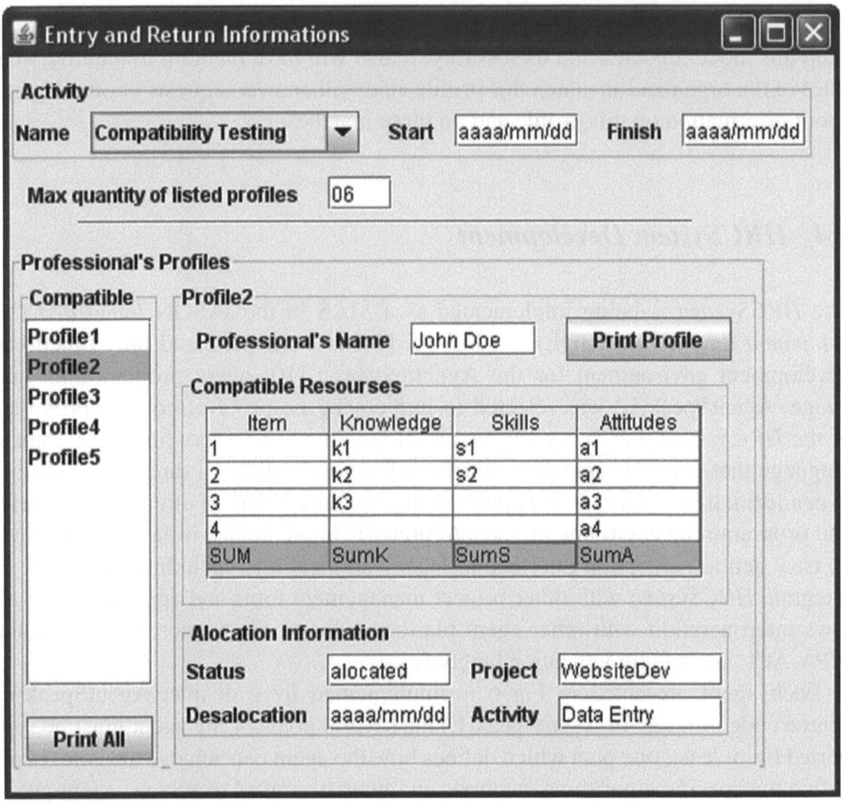

Fig. 9 Screenshot of Team Leaders and Project Manager user interface

The interface between *HRCSystem*'s and its project manager/team leader users is implemented through a dialog box similar to the screenshot showed in the Fig. 9.

This dialog box essentially implements an interview with these users, obtaining information for the *HRCSystem* databases, which cannot be obtained from the resource management databases of the company.

The initial validation of the CM model and the *HRCSystem* system will be made within the context of the website construction study case, applied to the organizational environment of a software development company.

Currently we are finalizing the mapping of competences for the activities and the professionals of the enterprise. When this task is finished we will start the experiments and tests of *HRCsystem.* The basic validation approach is to proceed with a series of controlled experiments, where *HRCSystem* selection of appropriate professionals to work in specific website development tasks, are compared with the selections made by company's managers and team leaders for the same task.

We already planned a series of experiments and tests that should check the functionality of the CM model. The basic idea is to apply the system to a series of simulated requests for project activities, registering the results obtained by it. The control check of these experiments will be formed by equivalent requests submitted to the project managers of the organization. After that we will compare the decisions suggested by *HRCsystem* with the decisions taken by company managers. In future we expect also to check the performance of the system, specifically when helping recently hired managers.

4 Final Remarks

This chapter presents a new agent-based cognitive model for human competencies management, and a new software tool, called *HRCSystem*, which is based on this model, and intends to help project managers and team leaders of software development projects to select the right professionals to integrate a development team.

The model and the application where developed through the integration between two different software engineering methodologies of multiagent systems analysis and design. The model can be applied to several software development situations. To show the abilities of this model we show its application to the specification of website software construction tasks. This study case is based on empirical data obtained with interviews made with website software development specialists from a software development company. The case is being elaborated in conformity with the mission and vision of the company, supporting its strategic planning, and the practices and methods proposed by the Personal Software Process (PSP) [8].

We believe that the main appeal of the *HRCSystem* is to serve as an assistant that can help recently hired (or promoted) managers and team leaders to assess their decisions. It can serve as a on-the-job learning tool for new managers. It also can be helpful to more experienced managers, when there are a huge base of developers and projects in the organization.

The long term goal of *HRCSystem* project is to actively incorporate in a single source the organization's knowledge about the competences and availability of software developers. The evolution of the knowledge base will occur through two main processes: (a) the direct adding of new knowledge through the human-machine frontend interface of *HRConsultant* agent and projects database backend interface of *ProjMgmnt* agent; (b) a conflict-solving mediation process that will solve differences between system's outputs and users (managers) expectations, and that is able to change/evolve the knowledge base when faced with discrepancies, mainly when these discrepancies arise from experienced users. There is a wide degree of latitude in the system's architecture to incorporate mediation and interface agents able to solve theses issues. This is an important future research direction of this work.

References

1. Bordini, R.H., et al.: JASON and the golden fleece of agent-oriented programming. In: Multi-Agent Programming: Languages, Platforms and Applications, Berlin: Springer, Vol. 15, pp. 3–37 (2005)
2. Bresciani, P. et al.: Tropos: An agent-oriented software development methodology. In: Autonomous Agents and Multi-Agent Systems, pp. 203–236 (2004)
3. Carbone, P. et. al.: Gestão por competências e gestão do conhecimento. Rio de Janeiro: FGV, 2 ed., pp. 40–78 (2006)
4. Dillenbourg, P., Self, J.A: A framework for learner modelling. Interactive Learning Environments, No.2, pp.111–137, (1992)
5. Dutra, J.: Competências: conceitos e instrumentos para a gestão de pessoas na empresa moderna. São Paulo: Atlas (2004)
6. Gluz, J.C. et al.: Formal Aspects of Pedagogical Negotiation in AMPLIA System. In: Intelligent Educational Machines. Intelligent Systems Engineering Series. Springer, (2006)
7. Gluz, J. C. et al.: Pedagogical Negotiation and Solidarity Assimilation Groups in Action: the Combined Efforts of E-M@T and Leibniz to Aid Calculus Students. In: Agent-Based Tutoring Systems by Cognitive and Affective Modeling. IDEA Group, Hershey:PA, USA (2008)
8. Humphrey, W.S.: A Discipline for Software Engineering – The Complete PSP Book. SEI Series in Software Engineering (2005)
9. Rao, A. S., Georgeff, M. P: Modeling rational agents within a BDI-architecture. In: Proceedings of Knowledge Representation and Reasoning (KR&R-91), San Mateo, CA: Morgan Kaufmann Publishers, pp. 473–484 (1991)
10. Rao, A.S.: AgentSpeak(L): BDI agents speak out in a logical computable language. In: Procs. of the 7th European Workshop on Modelling Autonomous Agents in a Multi-Agent World, Springer-Verlag, LNAI, Vol. 1038, pp. 42–55 (1996)
11. Tran, Q. N. et al.: A Preliminary Comparative Feature Analysis of Multiagent Systems Development Methodologies. In: Procs. of AOIS 2004 Workshop at AAMAS2004, NY, pp. 157–168 (2004)
12. Vicari, R., Gluz, J. C.: An Intelligent Tutoring System (ITS) view on AOSE. Int. J. Agent-Oriented Software Engineering, Vol. 1, Nos. 3/4, pp. 295–333 (2007)
13. Yu, E., Mylopoulos, J.: Towards modelling strategic actor relationships for information systems development - with examples from business process reengineering, In: Procs. of the 4th Workshop on Information Technologies and Systems (1994)
14. Gomes, B. Processos para Construção de Websistes. Porto Alegre, Records of the interview conceded to Stefan Oliveira about the website construction processes used in the company S1 Soluções Inteligentes e Consultoria LTDA, 20 april (2008)
15. Jennings, N.R. An Agent-Based Approach for Building Complex Software Systems. Comm. the ACM, v. 44, n. 4, pp. 35–41, Apr. (2001)
16. Cohen, P., Levesque, H. Intention Is Choice with Commitment. Artificial Intelligence, n. 42, pp. 213–261 (1990)
17. Glaser, N., Morignot, Ph. The reorganization of societies of autonomous agents. In: Proceedings of 8th MAAMAW, Ronneby, Sweden. LNAI Series (1997)
18. Müller, J. P., Pischel, M., M. Thiel. Modelling reactive behaviour in vertically layered agent architectures. Intelligent Agents: Theories, Architectures, and Languages, Lecture Notes in Artificial Intelligence LNAI 890, Heidelberg, Ger-many. Springer Verlag (1995)

Neural Accumulator Models of Decision Making in Eye Movements

Vassilis Cutsuridis

Abstract Humans and animals are constantly facing the problem of having to choose from a variety of possible actions as they interact with the environment. Both external and internal cues have to be used to guide their selection of a single action from many possible alternatives. Which action to choose in a given context may have important biological consequences to their survival. Decision making is regarded as an accumulation process of evidence about the state of the world and the utility of possible outcomes. Two well established neural accumulator models of decision making are presented to model the neural basis of decision making in behavioural paradigms such as the antisaccade task.

Keywords Superior colliculus · Antisaccade task · Decision making · Accumulator model · Eye movement

1 Introduction

Decision making is the process of selecting from sets of options based on current evidence about the state of the world and estimates of the value of different outcomes [1]. Decision making has been a topic of intense study by multiple disciplines such as economics, sociology, statistics, computer science, artificial intelligence, ethology, cognitive and behavioural neuroscience. Economists often investigate how decisions are formulated in the presence of uncertainty, whereas ethologists approach the problem of decision making in the context of foraging Psychologists frequently investigate a behavioural choice using a concurrent schedule of reinforcement. Sociologists investigate how the decision making processes of an individual are influenced by the decisions of others in the same group [11], whereas artificial

V. Cutsuridis (✉)
Department of Computing Science and Mathematics, University of Stirling, Stirling FK9 4LA, U.K.
e-mail: vcu@cs.stir.ac.uk

A. Hussain et al. (eds.), *Brain Inspired Cognitive Systems 2008*, Advances in Experimental Medicine and Biology 657, DOI 10.1007/978-0-387-79100-5_4,

intelligence (AI) and computer science analyze how an optimal decision-making strategy can be learned through experience [12].

In recent years, cognitive and behavioural neuroscientists have begun to investigate the neural basis of decision making using various behavioural paradigms. The behavioural paradigm often used is saccadic eye movements (i.e. rapid eye movements to bring the saccadic goal onto the fovea). Saccadic eye movements are important in understanding the neural basis of decision making, because (1) making a saccade, among a set of potential visual targets, one must be selected as the next end-point of a saccade and (2) initiating a saccade, the decision must be made to release the system from its previous state of fixation. The slowness, variability of response times (RT) and percentage of erroneous responses are some of the understudy variables in these behavioural paradigms.

In this paper, two neural accumulator models [2–7] of visually guided eye movements in the absence of distractors at various levels of abstraction (molecular, single neuron, population of neurons, multiple brain areas and behaviour) are summarized to provide functional roles to the neural substrates involved in preparation and execution of the saccadic eye movements and explain which neural mechanisms are responsible of the response variability and error rate in a well established oculomotor task (i.e. antisaccade task).

2 Brain Anatomy and Physiology of Saccade Eye Movements

Several brain areas are involved in the control of saccadic eye movements [13]. Visual information from the external world enters the brain from the eyes through two distinct anatomical pathways: (1) From the retino-geniculo-cortical pathway to the primary visual cortex and (2) from the retinotectal pathway to the superficial layers of the SC. Visual information is subsequently processed through several extrastriate visual areas before it arrives in the lateral intraprietal area (LIP) in the posterior parietal cortex. LIP is at the interface between sensory and motor processing. The LIP in turn projects to both the intermediate layers of the superior colliculus (SC) and the frontal cortical oculomotor areas including the frontal eye fields (FEF), the supplementary eye fields (SEF) and the dorsolateral prefrontal cortex (DLFPC). The FEF has a crucial role in executing voluntary saccades, whereas the SEF is important for internally guided decision-making and sequencing of saccades [14]. The DLPFC is involved in executive function, spatial working memory and suppressing automatic, reflexive responses [15]. All these frontal regions project then to the SC, which is a vital node in the premotor circuit where cortical and subcortical signals converge and are integrated.

Furthermore, the FEF, SEF and SC project directly to the reticular formation to provide the necessary input to the saccadic premotor circuit that a saccade is initiated or suppressed. Frontal and posterior cortical oculomotor areas also project indirectly to SC through the direct and indirect pathways of the basal ganglia. Cortical inputs to the direct pathway lead to disinhibition of the SC and thalamus, whereas cortical inputs to the indirect pathway lead to the inhibition of both SC and thalamus.

3 Empirical Signatures

A behavioural paradigm often used to investigate decision processes is the antisaccade task [10]. The antisaccade task is a choice reaction time task in which subjects perform eye movements in the opposite direction from the location of a peripheral stimulus [10]. Recently, a large epidemiological study was conducted [8, 9] testing the performance of a large population of young male subjects in the antisaccade oculomotor task. A population of 2075 conscripts performed 90 trials of the antisaccade task as fast as possible without any accuracy constraints (see Fig. 1). Each subject was seated in front of computer monitor and he/she was asked to fixate to a stimulus in the centre of the screen. After a variable period of 1–2 s, the central stimulus was extinguished and immediately after another stimulus appeared randomly at one of nine distances (2–10° at 1° intervals) either to left or to the right of the central fixation stimulus. The subjects were instructed to make an eye movement to the opposite direction from that of the peripheral stimulus as quickly as possible. The following indices of performance were measured:

- Percentage of errors
- Mean latency of the first eye movement regardless of whether this was an error prosaccade or a correct antisaccade eye movement
- Standard deviation of the latency of the first eye movement
- Mean latency of correct antisaccades
- Standard deviation of the latency of the correct antisaccades
- Mean latency of error prosaccades
- Standard deviation of the latency of the error prosaccades
- Mean latency of corrections
- Standard deviation of the latency of corrections

Saccade reaction time was defined as the time taken from the first appearance of the peripheral stimulus 'till the first detectable eye movement. Trials with reaction times <80 ms were excluded as anticipations and trials with reaction times >600 ms were excluded as no response trials. Only three eye movement behaviours were observed: (1) the subject made the correct antisaccade, (2) the subject made the error

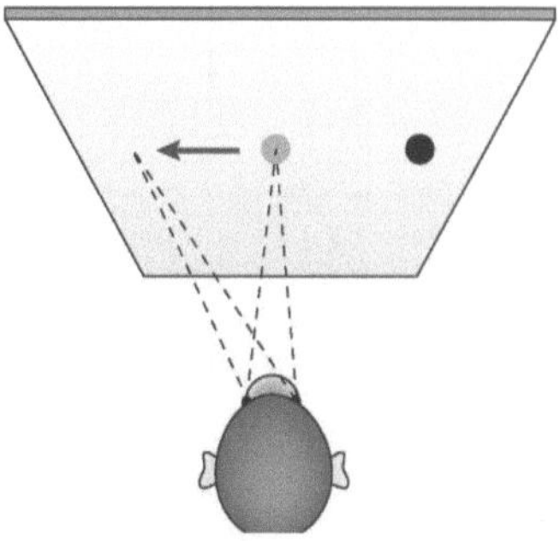

Fig. 1 Experimental setup of the antisaccade task (reproduced with permission from [14], Fig. 1,p. 221, Copyright © Nature publishing company)

prosaccade (very rare), and (3) the subject made an error prosaccade followed by a correct antisaccade. *At no time was ever observed a subject to make a correct antisaccade followed by an error prosaccade in the same trial.* A unimodal distribution of correct antisaccades and erroneous prosaccades were observed. The mean latency and standard deviation of the correct antisaccade from all subjects, respectively, were 270 ± 39ms and 56 ± 19ms. The mean latency and standard deviation of the error prosaccades from all subjects, respectively, were 208 ± 38ms and 46 ± 27ms. Finally, a 23% ± 17% of erroneous prosaccades of all subjects were reported.

These results raised some very important questions: (1) Why are the mean latencies of the correct antisaccades and error prosaccades so variable between trials in each subject and across all subjects? (2) Why the error rate is only 23%? Why not 50%? (3) What stops the error prosaccade from been expressed after the correct antisaccade has been released first? (4) Which are the neural mechanisms that justify these results?

These questions have been successfully addressed by the neural population accumulator model of the SC [2,6,7] summarized in the next section.

4 A Neural Population Accumulator Model of Decision Making Constrained by Antisaccade Data

The first model (see Fig. 2), which we will call the neural population accumulator model was a one-dimensional model of the intermediate layer of the superior colliculus (SC) [2,6,7]. The connectivity between neurons in the population was assumed to be on-centre off-surround. The internal state of each neuron i was given by

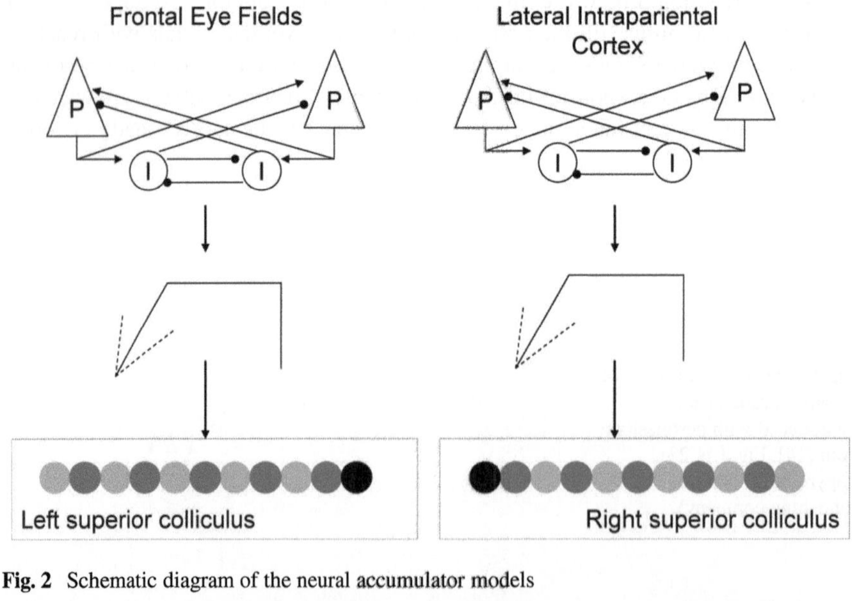

Fig. 2 Schematic diagram of the neural accumulator models

$$\tau \frac{dx_i(t)}{dt} = -x_i(t) + \sum_j w_{ij} A_j(t) + I_p(t) + I_r(t) - u_0 + I_n, \tag{1}$$

where I_n was the noise background input, u_0 was a global inhibition term, I_p and I_r were the two external inputs and τ was the integration constant. The average firing rate of each neuron was then given by

$$A_i(t) = \frac{1}{1 + \exp(-\beta u_i(t) + \theta)}, \tag{2}$$

where β was the sigmoid steepness and θ was the sigmoid offset. The interaction matrix between nodes was given by

$$w_{ij} = a \exp\left(\frac{-(j-i)^2}{2 \cdot \sigma_A^2}\right) - b \exp\left(\frac{-(j-i)^2}{2 \cdot \sigma_B^2}\right) - c, \tag{3}$$

where *a, b, c* were free parameters and σ_A, σ_B were spatial parameters. Three different types of SC neurons were modelled: fixation, buildup and burst neurons. Briefly, in the superior colliculus, fixation neurons discharge tonically when the subject is fixating and pause their activity when a saccade is initiated. On the other hand, buildup neurons discharge only when a saccade is initiated. Burst neurons discharge phasically and provide the final motor command to the brainstem neurons for the generation of an eye movement.

In the model, the two external inputs, which represented the FEF and LIP decision signals were modelled by

$$\begin{aligned} I &= A \cdot |slope \cdot t|, && \text{if } t_{on} + t_{delay} \le t \le t_{off} + t_{delay} \text{ and } I < I_{\max} \\ I &= A \cdot I_{\max}, && \text{if } t_{on} + t_{delay} \le t \le t_{off} + t_{delay} \text{ and } I \ge I_{\max} \\ I &= 0, && \\ A &= \exp\left(\frac{-(j-i)^2}{2 \cdot \sigma_A^2}\right), && \end{aligned} \tag{4}$$

where t_{delay} was the conduction delay from the retina to LIP (70 ms) and FEF (120ms), I_{max} was the theoretical maximum SC neuronal activity and *slope* was the slope of linearly rising phase of each input. The slope of each input varied from trial to trial from a different normal distribution with a certain mean, μ, and standard deviation, σ for each input. The value of the theoretical maximum SC neuronal activity of the FEF input was assumed to be larger from the theoretical maximum SC neuronal activity of the LIP input. This assumption reflected the instruction given to each subject in the beginning of each trial that they should always make the correct antisaccade even if their first eye movement was an error prosaccade.

In the model, decisions were formed via stochastic accumulating processes and contrast enhancement of the two decision signals. More specifically, the two cortically independent and spatially separated decision signals representing the reactive

(LIP) and planned (FEF) saccade signals, whose linearly rising phases were derived from two normal distributions with different means and standard deviations were integrated at opposite SC buildup cell populations, where they competed against each other via lateral excitation and remote inhibition. An ocular movement was initiated when the neuronal activities of the buildup cells reached a preset criterion level. The crossing of the preset criterion level in turn released the "brake" from the SC burst neurons and allowed them to discharge resulting in the initiation of an ocular movement.

To simulate the median reaction times, the shapes of the RT distributions of the correct antisaccades and the error prosaccades as well as the error rates of all 2075 subjects, we run the model for 1500 trials. In each trial, the slope of the reactive input took values for a normal distribution with mean μ_1 and standard deviation σ_1, whereas the slope of the planned input took values from another normal distribution with mean μ_2 and standard deviation σ_2. The mean values were estimated via a trial-and-error process, whereas the standard deviation values σ_1 and σ_2 were approximated so that the produced correct antisaccade and error prosaccade reaction times were greater than 80ms and less than 600ms. The threshold was adjusted so that the simulated error rate closely matched the observed one. Saccade reaction time (SRT) was estimated as the time taken from the first appearance of the peripheral stimulus till the time the burst activity started to deviate from zero. An additional 20ms efferent delay was also added.

To compare the SRT distributions of the experimental data with the simulated ones, we performed cluster analysis. The median RT and the inter-quartile range for antisaccades and error prosaccades of all 2075 conscripts were grouped into ten groups. The purpose of the cluster analysis was to partition the observations into groups ("clusters") so that the pairwise dissimilarities between those assigned to the same cluster tend to be smaller than those in a different cluster. We arbitrary chose ten clusters because we wanted each cluster to have a sufficiently large number of individuals (ranging from 30 individuals to 240 individuals in each cluster). We then normalized the SRT distribution of each subject data and then added the normalized distributions for all subjects belonging to the same group. For each category we calculated its percentage relative frequency of response times. More specifically, the time interval between 80 and 600ms was divided into twenty-six categories, each lasting 20ms. For each time bin, we added the SRTs. Plots of the simulated and experimental correct antisaccade and error prosaccade % density distributions of response times for all ten groups are displayed in Figs. 3 and 4.

The mean frequency for all subjects in a group was then calculated. The discrepancy in each category between the simulated and experimental correct and error distributions was measured by the squared difference between the observed (simulated) and the expected (experimental) frequencies divided by the expected frequency ((Observed-Expected)^2/Expected). The χ^2 value was the sum of these quantities for all categories. The rejection region was set at $\chi^2 \geq \chi^2_{0.05}$. The χ^2 test of homogeneity showed a significant difference in two of the ten comparisons for antisaccade RT distributions and two of the ten comparisons for the error prosaccade RT distributions (see Fig. 5).

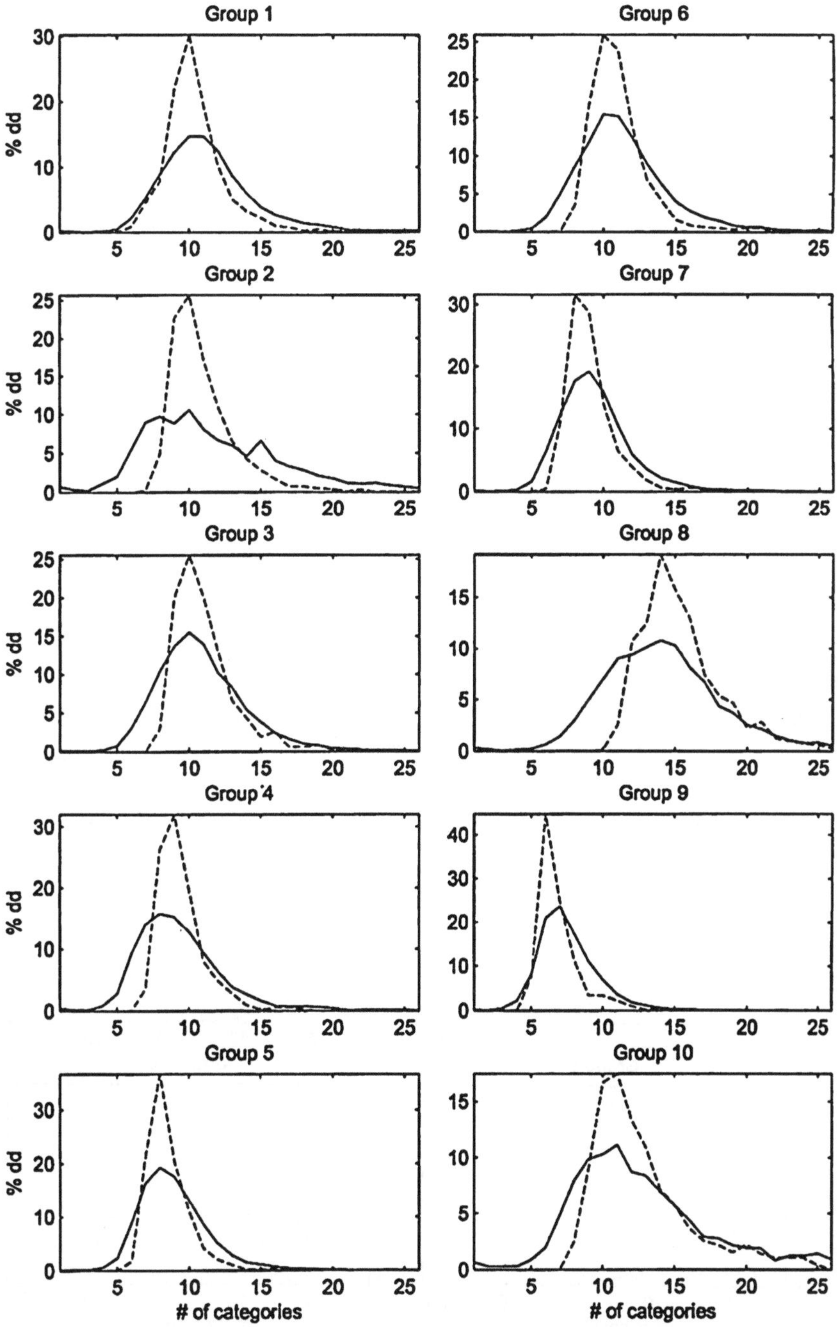

Fig. 3 Plots of correct percentage density distribution (y-axis) vs number of categories (x-axis) for all ten virtual subjects. Dashed lines: simulated correct percentage density distribution plots for all ten virtual subjects. Solid lines: experimental correct percentage density distribution plots for all ten virtual subjects. Reproduced with permission from [2], Fig. 5, p. 698, Copyright © Elsevier

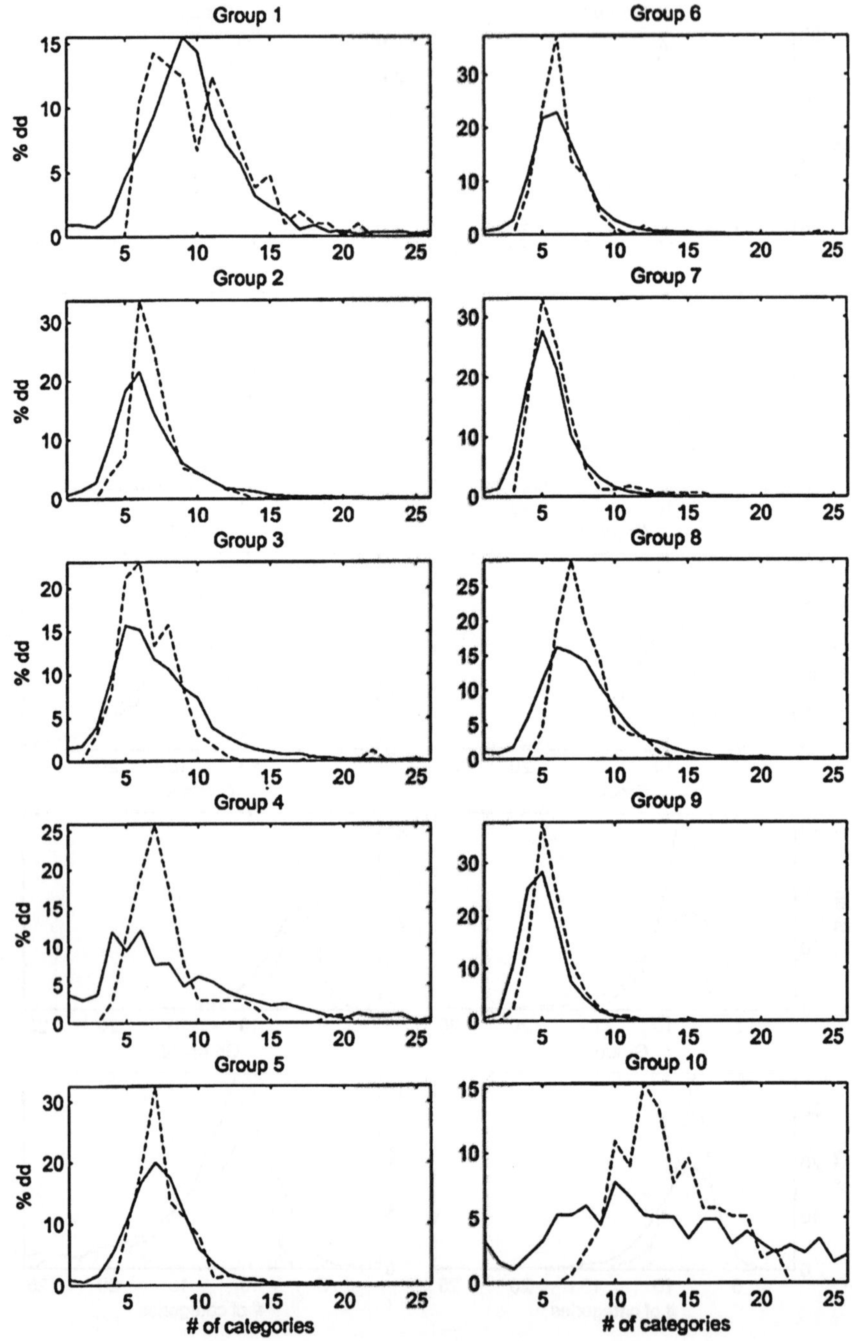

Fig. 4 Plots of error percentage density distribution (y-axis) vs number of categories (x-axis) for all ten virtual subjects. Dashed lines: simulated error percentage density distribution plots for all ten virtual subjects. Solid lines: experimental error percentage density distribution plots for all ten virtual subjects. Reproduced with permission from [2], Fig. 6, p. 699, Copyright © Elsevier

	Median correct antisaccade RT	Median error prosaccade RT	Percent antisaccade error rate	χ^2 correct antisaccades	χ^2 error prosaccades
Group 1	294.174 (288.16)	279.541 (265.20)	13.04 (16.15)	36.15	34.92
Group 2	276.50 (279.21)	202.97 (201.96)	38.62 (39.07)	90.5*	33.56
Group 3	281.89 (280.91)	212.54 (201.92)	20.15 (23.73)	32.16	32.89
Group 4	251.30 (249.27)	209.90 (211.65)	12.41 (12.02)	56.06*	96.24*
Group 5	254.80 (242.40)	212.99 (216.66)	24.27 (17.02)	35.21	24.18
Group 6	282.38 (288.44)	188.19 (193.66)	23.93 (28.86)	31.82	27.97
Group 7	263.10 (251.79)	180.63 (175.53)	20.87 (24.79)	30.34	21.82
Group 8	365.69 (349.42)	218.99 (221.36)	37.00 (34.58)	36.46	35.67
Group 9	218.20 (213.58)	177.85 (172.77)	27.36 (24.92)	36.99	23.15
Group 10	327.56 (307.5)	331.07 (326.99)	20.05 (21.81)	33.88	83.57*

Fig. 5 Simulated correct median, error median, error rate and values of χ^2 test of homogeneity between correct and error experimental and simulated percent density distributions for correct antisaccades and error prosaccades. χ^2 values marked with an asterisk indicate a significant difference between the simulated and the observed RT distributions. *Rejection region*: $\chi^2 \geq \chi^2_{0.05}$ (37.65). The degrees of freedom were 25. Units: correct SRT (ms); error SRT (ms). Values in parentheses stand for experimental values

The model was successful at explaining why the response times in the antisaccade task are so long and variable and at predicting accurately the shapes of correct and error RT distributions as well as the response probabilities of a large 2006 sample of subjects. The wealth of simulated results made the model unique in comparison to other models. The model predicted that there is no need of a top down inhibitory signal that prevented the error prosaccade from being expressed, thus allowing the correct antisaccade to be released. This finding challenged the currently accepted view of saccade generation in the antisaccade task, which requires a top-down inhibitory signal to suppress the erroneous saccade after the correct saccade has been expressed [14].

These results raised some additional important questions: (1) What are the biophysical mechanisms that produced the slowly varying climbing activity of the decision signals? (2) What are the biophysical mechanisms that produced the small varying threshold level (450 ± 50 Hz) across virtual subjects?

These questions were addressed successfully by the biophysical accumulator model summarized in the next section.

5 A Biophysical Accumulator Model

The second model [3–5], which I will call *biophysical accumulator model*, extended the previous neural SC population model of the antisaccade task by addressing the question of what were the biophysical mechanisms underlying the generation of the slowly varying accumulator like activity of the decision signals. The biophysical accumulator model was a multi-modular neural network model consisting of two cortical modules, each representing the population activity of FEF and LIP cortical neurons that drove the SC population rate model to produce saccade reaction times (SRT) and response probabilities in the antisaccade task.

The neuronal firing rates of both cortical modules were derived from the interplay of a wealth of ionic and synaptic currents. Hodgkin-Huxley mathematical formulations were employed to model these currents and the current balance equations of pyramidal neurons and inhibitory interneurons in the networks. The current balance equation of each pyramidal neuron was given by

$$C_p \frac{dV_p}{dt} = -I_L - I_{Na} - I_{Kd} - I_{HVA} - I_{NaP} - I_C - I_{Ks} - I_{AHP} - I_{AMPA} - I_{NMDA} - I_{GABAA} + I_{inj}, \quad (5)$$

whereas the current balance equation of each inhibitory interneuron was

$$C_{inh} \frac{dV_{inh}}{dt} = -I_L - I_{Na} - I_{Kd} - I_{AMPA} - I_{NMDA} - I_{GABAA} + I_{inj}, \quad (6)$$

Each ionic current followed the general ohmic relationship

$$I_{ionic} = g_{ionic} \cdot x \cdot (V - E_{ionic}), \quad (7)$$

where g_{ionic} was the maximal conductance of the particular ion channel, and E_{ionic} was the ionic reversal potential given by the Nernst equation for the particular ionic species. x was an activation or an inactivation variable (or a combination of variables depending on the particular current being modelled, and which can be raised to a non-unity power for a better fit to the data) that determined the fraction of open channels at a given time. These variables followed first-order kinetics:

$$\frac{dx}{dt} = \alpha_x \cdot (1 - x) - \beta_x \cdot x, \quad (8)$$

where α_x and β_x were voltage-dependent rate constants. Using a voltage-dependent time constant, τ_x, and a steady-state value, x_{-}, the differential equation was rewritten as

$$\frac{dx}{dt} = \frac{x_\infty - x}{\tau_x}, \quad (9)$$

where

$$\tau_x = \frac{1}{\alpha_x + \beta_x} \text{ and } x_\infty = \frac{\alpha_x}{\alpha_x + \beta_x}, \quad (10)$$

These types of equations were used to describe a variety of different voltage-gated ion channels. Experimentally, the steady-state activation variable could be measured using the voltage clamp protocol and fit to a Boltzmann function

$$x_\infty = \frac{1}{1 + \exp(-(V - V_{1/2})/k)}, \quad (11)$$

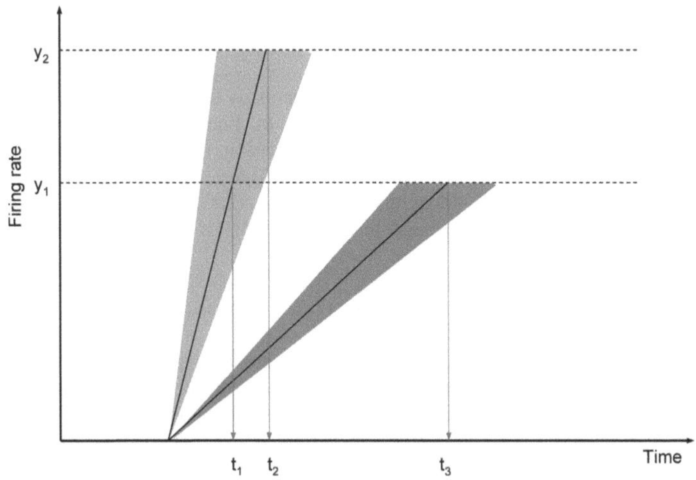

Fig. 6 Schematic diagram of the firing rate vs. time. Horizontal dashed lines depict two different threshold levels. Small increases in threshold level ($y_2 > y_1$) can result in large increases in the mean and standard deviation. Shaded area depict the variability (standard deviation) in response times

Complete mathematical formalism of the model and its parameters can be found in [3]. Both symmetric and asymmetric types of neuronal connectivities as well as homogeneous and heterogeneous neuronal firings were tested.

Detailed parametric analysis of all ionic and synaptic conductances in the model was performed to estimate which current(s) and what range of values could reproduce the full range of slope values (see Table 4 in [3]) of the planned and reactive inputs of the SC population rate model [2], while keeping the preset criterion level fixed.

It is important here to emphasize the need of keeping the criterion level fixed throughout all trials (see figure 6). Assume the criterion level is held fixed at a value y_1. Then, the resulting distribution has a mean at t_1 and a variance depicted by the light gray area. If we now move the criterion level by a small amount to y_2, then the new mean of the distribution shifts to a new value t_2 and its variance becomes much larger. That means that we have moved to a new category (i.e. a new virtual subject) with different mean and std.

The model predicted that only certain ionic and synaptic currents, namely the I_{NaP}, I_{NMDA}, and I_{AMPA} currents can produce the observed variability in the climbing activities of cortical decision signals, while keeping the preset criterion level fixed. We concluded that indirectly the model predicted the range of values of these currents' conductances' values that reproduced the correct antisaccade and error prosaccade reaction time (RT) distributions as well as response probabilities of a large group of 2006 subjects.

Acknowledgments VC was supported by the EPSRC project grant EP/D04281X/1.

References

[1] P Dayan, S Dehaene, K McCabe, R Menzel, E Phelps, H Plassmann, R Ratclif, M Shadlen, W Singer. Neuronal correlates of decision making. In: Engel C, Singer W, editors. Strungmann Forum Report: Better than conscious? Decision making, the human mind, and implications for institutions. MIT Press: Cambridge, MA, USA

[2] V Cutsuridis, N Smyrnis, I Evdokimidis, S Perantonis (2007b), "A Neural Network Model of Decision Making in an Antisaccade Task by the Superior Colliculus", *Neural Networks*, 20(6): 690–704

[3] V Cutsuridis, I Kahramanoglou, N Smyrnis, I Evdokimidis, S Perantonis (2007a), "A Neural Variable Integrator Model of Decision Making in an Antisaccade Task", *Neurocomputing*, 70(7–9): 1390–1402

[4] V Cutsuridis, I Kahramanoglou, S Perantonis, I Evdokimidis, N Smyrnis, "A Biophysical Neural Model of Decision Making in an Antisaccade Task Through Variable Climbing Activity", In: Artificial Neural Networks: Biological Inspirations – ICANN '05, Lecture Notes in Computer Science, LNCS 3696 (Springer-Verlag, Berlin, 2005a) 205–210

[5] V Cutsuridis, I Kahramanoglou, N Smyrnis, I Evdokimidis, S Perantonis, "Parametric Analysis of Ionic and Synaptic Current Conductances in a Neural Accumulator Model with Variable Climbing Activity", *19th Conference of Hellenic Society for Neuroscience*, Patra, Greece, September 30 - October 2, 2005b

[6] V Cutsuridis, N Smyrnis, I Evdokimidis, I Kahramanoglou, S Perantonis, "Neural network modeling of eye movement behavior in the antisaccade task: validation by comparison with data from 2006 normal individuals", Program No. 72.13. 2003 Abstract Viewer/Itinerary Planner. Washington, DC: Society for Neuroscience, 2003b

[7] V Cutsuridis, I Evdokimidis, I Kahramanoglou, S Perantonis, N Smyrnis, "Neural network model of eye movement behavior in an antisaccade task", *18th Conference of Hellenic Society for Neuroscience*, Athens, Greece, October 17–19, 2003a

[8] Evdokimidis, I., Smyrnis, N., Constantinidis, T. S., Stefanis, N. C., Avramopoulos, D., Paximadis, C., et al. (2002). The antisaccade task in a sample of 2006 young men I. Normal population characteristics. Experimental Brain Research, 147, 45–52

[9] Smyrnis, N., Evdokimidis, I., Stefanis, N. C., Constantinidis, T. S., Avramopoulos, D., et al. (2002). The antisaccade task in a sample of 2006 young males II. Effects of task parameters. Experimental Brain Research, 147, 53–63

[10] Hallett, P. R. (1978). Primary and secondary saccades to goals defined by instructions. Vision Research, 18, 1279–1296

[11] von Neumann, J., & Morgenstern, O. (1944). Theory of games and economic behaviour. Princeton: Princeton Univ. Press

[12] Sutton, R. S., & Barto, A. G. (1998). Reinforcement learning: an introduction. Cambridge: MIT Press

[13] Wurtz, R. H., & Goldberg, M. E. (1989). The neurobiology of saccadic eye movements. Amsterdam: Elsevier

[14] Munoz, D. P. & Everling, S. (2004). Look away: the antisaccade task and the voluntary control of eye movement. Nat Neurosci, 5, 218–228

[15] Guitton, D., Buchtel, H. A., Douglas, R. M. (1985). Frontal lobe lesions in man cause difficulties in suppressing reflexive glances and in generating goal-directed saccades. Exp Brain Res, 58, 455–472

Part II
Biologically Inspired Systems

Preface

Biologically inspired systems have come to refer to a wide range of systems which draw their inspiration from some aspect of biological systems. The reason for the interest in such systems is that biological systems are capable of feats simply not attainable by electronic systems. Biological inspiration may be from the actual physical form of the biological systems, as in the paper by Smith, or from a much more abstract view of biological systems, as in the papers by Windridge and Kittler, and by Loula et al. In both these cases, it is the capabilities of biological systems that the authors are hoping to emulate.

Another view of biologically inspired systems is that they are about attempting to understand the nature of what really is happening inside animal brains, and the paper by Estombelo-Montesco et al. takes this approach. Lastly, the term may be used to discuss machines which try to join the biological systems and the computer based systems, and this area is tackled in Bassani and Nievola paper.

As a result, the papers in this section cover a wide range, from learning for meaning and generation of representations, to investigating cortical activity using FMRI and Brain-Computer interfacing, to a review of Neuromorphic systems. What joins these chapters together is the biological inspiration of the techniques that each is using.

Loula et al. use a Peircean philosophy based approach to develop a computational experiment in which, under quite a carefully determined set of constraints, an intercommunication system develops which is essentially symbolic. Windridge and Kittler present a detailed and carefully thought out paper on learning which produces a self-updating cognitive representation and which is grounded in the perception/action loop. Their paper includes an experiment which validates their technique.

Estombelo-Montesco et al. use a novel version of ICA, namely DCA, to attempt to improve the localisation of the fMRI signals. Their technique does appear to be successful when applied to signals in the auditory cortex. Bassani and Nievola use a mixture of wavelet based transforms and Naïve Bayes classifiers in the interpretations of EEG signals, assessing this on an image classification task.

Lastly, Smith reviews work using primarily analogue techniques which are biologically inspired but are actually directly implemented in electronics. Tracing this back to early work with discrete components, he discusses when and how they might come out of the lab into more general use.

Leslie Smith

On Building Meaning: A Biologically-Inspired Experiment on Symbol-Based Communication

Angelo Loula, Ricardo Gudwin, Sidarta Ribeiro, and João Queiroz

Abstract The use of an appropriate set of empirical and theoretical constraints to guide the construction of synthetic experiments leads to a better understanding of the natural phenomena under study, and allows for a greater understanding of the experimental results. We begin this chapter with a description of a general approach for conducting experiments with artificial creatures within a synthetic ethological context. Next, we describe how this approach was used to build a computational experiment regarding the emergence of self-organized symbols. Our experiment simulated a community of artificial creatures undergoing complex intra and inter-specific interactions in which meaning evolved over time, from a *tabula rasa* repertoire of random alarm-calls to a specific set of optimal referential alarm-calls. To design different kinds of creatures as well as innanimate elements of the environment, we applied theoretical constraints from the Peircean philosophy of sign and empirical constraints from neuroethology. Our results suggest that the constraints chosen were both necessary and sufficient to produce symbolic communication.

Keywords Communication · Meaning · Semiosis · Symbol process · Self-organization · Emergence · Computer simulation.

A. Loula
Department of Exact Sciences, State University of Feira de Santana, Brazil
e-mail: angelocl@ecomp.uefs.br

A. Loula and R. Gudwin
Department of Computer Engineering and Industrial Automation, FEEC, State University of Campinas, Brazil
e-mail: angelocl@ecomp.uefs.br; gudwin@dca.fee.unicamp.br

S. Ribeiro
Edmond and Lily Safra International Institute of Neuroscience of Natal (ELS-IINN), Brazil
and
Department of Neuroscience, Federal University of Rio Grande do Norte, Brazil
e-mail: ribeiro@natalneuro.org.br

J. Queiroz (✉)
Research Group on Cognitive Science and Semiotics,
Federal University of Juiz de Fora, UFJF, Brazil
e-mail: queirozj@semiotics.pro.br

A. Hussain et al. (eds.), *Brain Inspired Cognitive Systems 2008*, Advances in Experimental Medicine and Biology 657, DOI 10.1007/978-0-387-79100-5_5,

1 Introduction

Synthetic approaches (as opposed to analytical ones) (cf. Braitenberg 1984) in Cognitive Sciences are advocated by many researchers as a promising pathway for a better understanding of higher-level cognitive functions as e.g. learning, memory, attention, emotions, communication and language. Among other synthetic strategies, one which drives attention of many different researchers comprises the field of 'artificial creatures'. An artificial creature is a special kind of agent which inhabits a given environment, where it lives and performs actions, based on some sort of perception of the surroundings. The main goal of building artificial creatures is to have a simple and controllable framework in which to study the evolution and development of low-level and higher-level cognitive functions, testing different theories and eventually creating new ones. Even though there are many successful examples of experiments with artificial creatures, in many different levels, there is one problem that still haunts the field, being a great shortcoming and sometimes jeopardizing the appreciation of these studies. This shortcoming relates to the way in which are conceived the design of the environment and the morphological definitions of sensors, effectors, cognitive architecture and processes of the conceived creatures. In many experiments, these decisions are somewhat naïve or arbitrary. In other experiments, despite being influenced by either meta-principles (formal theoretical constraints) and/or biological motivations (empirical constraints), these constraints are not explicitly stated, giving rise to fair criticism on the kinds of conclusions that can be derived from these experiments.

Here we argue that the successful conception of an experiment with artificial creatures requires being explicit about the theoretical and experimental constraints that will drive the experiment. This theoretical basis influences modeling on different degrees depending on how it constrains the model being built and what decisions it leaves to the experimenter. Constraints entail a reduction in the degrees of freedom that we can assume while building the experimental setup, by 'setting values to experimental parameters' following definitions and motivations from more reliable sources than naïve or arbitrary *ad-hoc* decisions. If theoretical foundations and constraints are used to develop computational experiments, these experiments may also provide contributions back to the theories and studies they were based upon. Simulations test hypotheses, the internal consistency of their theoretical background, and offer the opportunity to implement experiments that would be more/too costly or even impossible otherwise.

The general approach described above may be used widely in the field of Artificial Life in order to inspire and guide the design and construction of experiments. Here we focus our attention on a particular subfield of Artificial Life which comprises the study on the emergence of meaning among artificial creatures. Different computational approaches have been used to model and simulate *meaning processes (semiosis)*, including Evolutionary Robotics, Artificial Life, Synthetic Ethology, and Computational Semiotics (for some examples, see section 4). The understanding and modeling of 'Meaning' is certainly a great challenge to computer scientists. It is also related to two classical problems regarding the construction of artificial systems: the symbol-grounding problem and the frame problem. According to Deb Roy (2005), 'developing a computationally precise and tractable theory of language use which simultaneously addresses both referential and functional meaning is a grand challenge for the cognitive sciences'. A somewhat established approach describes that meaning process should be contextually grounded and acquired during local interactions among artificial distributed agents.

In the next sections, we present our specific methodology for the investigation of the emergence of self-organized symbol-based communication involving distributed interactions between artificial creatures. In the section "Theoretical and Empirical Constraints", we describe the formal constraints which we used to derive our experiment. Basically, the setup, design and synthesis of our creatures, along with their digital ecosystem, are theoretically based on the Peircean pragmatic philosophy of sign and empirically informed by neuroethological evidence. In order to infer the minimum organizational constraints for the design of our creatures, we also examined the well-studied case of semiosis in East African vervet monkeys (*Cercopithecus aethiops*), and its possible neuroanatomical substrates.

In the section "From Constraints to a Synthetic Experiment" we show how these constraints shaped our experiment. We view the emergence of communication as a self-organized process in a complex system of sign users interacting locally and mutually affecting each other, leading to an ordered state. Our methodology simulates the emergence of symbolic predator-warning communication among artificial creatures in a virtual world of predatory events, where these creatures continuously interact with each other.

Finally, we outline our conclusions and list the main advantages of using our methodology in the specific case of the study of emergence of meaning among artificial creatures.

2 Theoretical and Empirical Constraints

2.1 Constraint A. Semiosis and Communication in Semiotics

Which are the relevant attributes and properties to be considered to computationally simulate meaning processes? Scientists have been adopting different frameworks about the meaning phenomena, and produced computational models based in backgrounds as different as the more internalist versions and those more compromised with extended mind theory and distributed cognitive framework (see Queiroz and Merrell 2009). The main constraints considered here are derived from Peirce's pragmatic philosophy of sign. Peirce's model of meaning as the 'action of signs' (semiosis) has had a deep impact (besides all branches of semiotics) on philosophy, psychology, theoretical biology, and cognitive sciences (see Freeman 1983, Fetzer 1997, Colapietro 1989, Tiercelin 1995, Hoffmeyer 1996, Deacon 1997, Freadman 2004, Hookway 2002).

Peirce is often considered the founder of modern semiotics (Weiss and Burks 1945: 386). Semiotics was defined by Peirce (1967 §5.484) as "the doctrine of the essential and fundamental nature of all varieties of possible semiosis". In other words, semiotics describes and analyzes the structure of semiotic processes independently of their occurrence, or of the conditions under which they can be observed – inside cells (cytosemiosis), among tissues and cell populations (vegetative semiosis), in animal communication (zoosemiosis), or in typically human activities (production of notations, meta-representations, etc.).

According to Peirce's pragmatic approach, semiosis (meaning process) is an interpreter-dependent process that cannot be dissociated from the notion of a situated (and actively distributed) communicational agent. It is an interpreter-dependent process in the sense that it triadically connects sign (representation), object, and an effect on the interpreter (interpretant). The object is a form (habit, regularity, or a 'pattern of constraints') embodied as a constraining factor for interpretative behavior – a logically 'would be' fact of response. The notion of semiosis as a form communicated from object to interpreter through mediation of a sign allows us to conceive meaning, and meaning change, in a processual (non-substantive) way, as a constraining factor of possible patterns of interpretative behavior through habit and change of habit.

Semiosis is also pragmatically characterized as a behavioral pattern that emerges through the intra/inter-cooperation between agents in a communication act, which concerns an utterer, a sign, and an interpreter (Peirce 1958 §11, §318). Meaning and communication processes are thus defined in terms of the same "basic theoretical relationships" (Ransdell 1977), i.e., in terms of a self-corrective process whose structure exhibits an irreducible relation between three elements. In a communication process, "[i]t is convenient to speak as if the sign originated with an utterer and determined its interpretant in the mind of an interpreter" (Peirce 1958 §11).

2.2 *Constraint B. Sign Model and Classes*

In his "most fundamental division of signs", Peirce characterized icons, indexes, and symbols as matching, respectively, relations of similarity, contiguity, and law between sign and object. Icons are signs which stand for their objects through intrinsic similarity or resemblance irrespective of any spatio-temporal physical correlation that the sign has with an existent object. In contrast, indexes can only occur when the sign is really determined by the object, in such a way that both must exist as concurrent events. Finally, in a symbolic relationship, the sign refers to the object by a determinative relation of law or convention, a "habit (acquired or inborn)", regardless of "the motives which originally governed its selection." In terms of cognitive processes, icons are associated with sensory tasks. They are present in the sensory recognition of external stimuli of any modality, and in the cognitive relation of analogy. By contrast, the notion of spatio-temporal co-variation between sign and object is the most characteristic property of indexical processes.

The examples range from a demonstrative or relative pronoun, which "forces the attention to the particular object intended without describing it" (Peirce 1958 §1369), to physical symptoms of diseases, weathercocks, thermometers. We have claimed elsewhere that the alarm-call system used by African vervet monkeys (*Cercopithecus aethiops*), a well-known case of vocal communication in non-human primates, logically satisfies the Peircean definition of symbol (Ribeiro et al. 2007, Queiroz and Ribeiro 2002). Generally speaking, a symbolic sign communicates a habit embodied in an object to the interpretant as a result of regularity in the relationship between sign and object.

2.3 Constraint C. Referential Communication in Non-human Animals

An analysis of semiotic behavior we have made point out that some non-human animals can be seen as communicating using symbols as defined by Peirce's theory (Ribeiro et al. 2007, Queiroz 2003). They mostly constitute simple cases of symbol usage without further symbol-related properties, such as recursion or compositionality. The case of predator-warning alarm-calls in vervet monkeys constitutes a well-characterized example of referential communication. Field studies (Seyfarth and Cheney 1980, Struhsaker 1967) have revealed three main kinds of alarm-calls used to warn about the presence of (a) terrestrial stalking predators, (b) aerial raptors, and (c) ground predators. The correct use of alarms calls depends on some sort of learning processes since adult vervets are able to do so, while infant vervets initially do not, but gradually develop this ability (Seyfarth and Cheney 1986). The assumption that the mapping between signs and objects can be learned is also supported by the observation that cross-fostered macaques, although unable to modify their call production, "did learn to recognize and respond to their adoptive mothers' calls, and vice versa" (Cheney and Seyfarth 1998). The alarm-call system in vervet monkeys is a useful example of a symbolic semiotic system, which can be simulated through a community of agents that implement the 'minimum brain model' presented below (see Ribeiro et al. 2007).

2.4 Constraint D. Neural Representation Domains and Association Rules

For an adequate development of our semiotic creatures, it was crucial to determine the minimum set of neurobiological constraints to be implemented in programming code in order to generate the desired final behaviors. A minimum vertebrate brain was modeled as being composed by three major representational relays or domains: the sensory, the associative and the motor. According to such minimalist design, different first-order sensory representational domains (RD1s) receive unimodal stimuli, which are then associated in a second-order multimodal representation domain (RD2) so as to elicit symbolic responses to alarm-calls by means of a first-order motor representation domain (RD1m). The process by which a virtual creature learns to associate representations was modeled to follow the rules first postulated by Donald Hebb (Hebb 1949), by which synchronous pre-synaptic inputs generate synaptic reinforcement. The functions performed by associative representational domains include the combinatorial association of sensory and motor representations (e.g. cross-modal perceptual processing in the cerebral cortex (Calvert 2001, Andersen and Buneo 2002, Lloyd et al. 2003), the attribution of adaptive value to sensorimotor representations (e.g., emotional processing in the amygdala (Rodrigues et al. 2004, McGaugh 2004) and the implementation of short-term, fast-retrieve, erasable memory (e.g., working memory in the orbitofrontal cortex and hippocampus (Suzuki 1999, Rolls 2000). As discussed below, a neurosemiotic model of the alarm-call system in vervet monkeys assuming just such minimum neural constraints reveals the emergence of symbol-based referential communication (Ribeiro et al. 2007, Queiroz and Ribeiro 2002), and allows for the investigation of different semiotic stages of behavior ontogenesis.

2.5 *Constraint E. Self-Organization and Emergence of Communication Processes*

Self-organization is a process that mainly occurs in complex systems composed of many interacting entities that mutually affect each other's state, leading the system to an 'ordered' state, i.e. a state of reduced variability and ambiguity, with increased redundancy. Communication processes can be viewed as self-organizing if utterers and interpreters mutually affect each other, through local interactions in communicative acts, such that their future communication interactions are dependent of the past ones. In fact, sign users capable of learning though communicative interactions with others, correspond, in self-organizing systems, to entities capable of affecting others (as utterers) and of being affected (as interpreters) in a self-correcting process. By means of these ongoing processes, an ordered stated can be produced such that communicative variability (such as sign usage repertoire) or ambiguity is reduced, without any external or central control.[1]

We claim that the digital scenario we developed in our experiment leads to the emergence of self-organized symbol-based communication among artificial creatures. In the context of the sciences of complexity, the concept of 'emergence' has become very popular, to the extent that these fields are often described as dealing with 'emergent computation'. We employ the analysis of emergence applied to semiotics put forward by Queiroz and El-Hani (2006) and extended in Loula et al. (in press).

[1] The idea of communication/language as a self-organizing process have been presented also by other authors, e.g. (Steels 2003, Keller 1994).

Applying the hierarchical model for semiotic systems developed by Queiroz and El-Hani (2006) to explain emergent semiotics processes, we should consider (i) a focal level, where an entity or process we want to investigate is observed in the context of a hierarchy of levels; (ii) a lower level, where we find the parts composing that entity or process; and (iii) a higher level, in which the entities or processes observed at the focal level are embedded. Both the lower and the higher levels have constraining influences over the dynamics of the processes at the focal level. The emergence of processes (e.g., symbol-based communication) at the focal level can be explained by means of the interaction between these higher- and lower-level constraints so as to generate its dynamics. At the lower level, constraints amount to initial conditions and the limited set of possibilities arising from the emergent process. On the other hand, constraints at the higher level are related to the selective role played by the environment, establishing boundary conditions that coordinate or regulate the dynamics at the focal level.

Semiotic processes at the focal level are described here as communication events. Accordingly, what emerges at the focal level is the product of an interaction between processes taking place at lower and higher levels, i.e., between the relations within each sign-object-interpretant triad established by individual utterers or interpreters and the embedding of each individual communicative event, involving an utterer, a sign and an interpreter, in a whole network of communication processes corresponding to a semiotic environment or context.

The macro-semiotic (or higher) level regulates the behavior of potential S-O-I relations; it establishes the patterns of interpretive behavior that will be actualized by an interpreter, among the possible patterns it might elicit when exposed to specific signs, and the patterns of uttering behavior that will be actualized by an utterer, among the possible patterns it might elicit when vocalizing about specific objects. This macro-semiotic level is composed of a whole network of communicative events that already occurred, are occurring and will occur; it characterizes the past, present, and future history of semiotic interactions, where utterers are related to one or more interpreters mediated by communicated signs, interpreters are related to one or more utterers, and interpreters turn into utterers. We can talk about a micro-semiotic (or lower) level when we refer to a repertoire of potential sign, object, and interpretant relations available to each interpreter or utterer. Thus, in the micro-semiotic level we structurally describe the sign production and interpretation processes going on for an individual involved in a communicative act and, therefore, we talk about S-O-I triads instead of sign-utterer-interpreter relations. When an utterer, mediated by a sign, is connected to an interpreter, and thus a communication process is established, we can talk about a focal level, which necessarily involves individual S-O-I triads being effectively formed by utterer and interpreter. But in a communicative event, the actualization of a triad depends (1) on the repertoire of potential sign, object, and interpretant and (2) on macro-semiotic networks of communication processes, which define a context for communicative processes with boundary conditions that restrict possibilities to actual occurrences (for more details, see (Queiroz and El-Hani 2006)).

3 From Constraints to a Synthetic Experiment

3.1 An Experiment on Symbol-Based Communication Emergence

The creatures are autonomous agents inhabiting a virtual bi-dimensional environment. This virtual world is composed of prey and predators (terrestrial, aerial and ground predators), and of things such as trees (climbable objects) and bushes (used to hide). Prey produce vocalizations when they see predators, indicating the presence of a predator image in their visual systems. A vocalization immediately becomes available to other prey by way of audition, raising their arousal and increasing their chance to evade predation. In the present work, prey were not initially divided into apprentices and tutors, as seen in the contrast between infant and adult vervet monkeys, and as we have previously simulated (Loula et al. 2004). Rather, we sought to generalize the vervet monkey case of symbol-emergence by investigating a *tabula rasa* scenario in which no previous repertoire of alarms calls is known, and no symbol-based communication occurs yet.

Creatures are equipped with sensors (visual and auditory) and actuators that allow for specific actions (e.g. move, vocalize, change gaze direction), controlled by a behavior-based architecture (Mataric 1998), with multiple parallel behavior modules such as wandering, visual scanning, fleeing or chasing. This control architecture allows the creature to choose between different conflicting actions, given the state of the environment and the internal state of the creature.

Associative learning is the mechanism used by prey to gradually establish connections between auditory and visual data. The development of these associations relies on no explicit indication of the correct referent to be associated with alarms or whether the connection made was mistaken. The constitution of alarm-predator association mainly depends on the statistical co-occurrence of events, such as alarms being vocalized in the presence of nearby predators.

Working and associative memories were implemented in prey creatures, to allow them to learn temporal and spatial relations from the external stimuli and thus acquire association rules necessary to interpret signs as symbols. In working memory, stimulus-related information is kept for a few instants, allowing different stimuli received in close instants to co-occur in memory. Associative memory holds associations (with strength values between 0 and 1) that are created, reinforced and weakened according to the co-occurrence of stimuli in the working memories. In our model, associative memory formation follows Hebbian learning principles (Hebb 1949).

When a prey hears a nearby creature vocalize a specific alarm-call, it initially scans the surroundings, searching for possible co-occurring events and helping associations to be learned. A feedback may also be provided by the associative memory to the control mechanism, if the vocalization heard is already associated with a specific predator type. Depending on the association strength, it can influence the creature's behavior as if the related predator was actually seen, eliciting an escape response.

As associations are learned, prey use them to emit specific alarms when a predator is seen (when multiple alarms arise, the strongest one is eventually selected); if no alarm is known for the predator, a new one is randomly created and associated with the predator. More technical details about the experiment can be found in (Loula et al. 2004).

3.2 The constraints of the experiment

We simulate an ecosystem that allows the cooperative interaction of agents, including intra-specific communication by alarm calls to alert about the presence of predators (Constraint C). For the emergence of symbol-based communication and a global coherent repertoire of alarms, prey should rely only in local communicative interactions that affect individuals through learning, setting conditions for self-organization (Constraint E). Associative learning is the mechanism used by prey to gradually establish connections between auditory and visual data, in line with the evidence that alarm-calls are learned (Constraint C).

The associative learning conception was aided by several constraints. First, symbols are an interpretant-mediated sign-object relation, i.e. a mental association or a habit that has to be built in the creature to associate sign-object, in such a way that no external clue is needed for the creature to connect that sign to that object (Constraint B). Communication can play a major role in the constitution of such habit, as habits are transmitted from an utterer to an interpreter in a communication event (Constraint A). This process of habit transmission is aided by an indexical relation between each alarm call uttered and all the possible scanned referents at any given episode. When an alarm is heard, arousal raises and the information gathered by the sensors becomes available to associative memory. This indexical intermediate-step greatly helps to build symbols. Prey, however, must still be able to find out which referents are suitable, i.e., they must be able to generalize a useful association for future occurrences.

An architecture relying on two separate unimodal representation domains and a higher order multimodal representation domain where associations are established must be involved, thus our architecture follows this general model as a plausible scheme (Constraint D). The memory architectures of the artificial creatures were in essence the same as the minimum brain described before: creatures have working memories (RD1s) and an associative memory (RD2). Hebbian associative learning principles are a simple mechanism widely found in non-human animals (Constraint C). Associations established in RD2 may produce effects in the motor control architecture (RD1m), producing an immediate escape response after alarm hearing.

Self-organization is the process that describes the underlying dynamics of the emergence of symbol-based communication as much as a global pattern for a common repertoire of symbols (Constraint E). By communicating, a vocalizing prey affects the sign repertoire of the hearing prey, which will adjust their own repertoire to adapt to the vocalized alarm and the context in which it is emitted. Thus, the vocal competence will also be affected as it relies on the learned sign associations. This implies an internal circularity among the communicative creatures, which leads to the self-organization of their repertoires. This circularity is characterized by positive and negative feedback loops: the more a sign is used the more the creatures reinforce it, and, as a result, the frequency of usage of that sign increases; in turn, the less a sign is used the less it is reinforced, and, consequently, its usage is decreased.

In this self-organizing system, a systemic process (symbol-based communication), as much as a global pattern (a common repertoire of symbols), emerge from local communicative interactions, without any external or central control. This complex system of communicative creatures can be viewed as a semiotic system of symbol-based communication with three different hierarchical levels (Constraint E). The semiotic processes of symbol-based communication emerge at the focal level through the interaction of a micro-semiotic level, containing a repertoire of potential sign, object, and interpretant relations within an interpreter or an utterer, and a macro-semiotic level, amounting to a self-organized network of all communication processes that occurred and are occurring, involving vocalizing and hearing prey and their predators. It is in this hierarchical system that things in the environment become elements in triadic-dependent processes, i.e., alarms (signs) come to be associated with predators (objects) in such a manner that their relationship depends on the mediation of a learned association (i.e., they become symbols). In order to give a precise meaning to the idea that symbol-based communication emerges in the simulations we implemented, we argue that the semiotic processes at stake are emergent in the sense that they constitute a class of processes in which the behavior of signs, objects, and interpretants in the triadic relations actualized in communication processes cannot be deduced from their possible behaviors in simpler relations. Their behaviors, and, consequently, the semiotic process these behaviors realize, are irreducible due to their non-deducibility from simpler relations.

The system can be seen as moving in a state space defined by all individual sign repertoires. The system moves from point to point each time a creature adjusts its repertoire, i.e. when learning takes place. In this search space, attractors are defined as points where all individual repertoires converge to a common one, thus stabilizing the system. When the system stabilizes, creatures will be relating predators and alarms in the same way, and vocalizing and interpreting signs in the same manner.

3.3 *Simulations of Interactions Among Creatures*

In order to study the self-organization dynamics of communicative acts, we placed prey and predators in the same environment. During the execution of the simulations, we assessed the associative memory items and the behavior responses of the prey to alarm calls. At first, prey vocalize random alarms when predators are spotted by, and since no associations have been established yet, the hearing prey responds indexically to an alarm call through the visual scanning behavior that allows them to search for co-occurring events, triggering the learning process. After a while, no more new alarms are created since every prey already knows at least one alarm for each predator. After many iterations of this process, creatures begin to engage in symbol-based communication, which is characterized by faster responses, as previously verified in the specific case of tutor/apprentice populations (Ribeiro et al. 2007, Loula et al. 2004). The symbolic threshold is reached when the association between alarm and predator gets near maximum, leading to the consistent use of that association and eventually to the direct activation of the fleeing behavior, without the need to visually search for a predator. Hence, at this optimum value, the prey stops scanning after an alarm is heard, and flees right away; at this point, the communicative behavior can be interpreted as a symbol-based one. Now, the interpretation of a sign (alarm), i.e., the establishment of its relation to a specific object (a predator type) depends solely upon an acquired habit, and not on a physical correlation between sign and object, a property that qualifies the alarm sign to be interpreted as a symbol.

Simulation results show that there was a convergence to a common repertoire of associations between alarms and predators. This is a repertoire of symbols that make the prey engage in escape responses when an alarm is heard, even in the absence of visual cues. Here we present results from a typical simulation run, using 4 prey and 3 predators, together with various bushes and trees. We let the simulation run until the population of prey converged to a common sign repertoire for the predators. Initially none of the prey have alarms associated with predators. As previously described, when no alarm is known for a seen predator, prey create alarms by randomly selecting one out of 100 possible alarms, vocalizing it and establishing an initial association in its memory. Therefore, at the beginning of the simulation, new alarms are randomly created when prey meet predators. This creates an initial explosion in the amount of available alarms, which tend to be in greater number than the existing predator types. In Figure 1, we see that various alarms were created to refer to each predator at first, but soon they stop appearing because every prey will know at least one alarm for each predator. In the graph of figure 1a, the terrestrial predator is associated with alarms 12, 14, 32, 38, 58 and 59, but only alarm 32 reaches the maximum value of 1.0, and the competing alarms are not able to overcome it at any time. Similar results were found in the case of alarms 14, 32, 58 and 59 associated with the aerial predator (figure 1b): only alarm 58 reached a maximum value. But among the alarms for the ground predator (figure 1c), there was a more intense competition that led to the inversion of positions between alarms 38 and 59. They were created almost at the same time in the population, and initially alarm 38 had a greater mean value than alarm 59. But between iteration 1000 and 2000, the association value of alarm 59 overcame the value of alarm 38, which slowly decayed, reaching the minimum value after iteration 9000. This 'competition' between signs and the convergence to a unique one mainly stem from the self-organization dynamics.

As prey are both sign users and sign learners, they work as media for signs to compete. In a successful interaction, the interpreter associates the sign with the

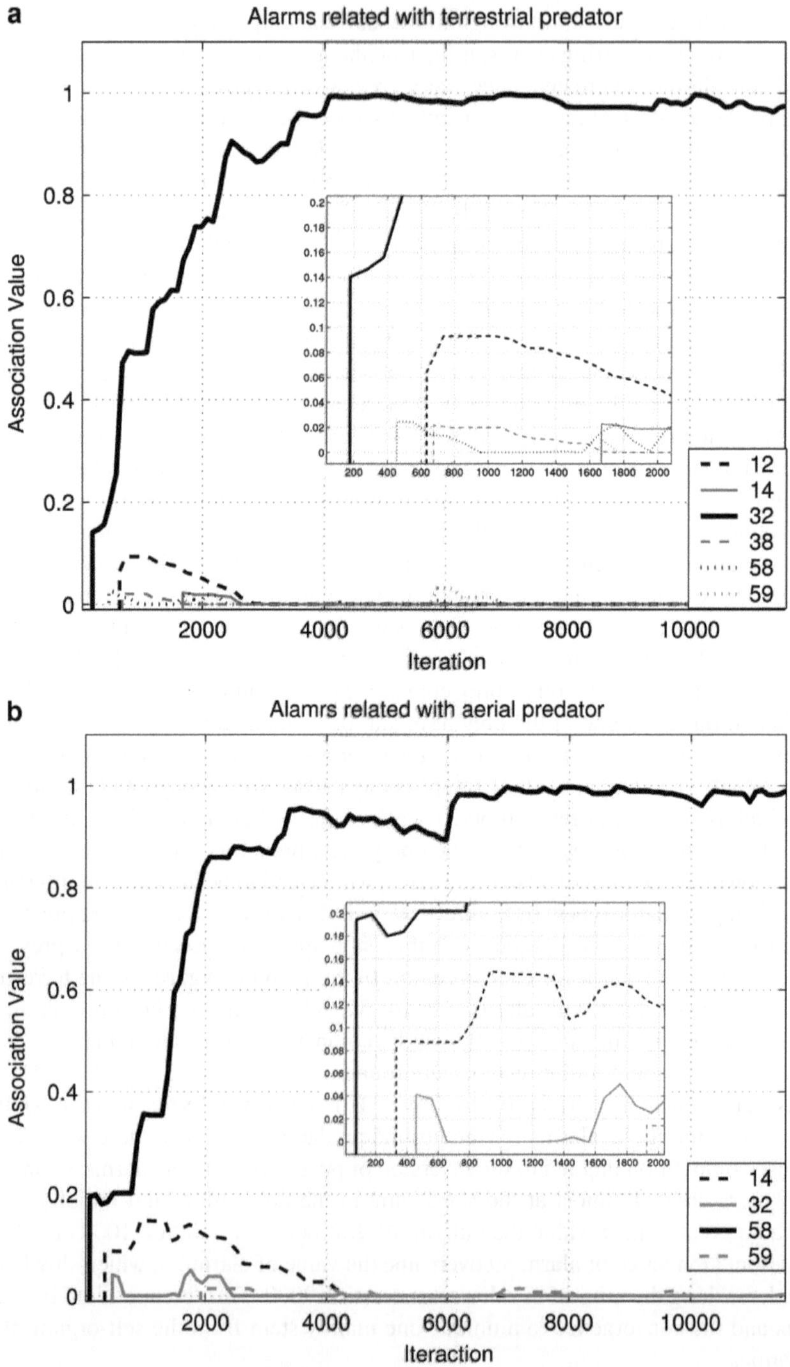

Fig. 1 The mean association values of the alarm-referent associations for 4 self-organizers: (**a**) terrestrial predator, (**b**) aerial predator

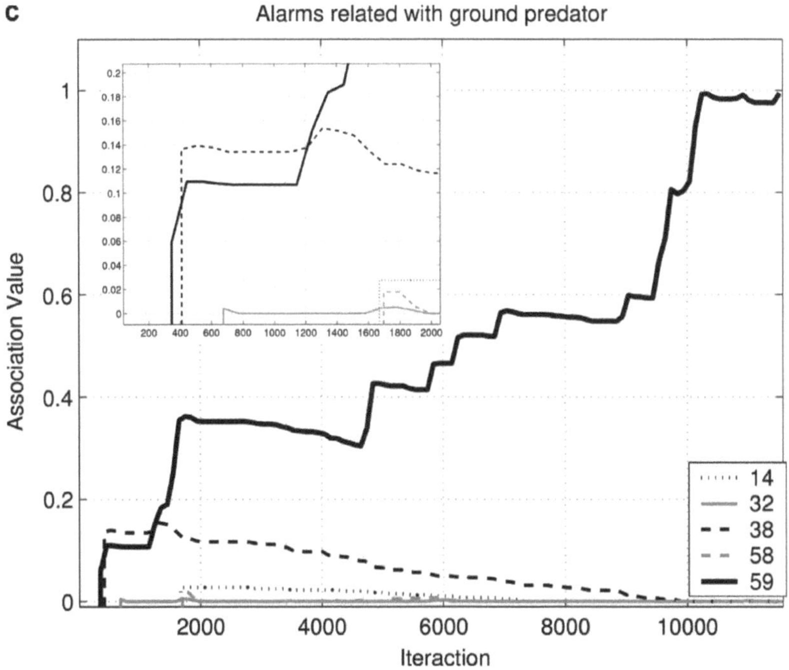

Fig. 1 (continued) (**c**) ground predator

same referent the utterer used it for. In this case, the associations employed by both interpreter and utterer will be reinforced. In the opposite case, the associations will be weakened. The stronger the sign association is, the more it will be used, and the more it is used, the more it will be reinforced. This positive feedback loop allows the self-organization of the sign repertoire of the prey population, with alarm-referent associations getting ever stronger so as to eventually lead to the transformation of indexes into symbols, i.e., a habit shared by the entire population of prey.

4 Related Work

There are connections of our work with experiments concerning the symbol grounding problem, and the self-organization and emergence of shared vocabularies and language in simple (real or virtual) worlds (Steels 2003, Sun 2000) (for a recent review of other works, see (Wagner et al. 2003)). As a typical project in ALife, we simulate an ecosystem that allows cooperative interaction between distributed agents, including intra-specific communication, a process that can raise the fitness of individuals in the face of predatory events.

Some of the related work follows empirical constrains as biological motivations (MacLennan 2002, Noble 1998), but none applies neurobiological constraints at the same time. MacLennan (2002) proposes an approach called synthetic ethology, which can be viewed as a sub-field inside Artificial Life. In synthetic ethology, simple worlds and creatures are created to study anima behavior, but in the experiment proposed for studying the evolution of communication no ethological case was analyzed and used to built the experiment, only general principles where applied, ending with a quite simplified experiment, far from real world study cases. Noble (1998) on the other hand relied on biological theories about the evolution of communication, such as the handicap principle and communication as manipulation, and evolutionary simulation models (close to a game theory approach) to test this theories, but no specific ethological case was used and there was no intend to model the cognitive apparatus of the creatures.

Other studies present theoretical foundations by referring to Peirce's work (Cangelosi et al. 2002, Jung and Zelinsky 2000, Sun 2000, Vogt 2002, Roy 2005), but they just borrow Peircean definitions of symbol or sign without generating any further consequences to the designed experiment. Sun (2000), Vogt (2002) and Roy (2005) bring forth definitions of signs and symbols from Peirce's work, but they end up applying them, changing them and mixing them with definitions from others, in such a way that we cannot conclude whether the experiments built were actually based on Peirce's theory or if it contributed, validating or not, Peirce's theory. Cangelosi et al. (2002) and Jung and Zelinsky (2000) presents Peirce's theory through a second hand reading of Deacon's work, which is at least a limited analysis of Peircean theory and particularly of his definition of a symbol. As a consequence, we can that they were not able to recognize a symbol when it first occurred in their experiments.

Deacon's reading of Peirce's theory is the most popular example at hand of such disconnection between theoretical framework and actual research (Deacon 1997). His depiction of humans as the only 'symbolic species' is based on the assumption that symbols necessarily have combinatory properties, and that only the human prefrontal cortex could possibly implement such properties. However, this proposal is incongruent with Peirce's theory and frontally collides with several empirical lines of evidence (for a discussion of this point, see Ribeiro et al. (2007), Queiroz and Ribeiro (2002)). Poeppel (1997) already recognized the 'problematic' and 'speculative' manner in which Deacon built his arguments using Peirce's theory, comparative and evolutionary approaches to language and even linguistic theories.

Just bringing forward a definition from Peirce's theory without deriving any consequence or constraint to the experimental setup certainly reduces the explanatory power of the proposed model. Recognizing the inter-dependence of Peirce's concepts at different levels, such as the sign model and its derived sign classification, substantially enrich computational experiments willing to simulate communication and its relationship to meaning.

5 Conclusion

The simulation of virtual ecological communities formed by synthetic cognitive creatures constitutes a powerful tool for the investigation of communication, allowing for the generation and testing of hypotheses. This approach, however, is extremely sensitive to *a priori* choices of theoretical and empirical constraints, which ultimately determine the occurrence of the phenomena of interest. The definition brought forth to describe, for example, what constitutes a symbol, may change the way the whole experiment is conceived; if an incorrect set of constraints is adopted, the experimenter may even fail to recognize the sought phenomena when it happens. A project that builds a simulation from theoretical and experimental constraints, is also an implementation of the underlying models and thus a test bed for hypotheses derived from these models. By the same token, such experiments constitute a way to falsify or corroborate the proposed models.

In our experiment, the simulation of a virtual community leads to the emergence of symbolic communication and representations, suggesting that the constraints adopted are sufficient to implement symbol-based communication. An analysis of cognitive processes observed in vervet monkeys suggests that symbol learning begins with the acquisition of indexical relations, which reproduce spatial-temporal regularities, detected during this process. Simulations indicate that the learning process will, eventually, result in law relations, which can be generalized to other contexts, particularly when a sign stands for a class of objects, formally satisfying the established conditions to describe symbolic semiosis. Symbols thus result from simple mechanisms of associative learning and self-organized interactions, which allow sign users to mutually affect each other communicative behavior but feedback loops (both positive and negative) conduct the system to an ordered state where symbol-based communication can be achieved.

Self-organizing principles are a common feature of many biological systems (see Morgavi et al. (2005)). Since we are simulating communication processes among biologically inspired creatures, it is expected that self-organization dynamics would play a major role in this process. Self-organization is also compatible with Peirce's theory, especially with his communication model accompanied by habit change processes, its self-correcting dynamics and the circular relations between interpreters and utterers. Self-organization can be seen as an important element in the emergence of new systemic processes in semiotic systems, where a hierarchy of levels can be described and used to better understand the generation of the phenomena. Emergence theory in the context of complexity sciences and semiotic systems, as well as computational experiments that simulate this process, are described in detail elsewhere (see Loula et al. (in press), Queiroz and El-Hani (2006)).

In summary, our synthetic experiment involves a virtual community characterized by random inter-specific predation and intra-specific cooperative referential communication among prey. The simulation was inspired on the case of referential communication provided by African vervet monkeys, which employ predator-specific alarm calls to warn against attacks. The simulation was also shaped by carefully selected semiotic and neurobiological constraints. Our present results generalize previous findings obtained in apprentice/tutor populations (Ribeiro et al. 2007) to groups of artificial creatures initially void of any symbolic competence. The results demonstrate the feasibility of implementing symbolic communication based on a very limited but well-chosen set of constraints. Virtual neurosemiotic communities constitute a flexible and fruitful tool for the generation and testing of hypotheses regarding the ontogeny and phylogeny of animal communication.

Acknowledgments This work was supported by FAPESB, CNPq and AASDAP.

References

Andersen RA, Buneo CA (2002) Intentional maps in posterior parietal cortex. Annual Review of Neuroscience 25:189–220.

Braitenberg V (1984) Vehicles: Experiments in Synthetic Psychology. MIT Press, Cambridge, Massachussets.

Calvert GA (2001) Crossmodal processing in the human brain: insights from functional neuroimaging studies. Cerebral Cortex 11(12):1110–23.

Cangelosi A, Greco A, Harnad S (2002) Symbol grounding and the symbolic theft hypothesis. In: Cangelosi A, Parisi D (ed) Simulating the Evolution of Language. Sprinter. London. chap. 9.

Cheney DL, Seyfarth RM (1998) Why monkeys don't have language. In: Petersen G (ed) The Tanner Lectures on Human Values, vol. 19.University of Utah Press, Salt Lake City. pp.173–210.

Colapietro V (1989) Peirce's Approach to the Self: A Semiotic Perspective on Human Subjectivity. State University of New York Press, New York.

Deacon TW (1997) The symbolic species: The co-evolution of language and brain. W.W. Norton Company, New York.

Fetzer JH (1990) Artificial Intelligence: Its Scope and Limits. Kluwer Academic Publishers, Dordrecht, The Netherlands.

Freadman A (2004) The Machinery of Talk — Charles Peirce and the Sign Hypothesis. Stanford University Press, Stanford.

Freeman E (1983) The Relevance of Charles Peirce. Monist Library of Philosophy, La Salle.

Hebb DO (1949) The Organization of Behavior: A Neuropsychological Theory. John Wiley & Sons, New York.

Hoffmeyer J (1996) Signs of Meaning in the Universe. Indiana University Press, Bloomington, IN.

Hookway C (1985) Peirce. Routledge & Kegan Paul, London.

Jung D, Zelinsky A (2000) Grounded symbolic communication between heterogeneous cooperating robots. Autonomous Robots journal 8(3):269–292.

Lloyd DM, Shore DI, Spence C, Calvert GA (2003) Multisensory representation of limb position in human premotor cortex. Nature Neuroscience 6(1):17–18.

Keller R (1994) On language change: The invisible hand in language. Routledge, London.

Loula A, Gudwin R, El-Hani CN, Queiroz J (in press) Emergence of Self-Organized Symbol-Based Communication in Artificial Creatures. Cognitive Systems Research.

Loula A, Gudwin R, Queiroz J (2004) Symbolic Communication in Artificial Creatures: an experiment in Artificial Life. In: Bazzan A, Labidi S (ed) 17th Brazilian Symposium on Artificial Intelligence – SBIA (Lecture Notes in Computer Science 3171:336–345). see also, www2.uefs.br/graco/symbcreatures/

MacLennan BJ (2002) Synthetic ethology: a new tool for investigating animal cognition. In: Bekoff M, Allen C, Burghardt GM (ed) The Cognitive Animal: Empirical and Theoretical Perspectives on Animal Cognition. MIT Press, Cambridge, Mass. chap. 20, pp. 151–156.

Mataric M (1998) Behavior-Based Robotics as a Tool for Synthesis of Artificial Behavior and Analysis of Natural Behavior. Trends in Cognitive Science 2(3):82–87.
McGaugh JL (2004) The amygdala modulates the consolidation of memories of emotionally arousing experiences. Annual Review of Neuroscience 27:1–28.
Morgavi G, Morando M, Biorci G, Caviglia D (2005) Growing up: emerging complexity in living being. Cybernetics and Systems 36(4):379–395.
Noble J (1998) The Evolution of Animal Communication Systems: Questions of Function Examined through Simulation. D. Phil. thesis, University of Sussex, November, 1998.
Peirce CS (1967) Annotated catalogue the papers of Charles S. Peirce. Robin RS (ed). University of Massachusetts Press, Amherst. §11, 318.
Peirce CS (1958) Collected Papers of Charles Sanders Peirce. Harvard University Press, Cambridge, Mass.
Poeppel D (1997) Mind over chatter. Nature 388:734.
Queiroz J (2003) Comunicação simbólica em primatas não-humanos: uma análise baseada na semiótica de C.S.Peirce. Rev Bras Psiquiatr 25 (Supl II): 2–5.
Queiroz J, El-Hani CN (2006) Semiosis as an Emergent Process. Transactions of the Charles Sanders Peirce Society 42(1):78–116.
Queiroz J, Merrell F (2009) On Peirce's pragmatic notion of semiosis – a contribution for the design of meaning machines. Minds & Machines 19: 129–143.
Queiroz J, Ribeiro S (2002) The biological substrate of icons, indexes and symbols in animal communication. In: Shapiro M (ed) The Peirce Seminar Papers 5.Berghahn Books, Oxford, UK. pp. 69–78.
Ransdell J (1977) Some leading ideas of Peirce's semiotic. Semiotica 19(3):157–178.
Ribeiro S, Loula A, Araújo I, Gudwin R, Queiroz J (2007) Symbols are not uniquely human. Biosystems 90:263–272.
Rodrigues SM, Schafe GE, Ledoux JE (2004) Molecular mechanisms underlying emotional learning and memory in the lateral amygdale. Neuron 44(1):75–91.
Rolls ET (2000) Memory systems in the brain. Annual Review of Physiology 51:599–630.
Roy D (2005a) Semiotic Schemas: A Framework for Grounding Language in Action and Perception. Artificial Intelligence 167(1–2):170–205.
Seyfarth RM, Cheney DL (1986) Vocal development in vervet monkeys. Animal Behaviour 34:1640–1658.
Seyfarth RM, Cheney DL, Marler P (1980) Monkey responses to three different alarm calls: Evidence for predator classification and semantic communication. Science 210:801–803.
Steels L (2003) Evolving grounded communication for robots. Trends in Cognitive Science 7(7):308–312.
Struhsaker TT (1967) Behavior of vervet monkeys and other cercopithecines. New data show structural uniformities in the gestures of semiarboreal and terrestrial cercopithecines. Science 156(779):1197–203.
Sun R (2000) Symbol grounding: A new look at an old idea. Philosofical Psychology 13(2): 149–172.
Suzuki WA (1999) The long and the short of it: memory signals in the medial temporal lobe. Neuron 24(2):295–8.
Tiercelin C (1995) The relevance of Peirce's semiotic for contemporary issues in cognitive science. In: Haaparanta L, Heinämaa S (ed), Mind and Cognition: Philosophical Perspectives on Cognitive Science and Artificial Intelligence. Acta Philosophica Fennica 58. pp. 37–74.
Vogt P (2002) The physical symbol grounding problem. Cognitive Systems Research 3(3): 429–457.
Wagner K, Reggia JA, Uriagereka J, Wilkinson GS (2003) Progress in the simulation of emergent communication and language. Adaptive Behavior 11(1):37–69.
Weiss P, Burks A (1945) Peirce's sixty-six signs. Journal of Philosophy XLII: 383–388.

Perception-Action Learning as an Epistemologically-Consistent Model for Self-Updating Cognitive Representation

David Windridge and Josef Kittler

Abstract As well as having the ability to formulate models of the world capable of experimental falsification, it is evident that human cognitive capability embraces some degree of representational plasticity, having the scope (at least in infancy) to modify the primitives in terms of which the world is delineated. We hence employ the term 'cognitive bootstrapping' to refer to the autonomous updating of an embodied agent's perceptual framework in response to the *perceived* requirements of the environment in such a way as to retain the ability to refine the environment model in a consistent fashion across perceptual changes.

We will thus argue that the concept of cognitive bootstrapping is epistemically ill-founded unless there exists an *a priori* percept/motor interrelation capable of maintaining an empirical distinction between the various possibilities of perceptual categorization and the inherent uncertainties of environment modeling.

As an instantiation of this idea, we shall specify a very general, logically-inductive model of perception-action learning capable of compact re-parameterization of the percept space. In consequence of the *a priori* percept/action coupling, the novel perceptual state transitions so generated always exist in bijective correlation with a set of novel action states, giving rise to the required empirical validation criterion for perceptual inferences. Environmental description is correspondingly accomplished in terms of progressively higher-level *affordance* conjectures which are likewise validated by exploratory action.

Application of this mechanism within simulated perception-action environments indicates that, as well as significantly reducing the size and specificity of the *a priori* perceptual parameter-space, the method can significantly reduce the number of iterations required for accurate convergence of the *world-model*. It does so by virtue of the active learning characteristics implicit in the notion of cognitive bootstrapping.

Keywords Cognitive representation · Perception-action learning · Affordance · Subsumption hierarchy · Cognitive agency

---

D. Windridge (✉) and J. Kittler
Centre for Vision, Speech and Signal Processing, Faculty of Engineering & Physical Sciences, University of Surrey, Guildford, UK
e-mail: d.windridge@surrey.ac.uk,j.kittler@surrey.ac.uk

A. Hussain et al. (eds.), *Brain Inspired Cognitive Systems 2008*, Advances in Experimental Medicine and Biology 657, DOI 10.1007/978-0-387-79100-5_6,

1 Introduction

An autonomous cognitive agent is one capable of functioning within an unstructured environment. Typically, this functionality takes the form of the updating of a cognitive agent's internal world-model in response to exploratory findings, and the subsequent specification of appropriate goals within this revised domain [35]. In the most general cases, an autonomous cognitive architecture has the ability gain new goal-setting capabilities on the basis of generalizations about its previous exploratory findings (eg [19, 44]). In principle, this ability to gain new capabilities should extend to the possibility of autonomously revising the *perceptual* basis on which the agent's world experience are predicated (that is to say, the *manner* in which the world is represented).

It is evident that (within certain important constraints) human cognitive agency must exhibit this capability. For instance, a human adult may, within low-capacity short-term memory, characterize an external scene (its world model) as consisting of the primitives *car*, *buildings*, *trees*, etc, in various states of spatial interrelationship. However, since none of these perceptual categories are implicit in a newborn infant, the adult human must have acquired the perceptual categorizations in the course of their life. To some extent, these perceptual categories are social acquisitions, learned via interaction with other humans (perhaps in the manner of [48]). However, their initial delineation would still require explanation. More subtly (and at an epistemologically-deeper level), we might wonder why we, for instance, categorize the individually perceivable components *hands*, *arms*, *legs*, *etc* as the singular percept *person*. Equally, we might wonder why we do not naturally amalgamate the such sub-percepts as *tree-branch* and *house-window* into a singular percept '*branch-window*' (cf Quine's notion of 'ontological relativity' [37]). It is not obviously the case that these category-amalgamations are based on the innate perceptual (or physical) distance between sub-categories: even if such were the case, it would not be clear that the parameters governing perceptual clustering and sub-clustering could be innately specified so as to constitute 'useful' categories such as '*person*'.

For a static world, it is hence evident that if an individual's ability to create perceptual categories is not constrained in some manner (so that we *were*, for instance, free to amalgamate the sub-percepts *tree-branch* and *house-window* into a single category), there could be no absolute distinction made between world-model and perceptual category. However, this distinction constitutes the fundamental basis of cognition as being the act of 'knowing the external world'.

Given this problem with static distinctions, it is evident, then, that the capacity for the world to undergo change is crucial to its perceptual characterization. Hence, it is logical to form the composite perceptual category *car* from its individual components *door*, *wheels*, *windscreen*, *etc*, because these constitute *co-moving* entities. We thus implicitly constitute at least some of our perceptual categorizations on the basis of perceptual *co-articulation* (that is, the mutual dependency existing between low-level translatable percepts). Independently-mobile entities (eg *cars* and *humans*) thus constitute separate perceptual categorizations (even during moments of interaction).

However, this only accounts for the *self-motive* aspects of the world (those entities capable of initiating co-movement of individual parts without any external input). How then are we to account for the distinct segmentation of static, background perceptual entities such as *tree* and *building*? To arrive at a meaningful basis for these perceptual categorizations, we need to consider the motive possibilities of the *agent*; specifically an active agent embodied within the scene.

The problem addressed in this paper is therefore that of unsupervized perceptual category creation in embodied agents. A central concern is thus how to render this process empirically meaningful and free of foundational paradoxes (in the sense of the distinction between world-model and perceptual category being lost). We propose the strategy of *cognitive bootstrapping* as a means to accomplishing this.

We have elsewhere [2, 47] defined cognitive bootstrapping among autonomous sensory agents as the act of spontaneously inferring new perceptual categories in a manner that *simultaneously* allows for the continuous refinement of models of the embodied agent's external environment described in terms of those categories. It is evident from the preceding discussion that such perceptual inference must be appropriately constrained in order to ensure that novel percepts are meaningful and non-arbitrary, and further that perceptual-framework refinement and world-model refinement are kept distinct. That is, (for some hypothetical [potentially surjective] function, $p : W \rightarrow 2^{|P|}$, capable of relating a particular world-datum W to a corresponding *set* of observed percepts, $\{P\}$), we require that errors of representation, (eg of the form $p_{\text{observer}}(\{W_{\text{partial}}\}) \neq p_{\text{optimal}}(\{W_{\text{partial}}\})$), can be readily distinguished from errors of world-modeling (of the form $p_{\text{observer}}(\{W_{\text{generalized}}\}) \neq p_{\text{observer}}(\{W_{\text{all}}\})$). (The function $p_{\text{observer}}(\{W_{\text{generalized}}\})$ is thus a generalized world model in the observer's 'frame' of perceptual reference constructed from the partial observations $p_{\text{observer}}(\{W_{\text{partial}}\})$; the set $\{W_{\text{all}}\}$ is consequently the totality of (potentially) measurable world data, and p_{optimal} is the frame of perceptual reference when optimized via some appropriate criterion).

The chief epistemological danger for a self-updating cognitive system is thus that any completely unconstrained capacity for perceptual updating (for instance, permitting p_{observer} to range freely over the entire functional space: $p : W \rightarrow 2^{|P|}$) would have the potential to accommodate *any* given error-configuration of the world-model (ie so that $p_{\text{observer}}(\{W_{\text{generalized}}\}) = p_{\text{optimal}}(\{W_{\text{all}}\})$ for $p_{\text{observer}} \neq p_{\text{optimal}}$ and $\{W_{\text{generalized}}\} \neq \{W_{\text{all}}\}$), and thus lack the ability to found the distinction between the objective world and its representation. An appropriate analogy for p here is an orthonormal basis of perceptual primitives embedded within a Hilbert space of world-data vectors (such that the orthonormal perceptual primitives represent independent attribute classes such as color and shape). Thus, while it is possible for *any* given set of world data and generalized world models to be represented in *any* given basis (including a correct world-model within an incorrect basis), there always remains an optimal basis in which the distribution of world vectors is most compactly represented (in this case when the marginal distributions of world-vectors are maximally non-Gaussian). Consequently, independent optimization of p and $W_{\text{generalized}}$ is possible (by minimum description-length encoding and empirical refinement, respectively) if there exists a percept-independent way of generating the distribution of world-data vectors. It is the argument of the current paper that *actions* precisely fulfill this criteria.

A significant aspect of the survey endeavor of [46] was therefore to identify a mechanism for empirically-meaningful cognitive bootstrapping reconcilable to the concerns of both cognitive science (eg [6,13,43]) and philosophy (eg [18,22,24,29]). The identified mechanism consists in an *a priori* perception-action coupling capable of progressive, open-ended abstraction, such that an embodied agent equipped with it is able to form higher-level symbolic generalizations that are always grounded via corresponding actions hierarchically connected to the *a priori* perception-action coupling. It is thus the aim of an agent employing cognitive-bootstrapping to maximize the general descriptivity of percepts *in so far as they are relevant to the agent's actions*.

Hence, any formalization of this notion will require perceptual inference to be driven by the attempt to form a generalized mapping between perceptual states and environmentally-legitimate actions commencing from an initial, generic sensor-motor complex (which, being generic, is *not* optimized for the specific environment in which it is placed). This generalization can be accomplished by a variety of machine-learning methods (an early approach to symbolic generalization of the perception-action coupling was demonstrated via stochastic clustering in [2], though without the system having the genericity required to suggest actions existing outside and agent's previous experience). The concern of the current paper is hence to present a method for achieving this generalization in the most universal logical terms, in such a way as to enable entirely novel action classes to be induced by the agent.

1.1 Cognitive Bootstrapping in a Perception-Action Context

The importance of a perception-action (P-A) learning framework to cognitive bootstrapping arises from its reversal of the classical approach to agent-based learning, in which perceptual inference typically *precedes* exploration. (Such an agent forms a perceptual model of the world prior to attempting to interact with it). Instead, the perception-action agent performs perceptual inference only *after* exploratory actions have taken place [10, 14, 15, 40]. Hence, the P-A agent's world model is constructed only from those percepts that change as the result of the agent's actions. Thus for a set of *a priori* percepts $\{P\}$, and a set of actions $\{A\}$, the agent's world model is some generalization, f_{gen}, of the partial set of mappings $f : \{A_{\text{exp}}\} \rightarrow \{P_{\text{exp}}\} \times \{P_{\text{exp}}\}$ brought about by the set of exploratory activities A_{exp} apparent in the percept space $\{P_{\text{exp}}\}$ (where $\{P_{\text{exp}}\} \in \{P\}$). (This mapping has the form of a Cartesian product in consequence of the fact that an individual action A links a percept P_{initial} [existing at the outset of the action] to a percept P_{final} [existing at the end of the action]). Note here that we have assumed for simplicity (in distinction to the more general Hilbert-space example given earlier) that the set of *a priori* percepts $\{P\}$ is such that world configurations have an unambiguous representation as a specific *singular* percept; that is, we assume that perceptual attribute classes that are fully correlated with each other with respect to actions exist in some unspecified perceptual sub domain so as to keep the focus here purely on the percept-action relation. The generalized world-model f_{gen} hence represents the P-A agent's beliefs about its potential to undertake actions within the world having *directly observable consequences*. The *a priori* percepts $\{P\}$ are thus very general; they might, for instance, represent all of the distinguishable configurations of a visual sensor, such as a camera. (Hence, for an n-state, $X \times Y$-sized monochrome camera array, $|\{P\}| = n^{X \times Y}$). However, while it is necessary that f_{gen} has a domain and range derived from the sets $\{A\}$ and $\{P\} \times \{P\}$, respectively, it is unlikely that the mapping will prove exhaustive: the actions available to the agent, $f_{gen}(A_{gen})$, will, in general, only be able to access a subset of the total range of perceivable states $\{P\}$.

It is hence very unlikely that the *a priori* percept set $\{P\}$ is the most efficient means of representing the world. If the non-redundant (ie intentionally accessible) set of percepts brought about by exploratory activity $\{A_{exp}\}$ is denoted $\{P'_{exp}\}$, then we would expect that the generalized function f_{gen} has an equally generalized range $\{P'_{gen}\}$ deriving from $\{P'_{exp}\}$[1]. The set of generalized percepts $\{P'_{gen}\}$ hence consists in an abstracted, higher-level set of perceptual categorizations that are predicated purely on the basis of the agent's active capability.

To see how this might apply in practice, consider a practical example in which an *a priori* perceptual description is given of a discretized Cartesian space, X, such that each location can have a particular label selected from a set, L. This corresponds to the typical assumptions underlying standard image processing systems; ie an intensity-labeled pixel-grid. Under a perception-action bijectivity constraint of the form, $\{P\} \times \{P\} \Leftrightarrow \{A\}$ (see section 1.2), the set of all *conceivable actions* corresponds to the set of possible perceptual transitions in the *a priori* space; ie the bijectivity has the form: $\{|L|^{|X|}\} \times \{|L|^{|X|}\} \Leftrightarrow \{A\}$. Now suppose that an embodied cognitive agent equipped with this visual system determines via randomized exploratory motor activity (that activates, say, a series of robot limbs) that one of these labels persists over the exploratory transitions, ie that $l \rightarrow X_{\text{initial}}$ & $l \rightarrow X_{\text{final}}$ for some $(l \in L)$, $(X_{\text{initial}}, X_{\text{final}} \in X)$. A natural generalization of this assumption is that *all* labels, l, persist for *all* transitions, ie $\forall (l, X_{\text{initial}}, X_{\text{final}}) : l \rightarrow X_{\text{initial}}$ & $l \rightarrow X_{\text{final}}$ $(l \in L)$, where $X_{\text{initial}}, X_{\text{final}} \in X$.

However, under this generalized assumption, the space of possible transitions is vastly reduced, being now of magnitude $\{|L| \times |X|\} \times \{|L| \times |X|\}$, since each of the $|L|$ objects can only occupy *one* of $|X|$ locations. Under the perception-action bijectivity constraint ($\{P'\} \times \{P'\} \Leftrightarrow \{A_{\text{generalized}}\}$) governing revisions $\{P'\}$ of the percept space $\{P\}$ in the light of experimental data that indicate the existence of a generalized constraint, $\{A_{gen}\}$, upon the action-space $\{A\}$ ($A_{gen} \subset A$), we now have instead that: $\{|L| \times |X|\} \times \{|L| \times |X|\} \Leftrightarrow \{A\}$.

That is to say, the new percept-space, $\{P'\} = \{|L| \times |X|\}$, has the form of a polynomial of first order in $|X|$, while the former percept-space, $\{P'\} = \{|L|^{|X|}\}$,

[1] $\{P'_{exp}\}$ is a subset of $\{P_{exp}\}$ since, in general, not all of the exploratory actions will result in percepts that could be regarded as intentional; for instance, if an action end-point is unstable.

is exponential in $|X|$. Given that $|X|$ is typically large, this is a very significant economization of the percept domain. Furthermore, this alternative percept domain is a subset of the original domain (since the *a priori* space, $\{P\}$, must be chosen so as to *overpopulate* the space of action possibilities, in order that environmental constraints can be determined at all).

In general, following any such perceptual updating, we would still wish to retain the original *a priori* percept space so that any perceptual transitions not accommodated by the new perceptual space could still be perceived, enabling the proposed perceptual transition to be reversed if unsuitable. We thus make explicit the implicit subsetting $P' \subset P$ and consider, instead of complete transitions from one perceptual domain to another, rather the *hierarchical extension* of the percept space: $P' \supset P' \supset P'' \supset P''' \supset \ldots$. Hence, exploratory activity takes places at the *apex* of a recursively-inferred perceptual hierarchy, rather than within the original *a priori* space; lower levels of the perceptual hierarchy thus act to progressively contextualize the increasingly abstract action-specification occurring at higher levels of the hierarchy. In this way, (ie via recursive hierarchical abstraction of the percept space), we automatically construct a grounded symbol-generating and manipulating agent. Critically, in ascending the perceptual hierarchy, we still retain the ability to falsify *specific* action transitions, $a \in \{A'\}$, and thus retain the ability to construct a falsifiable model of the world in accordance with the standard (ie *non-perceptually updating*) conception of a learning agent.

Autonomous, exploration-based abstraction of an *a priori* perception-action coupling hence provides a means of updating both an agent's perceptual framework as well its model of the external world, without incurring the foundational paradoxes we might expect in attempting to employ a perceptual framework to postulate the existence external of objects, and those same objects to postulate the existence of alternative perceptual frameworks. Hence, by constraining proposed perceptual modifications to be hierarchical abstractions of the *a priori* perceptual framework $\{P\}$ (which cannot themselves be subject to empirical confirmation or refutation), we always ensure that both the perceptual framework and the objects perceived in terms of it (ie f_{gen}) are subject to empirical verifiability.

A further advantage of this strategy is that issues of framing [8,27] are largely resolved, since the agent eliminates much of the operationally-redundant environment descriptivity that occurs in classical approaches to autonomous cognition.

In an agent that employs perception-action learning *actively*[2] we might therefore expect that the generalized perceptual hypotheses built on exploratory actions could *themselves* serve to guide the exploration process. They would do this by suggesting novel percept states to which hypothesized agent actions could transit (ie the set of actions $\{A'_{\text{gen}}\}$ that connect the novel percepts $\{P'_{\text{gen}}\}$ together via f_{gen}). Exploratory actions in this scenario would hence serve to refine both object and percept hypotheses *at the same time*, bootstrapping object and percept models in response to environmental demands.

[2] Cf eg [3,32] for an illustration of differing forms of active learning.

It is hence the endeavor of this paper to exploit perception-action theory to set out a *cognitive bootstrapping mechanism* for autonomous logical agents capable of spontaneously inferring both the most appropriate perceptual model by which to interpret the agent's external surroundings, as well as the most accurate model of those surroundings *in terms* of the chosen perceptual framework. This environmental model, for a first-order-logic-based perception-action learner, then consists in a parameterized, clause-based description of the subset of the total motor-space that constitutes the set of legitimately performable actions, according to some set of very generalized physical criteria (in our case the performability and stability of the proposed action). Updating the perceptual framework will consequently involve the *reparameterization* of the perceptual variables in this action-based first-order-logic model in order to allow representation of the environment in the most action-relevant manner, permitting simultaneous empirical refinement of both the agent's world-model and its perceptual framework. Typically, because perceptual redundancy is progressively eliminated, actions are carried out at *increasingly higher levels of symbolic abstraction* as the mechanism proceeds.

1.2 Implementing Cognitive Bootstrapping in the Logical Domain

The central motivating factor of this work is thus the notion that specifying a simultaneous criterion for the inference of *both* objects and percepts is illogical, or even paradoxical, within classically dualistic cognitive models (eg [4]), since object-model errors can be arbitrarily subsumed within perceptual states and vice-versa[3].

The potential for unintentional 'cognitive tautology' through reciprocally defining percept and object states hence always exists in *fully* autonomous cognitive agents capable of updating their cognitive functionality in response to their external world model. We have hence argued that this can be overcome if there exists a strong *a priori* interconnection between percept and motor states. However, in addition to this percept-motor interconnectivity, the fully autonomous cognitive agent must, by definition, be capable of *generalized representation* of its experience, since the actions triggered in relation to particular perceptions would otherwise have to be specified individually, in advance. They would hence not be constitutive of the *intentionality* of action required for true autonomy. Various approaches to the generalization of experience within artificial cognitive agents are possible; for instance, clustering [26] and subspace representation [16]. We focus here exclusively on first-order logical representation, the simplest logic that permits *quantification* over instances. *Variables* are thus the entities within first-order logic that serve to unify possibilities of *instantiation*. Specific perceptions can hence be represented as

[3] Indeed, for a non-interacting cognitive agent, there are effectively no meaningful autonomously-derived criteria that can be applied to perceptual inferences in consequence of the fact that no non-arbitrary cost function capable of distinguishing between rival perceptual hypotheses exists *a priori* (cf Quine [37] on the ontological underdetermination problem).

instantiations of perceptual variables, and similarly, specific configurations of the objective domain can be considered as instantiations of the environmental variables.

Obviously, this implies that the environment is susceptible to characterization via first-order logical clauses. We contend that this will always be true at some level of the perception-action hierarchy, typically the very highest levels (where the level of abstraction of generalized percepts become such that formalized symbolic manipulation can be applied). Thus, although the human perceptual environment might be highly stochastic and contingent in the lower-levels (for instance, saliency clusters in the visual field), higher level representations (eg 'ladder', 'car') will, in general, be susceptible to rule-based reasoning of the kind: 'the object ahead is a ladder and could therefore be climbed', 'the object over there is a car and could thus be driven', etc. The current investigation is hence an attempt to demonstrate a mechanism for achieving cognitive bootstrapping at the formal and relational level, with the test environment thus characterized as entirely relational and without stochastic complication. As such, the method is intended to demonstrate the top-level of activity of an open-ended P-A learning agent (the relationship between the formal and stochastic learning is illustrated in [26], for a non-perceptually bootstrapping cognitive agent).

By attempting the implementation of active perception-action agency within a protocol-driven (ie *rule-based*) environment, we will hence show that in carrying-out first-order logic *induction* of the protocols that govern permissible actions it is possible to perform a remapping of the variable input/output structure of inferred clauses such that a *maximally compact* set of variables governing action-legitimacy can be obtained. Because of the perception-action linkage implicit in the agent's design, this variable remapping also serves to define a maximally compact set of perceptual variables through which the agent can reinterpret its external environment, with important consequences for the learning algorithm as indicated below.

This remapping is carried-out in the manner indicated earlier. Since, in perception-action theory, actions serve as the linkages or *relations* between individual perceptual states, one criterion for optimal perceptual representation is immediately suggested. This is the notion that perceptual states should be both *exhaustively* and *unambiguously* delineated by the hypothesized set of *legitimate* actions, $\{A_{\text{gen}}\}$, that have been previously determined by first-order logical inference. It is hence possible to define a *generalized* set of perceptual states that are ideally correlated with the inferred generalized action states through the application of a condition of *bijectivity* of action states with respect to pairings of the proposed perceptual states. (Recall that these pairings are necessarily temporal in nature, being between initial perceptual states, P_{initial}, and final perceptual states, P_{final}). Hence, we have that $\{A_{\text{gen}}\} \Leftrightarrow \{P'_{\text{gen}}\}_{\text{initial}} \times \{P'_{\text{gen}}\}_{\text{final}}$. This bijectivity criterion is applicable *at all stage of the perception-action hierarchy*, serving as a guiding constraint throughout the open-ended inference of novel perceptual frameworks and object-entities.

The notion of bijectivity between perception and action can be thought of as a formalization of the phenomenological theory that perception directly relates to, and indeed, *expresses* action possibilities (cf in particular [18]). Similarly, the notion of *affordance* (eg [13, 28]), currently the subject of much attention within cognitive science, proposes that what is perceived in the external world is not simply the inert set of spatial relations that characterize geometric objects. Rather, what is perceived is a set of entities characterized in terms of the activities and manipulations that an observing subject can perform with them. Both of these schools of thought hence attempt to define a middle ground within the classical distinction between subjects and objects by introducing the notion of embodied, active agents whose perceptual apparatus is an intrinsic part of the environment that they seek to describe[4].

The *maximally-compact descriptivity* of legitimate actions in the objective domain consistent with the bijectivity criterion thus constitutes one half of the proposed strategy for the optimization of perceptual representation. Acting in parallel with this is the opposing perceptual constraint of retaining maximal *agency* within the environment, such that the agent must be capable of perceptually distinguishing the consequences of *all* of the possible legitimate actions determined by first-order logical inference. The result of these two constraints is hence a criterion capable of overcoming the Quinean [37] problem of the arbitrariness of *perceptual* inference with respect to observed data, which, unlike the standard problem of object-model inference with respect to observed data, is underdetermined for non-interactive cognitive agents. In this scenario, optimal perceptual representation thus explicitly maintains the link with optimal *objective* representation, yet retains sufficient distinction from it to enable relatively independent cognitive updating without compromising the 'objectivity' of the perceptually-derived environment model.

Implementing cognitive bootstrapping in the logical domain will hence firstly involve the use of first-order logical induction to derive a generalized model of legitimate actions, f_{gen}, from exploratory samplings of the *a priori* action space. f_{gen} thus has the form of a set of logical *clauses* defined over the set of *a priori* perceptual variables $\{P\} = \{P_1, P_2, P_3 \ldots\}$. Secondly, a reparameterization, $(\{P'_{\text{gen}}\} = \{P'_1, P'_2, \ldots\}$ s.t. $|P'_{\text{gen}}| \subset |P|)$, of the perceptual variables present in the inferred clauses will take place in order to ensure maximally-compact perceptual descriptivity of the permissible actions via the bijectivity criterion. This latter aspect of the approach (occupying the whole of Section 4) is thus the principle methodological contribution of the current investigation.

1.3 Generalized Object Classes in Cognitive Bootstrapping: The Affordance/Schemata Distinction

The affordance/schemata distinction lies at the heart of cognitive agency, and can be understood in this investigation in the following way: Affordances are those possibilities offered by objects in the world *in relation to* the agent's motor capabilities. As such, they occupy an intermediate position between the classical Cartesian poles of *subject* and *object*. However, they are unquestionably empirical facts (facts

[4] More general arguments for the importance of embodiment to cognition can be found in [1,9,12, 24,25].

relating to the interaction of agent and environment), in the sense of being falsifiable according to the standard Popperian [36] criterion of empiricism.

Schemata are high-level representations of action possibilities predicated on *assumptions* about the affordances offered by the world (the agent's world-model is characterized as the total set of assumed affordances). Hence, because the cognitive bootstrapping agent is intended to produce novel percepts in relation to its exploratory findings about the world, schemata for employing these inferred percepts must also be capable of falsifying them.

An example of this falsifiability might hence be the postulation of a novel perceptual category, 'ladder', on the basis of certain saliencies, or aggregate saliencies in the agent's visual field. In a P-A framework, however, this percept is further characterized by its assumed affordance of agent action possibilities (such as the agent's being able to reach-out and grasp the individual rungs of the ladder). But in order to justify its characterization as a *single* perceptual entity within a perception-action framework, the ladder must also have a singular function in terms of the the schemata for its employment. In this case, the singular function is 'climbability', in the sense that an entire ladder must be utilized to achieve this. The perceptual category 'ladder' is thus falsifiable if it turns out that the 'climbing' schemata is unfulfillable (perhaps the rung materials in agent's world are too weak to support climbing), in which case the P-A-based segmentation of 'ladders' in the world is redundant. All postulated perceptual categories in a P-A framework must be similarly falsifiable via their associated schema.

In cognitive bootstrapping terms, schemata are thus abstractions of the existing perception-action hierarchy that (if they achieve a sufficient level of empirical confirmation) can themselves *become* affordances by virtue of being linked to perceptual inference. (The identification of the percept category 'ladder' is directly correlated with the schemata 'climb'). The perception-action hierarchy is thus endlessly extensible, with the inferred schemata becoming increasingly abstract (such that, for instance, they can be treated in formal logical terms).

Inferred high-level level perceptual representations in the cognitive bootstrapping agent must therefore always have this characteristic of hierarchical dependency on lower-level percepts. The percept-action link always ensures that these perceptual representations are correlated with high-level actions (such as climbing), that are in turn built on lower-level action capabilities (in this case, the basic object manipulation capability): The following paper can be seen as an attempt to formalize this idea at the level of logical induction. The formation of generalized object classes via perceptual reparameterization can hence be seen as a means for forming high-level schemata in relation to the affordance possibilities of the environment.

1.4 Active Learning and Cognitive Bootstrapping

Perceptual inference of the type detailed above, while significant in overcoming a number of conceptual difficulties associated with autonomous, self-updating cognitive agency, would be of only abstract interest unless it can be shown to have useful consequences in the objective domain. However, it is clear from the above that, in a perception-action context, the postulation of a novel set of *generalized* perceptual variables also implicitly suggests a set of generalized action states. In performing the perceptual remapping ($\{P\} \rightarrow \{P'_{\text{gen}}\}$), the agent hence *narrows-down* its choice of potential exploratory actions to those consistent with the notion of testing its inferred model of the environment.

Thus, while the non-bootstrapping perception-action agent must carry out exploration via random sampling of the *a priori* motor-space, the same agent employing perceptual updating of the type indicated instead samples the *inferred* action space randomly. This has the natural corollary that *a priori* motor-space is sampled *actively* (cf eg [3, 32]), exploration being conducted on the basis of the agent's perceptual inferences. Active learning is characterized by the intentional selection of learning examples: it is typically used in environments in which there is an abundance of unlabeled data and where labeling data is costly. This is hence true of the P-A learning environment in that the determination of the success of an exploratory action (ie, its labeling as 'successful' or 'unsuccessful') is intrinsically expensive in temporal terms. If we know that a broad class of actions are likely to be unsuccessful, it is hence not necessary to sample them at the same rate as the generally much smaller class of successful actions. Consequently, the fact that cognitive bootstrapping renders unperceivable (at the highest level of representation) those actions that are not consistent with the agent's estimation of what is possible in action terms, implies that exploration acquires an active aspect.

Thus, although it is possible for the agent to achieve a similar sort of active learning without perceptual inference by randomly sampling the *a priori* motor-space and checking for consistency with the inferred model of legitimate actions[5], the fact of having reparameterized the perceptual space in a more compact form means that we can *directly* suggest exploratory actions without the computational inefficiencies of consistency checking the samples. This is then an *active* learning strategy similar to that of [39], but arising naturally within the context of perception-action learning as a result of having determined an empirically consistent and non-paradoxical model for perceptual updating. Thus, a robotic learning agent that initially sets out to stochastically sample the action-domain 'move the gripper from position X to position Y', would, under an appropriate perceptual reparameterization, instead stochastically sample the action-space 'move the object A onto the object B'. It thereby 'intentionally' selects hypothesis-specific learning examples in the manner required of the active learning process. These samples will then be such as to more readily confirm or refute the perceptual hypothesis that the agent's world consists in 'movable objects'. They will also more readily refute the external *object* hypothesis (namely, that the specific percept experienced by the agent is in fact a movable object within the world). Hence, a cognitive bootstrapping system is simultaneously an active learning system with respect to both the object and percept hypotheses.

[5] Though, importantly, without the same guarantee of uniform sampling of the legitimate actions (see Conclusions).

We hence envisage an iterative, three-stage process of active-learning consisting of: (1) Randomized exploratory activity carried-out in terms of the currently-assumed perception-action model; (2) Induction of a novel action legitimacy model in terms of the current perceptual framework; (3) Reparameterization of the perceptual framework in order to most appropriately (ie most compactly) represent the inferred action-model.

Note that this active exploration by random sampling of the remapped perceptual variables is only possible because of the *hierarchical link* that exists between the inferred percept space and the *a priori* percept space (the hierarchicality arising from the fact that the former space is a *subset* of the latter). As such, it represents a 'higher', more *symbolic* level of representation of the sensory data (we shall later give an intuitive example of this). Cognitive bootstrapping in the manner described can therefore be considered a method for the spontaneous generation and grounding of symbols in the manner of Harnad [17]. Active exploration, in driving exploration from the highest, most symbolic level of the hierarchy, hence serves to continuously re-train *all* of the hierarchical levels existing beneath it. Though beyond the scope of this paper to demonstrate, the method thus has, as a hierarchical reinforcement learner [7], the capacity to update itself in a manner capable of robustly accommodating environmental changes.

The most prominent aspect of the hierarchicality inherent in cognitive bootstrapping demonstrated in the following paper, however, is that brought about by the progressively higher levels of perceptual representation. Thus, despite the fact that exploration is always undertaken stochastically, randomly-chosen actions within the abstracted, top-level percept domain increasingly take the form of *intentional actions* when considered in terms of the *a priori* motor space.

1.5 Structure of Investigation

In setting out our approach to cognitive bootstrapping within the relational domain and demonstrating its utility within a simulated logical environment, we shall structure the investigation as follows: Section 2 is concerned with the nature of the test-environment and the agent's relation to it in terms of its *a priori* perception-action capabilities. It also sets out the means by which these are described in first-order logical terms. Section 3 deals with mechanism by which the generic cognitive agent is to infer the specific logical rules underlying its environment given only the outcomes of exploratory actions. Section 4 then describes how the cognitive bootstrapping agent can utilize this logical inference to perform a remapping of its perceptual space, such that *apparently* random exploration of its environment constitutes an *active* learning approach with respect to environmental rule inference. Results of the application of this technique with respect to a benchmark passive-learner then constitutes Section 5 of the paper. Section 6 finishes by discussing the implication of these results both in practical terms, and in terms of the implications for autonomous cognitive agents capable of *meaningfully* updating their own cognitive apparatus at the same time as their models of the external world.

2 A Simulated Environment for Active Perception-Action Learning

Following the preceding description of P-A learning in abstract terms, we shall henceforth find it convenient to introduce the proposed cognitive-bootstrapping method within a concrete context, and only after which will the generalization to arbitrary environments be discussed. We are hence here interested in cognitive bootstrapping only at the relational-levels of the perception-action hierarchy, in which an idealized, discrete representation of the environment is already assumed to exist. It is hence clear that the domain in which we seek to implement the preceding ideas should be one for which there exists a clear, protocol-driven criterion of action legitimacy. This legitimacy should obviously be reflective of the constraints of an actual physical environment, albeit at a sufficiently abstracted level. Hence, in order to generate a preliminary instantiation of the notion of cognitive bootstrapping, we select a simulated 'shape-sorter' puzzle as the domain of agent activity.

Within the simulated shape-sorter, variously shaped puzzle pieces may be arbitrarily transported around a three-dimensional volume via a 'gripping arm' (that hence constitutes the active component of the artificial agent). Other entities existing within the active arena include the surface of the puzzle and, within this, a series of holes that correspond uniquely to each of the shapes. The totality of these entities are assumed to rest on an impermeable surface. We further assume that the agent has an idealized perception of this environment (enacted, perhaps, via hypothetical cameras positioned outside of the immediately active domain). Low-level vision tasks such as object segmentation and three-dimensional reconstruction are hence assumed to have been flawlessly carried out at a hierarchical-level prior to that in which cognitive inference is to take place, such that no perceptual errors exist at this level (this artificial restriction is in no way a prerequisite of the method; rather it is a simplifying assumption).

2.1 *The A Priori Action Space*

Within this simplified environment the potential range of actions available to the agent (corresponding to the *a priori* motor space) are thus the positional translations of the gripping arm, which is assumed to perform a 'grasping' action at the initial stage of the attempted translation and a 'releasing' action at the final stage of the proposed translation. At this stage we do not, for simplicity of inference, consider the possibility of explicit object *rotation*, although the concept of object *orientation* is of relevance, as we shall see. Actions are hence specified via a six-tuple instruction '$move(x1, y1, z1, x2, y2, z2)$' corresponding to the transition from position-vector $(x1, y1, z1)$ to position-vector $(x2, y2, z2)$ within the active volume (which equates in physical terms to the three-dimensional spatial range of the gripping arm).

However, the existence of an action state within the *a priori* motor-space is not a guarantor of action *legitimacy*, a notion that is required if there is to be meaningful interaction between the agent and its environment[6]. Defining an appropriate criterion for the legitimacy of actions within the environment can, in a sense, be considered a form of agent *meta*-goal specification, though one that is minimally restrictive with respect to the agent's potential exploratory activities.

Action legitimacy in the shape-sorter environment is hence determined by the *feasibility*, *stability* and *utility* of the proposed transition. The first of these constraints, feasibility, is determined by the physical requirement of the non-coincidence of objects; one object cannot be legitimately moved into another. The *stability* of a proposed action is determined by whether its intended final state would undergo any further observable transition *not* initiated by the agent; situations in which a moved object is released without anything beneath it are hence illegitimate. The stability condition thus ensures the temporal reversibility of actions, such that the environment can be described in terms of relatively simple first-order relational predicates throughout the experimental process. The set of transitions hence has the closed mathematical structure of a *monoid*. To this end, we also require that positions and lengths are discretized to identical unit-lengths, such that partial overlaps between objects are not permitted. The final constraint, action *utility*, refers to the notion that the proposed transition should do actual physical work (that is, result in a perceived environment change) if it is to be considered legitimate. The gripper cannot therefore simply transit from one unoccupied position to another, though this is both *stable* and *feasible* under the preceding definitions.

Hence, if an instruction $move(x1, y1, z1, x2, y2, z2)$ is to be considered legitimate via the above criteria, it must involve the gripping of an object located at position $(x1, y1, z1)$, followed by the release of the same object at a location immediately above the supporting surface provided by an unencumbered solid entity located at $(x2, y2, z2-1)$. A supporting entity can be any of those we have defined: the puzzle-base, another shape, or a hole. However, the latter entity is only supportive if it *does not* match the morphology of the moved shape, or if it *does* match the shape's morphology, but has a differing orientation.

It is consequently possible to quantify the restrictions that this notion of action legitimacy places on the agent's generic, *a priori* motor space when it is embodied within the particular confines of the shape-sorter environment in the following way. The initial motor space has a numeric magnitude given by $(|x| \times |y| \times |z|)^2$, representing the combination of initial and final location possibilities, with $|x|$, $|y|$ and $|z|$ the respective cardinalities of the discretized ordinal vectors. Within this space there is a consistent quantity, $|shapes|$, of legitimately movable objects that can be placed on any suitable unencumbered surface. This supporting surface has to completely bisect the volume in the z direction, but must not include the position of the

[6] Indeed, in certain phenomenological models [45], this differential in actual and potential action capabilities *constitutes* the agent's environment.

object itself[7]. Quantized in unit areas, the surface therefore has an average numeric magnitude of $(|x| \times |y| - p \times |1|^2)$, where p is the probability that a given object is correctly oriented with respect to its corresponding hole (for random initial configurations this is determined by the level of discretization of the orientation parameter, specifically its *reciprocal*). The fraction of the agent's *a priori* action space that can be considered legitimate within the particular environmental context of the shape-sorter puzzle is hence:

$$\frac{|shapes| \times (|x| \times |y| - p)}{(|x| \times |y| \times |z|)^2} \tag{1}$$

2.2 The A Priori Percept Domain

As well as the *a priori* action-space detailed above, the agent is also equipped with an *a priori* perceptual 'space' through which it (at least initially) interprets its environment. The aim of cognitive bootstrapping is hence to determine the subset of this perceptual space that most efficiently delineates the current hypothesis as to what constitutes the legitimate *action* sub-space, but in such a way as to conserve the empirical falsifiability of this hypothesis. As such, the *a priori* percept categories employed for this purpose, like that of *shape*, do not, so far as the agent is concerned, yet have any action-determined meaning (which may vary throughout the bootstrapping procedure). The nominal designations of the *a priori* percept categories are thus, at this stage of the experimental process, labeled purely for the purposes of our comprehension. In full, these categories are hence as follows; *positional occupancy* (ie, the detection of the presence or absence of an entity at a particular location), *shape awareness*, *hole awareness*, *an awareness of hole-shape morphological equivalence*, *angular orientation* and *spatial adjacency*[8]. These base perceptual *categories* are hence distinct from the percepts we shall later infer in that they may be regarded, in phenomenological terms (cf [20]), as the in-analytic (non-separable) *attributes* of perception, or *qualia*. Thus, 'red circle' might constitute a singular perceptual category following perceptual inference, with the individual qualia 'red' and 'circular' the relevant intrinsic categories, or attributes, of perception. The perception of specific entities hence constitutes, in an imprecise sense, the *co-ordinatizing* of the intrinsic perceptual categories (we shall later utilize the

[7] Note that this supporting surface is different for each object, since differing objects slot into differing holes, with the holes acting as support-entities otherwise.

[8] In a more realistically complex cognitive environment it would be possible to eliminate the *a priori* awareness of the morphological correspondence between holes and shapes by resolving it into a suitable composite of the two perceptual categories of positional occupancy and spatial adjacency. In fact, these are more like the true Kantian *a priori* perceptual categories. (*A priori* in the sense of their being empirically undiscoverable, being rather the *conditions* of empirical discovery).

bijection principle to find an explicit coordinatization of the manifold underlying the *a priori* percepts that *is* precisely defined).

Thus, for this demonstration, we have elected to regard shapes and holes as *a priori*, rather than as *composite* perceptual entities. The agent's perception of these entities consequently consists (if there are no other attributes) in the returning of a particular *index* or *label* from the set of all entities of the same type. Shape and hole entities, are thus distinguished *extrinsically*, rather than via any intrinsic characteristics they might have for a more complex cognitive system, such as human perception.

Having thus given an indication how the *a priori* action and percept spaces are delineated, we can now turn to the means by which we are to implement this percept/motor environment in terms of first-order logic. In such a framework, it is *logical variables* that will serve as the means for generalizing over perceptual entities such that it becomes possible to express exploratory propositions directed at perceptions that have *not* yet been directly experienced by the agent. The corresponding mechanism for 'perception' within the first-order logical domain will hence be environmental *predication* carried out by the cognitive agent, in which the percepts are the full range of possible *co*-instantiations of the perceptual variables.

We thus lay the foundation for the use of inductive logic programming (ILP) methods in order to determine generalized environmental rule hypotheses from which novel perceptual variables can be extracted.

2.3 *Implementation of the Shape-Sorter Puzzle in First-Order Logic*

In having opted for a first-order logical description of the shape-sorter, it becomes possible to render the underlying constraints of the environment as a set of physical *protocols*. Attempted transitions within the *a priori* motor-space hence become logical *propositions* of the form $move(x1, y1, z1, x2, y2, z2)$, that are true or false according to their legitimacy in terms of these axioms. (Hence in a generic motor-space equipped with a set of *a priori* perceptual variables, $[L]$, the action propositions will be predicates of the form $action_n([L]_1, [L]_2)$, with the numeric subscripts denoting initial and final perceptual states, respectively). In choosing to model this system in *PROLOG*, we find that the agent's attempted actions become *goal states*, or theorems to be proved via first-order resolution. We are thus, in effect, seeking to construct a *semantic parser* for agent actions in terms of the shape-sorter axiom set.

Hence, describing, within PROLOG, the *a priori* cognitive categories listed above as *predicates with variable arguments*, we render the perceptual categories of *shapes*, *holes*, *hole-shape correspondence*, *angular orientation* and *positional occupancy* as, respectively: $is\_shape(A, L_1)$, $is\_hole(A, L_2)$, $hole\_shape\_match(A, B)$, $orientation(X, O)$ and $position(A, X, Y, Z)$ (quantification over variables is *implicit* in PROLOG).

The variables ranging over these perceptual categories are delineated as follows: A (and B) represent *positionally-determined entity labels* (ie holes, shapes, the puzzle-base itself) if the subgoal $position(A, X, Y, Z)$ is fulfilled. X, Y and Z represent *ordinal positions* along the three spatial axes; O is an *angle*. The variables L_1 and L_2 are label variables acting over the sets $\{shape_1, shape_2, \ldots\}$ and $\{hole_1, hole_2, \ldots\}$, and are thus subsets of the positionally-determined entity label class. (We shall later discuss the necessity of adopting this predicate form in general environments). Recall that notions like *object* and *angle* are not yet operationally defined for the agent; they are, so far, only perceptually differentiated with respect to each other. Hence predicates such as '*is_hole*' and '*is_shape*' should only be regarded as making elementary perceptual *distinctions* that will later acquire *meaning* via association with specific action-possibilities. (Thus, the predicate *is_hole* might correspond to an agglomeration of dark perceptual entities established as constituting a single class via unsupervized clustering at a lower level of the perception-action hierarchy). The predicate names we have adopted "*is_hole*", "*is_shape*", etc, are purely to assist reader comprehension.

The cognitive predicates associated with the *a priori* notion of spatial adjacency are rendered as $inc\_x(X1, X2)$, $inc\_y(Y1, Y2)$ and $inc\_z(Z1, Z2)$ for the three respective spatial directions. Hence, in consequence of the discretization of these axes, the term $inc\_x(X1, X2)$ is satisfied only when $X2 = X1 + 1$. (Since we do not yet, for simplicity, include the possibility of object rotation by the agent, there is no corresponding notion of *angular adjacency*; angles are represented by discrete tokens without relational content).

Higher-level concepts such as *the unoccupied location immediately above an object* can hence be represented by concatenations of the *a priori* categories: in this particular case the concatenation; $position(A, X, Y, Z)$, $inc\_z(Z, Z1)$, $not(position(\_, X, Y, Z1))$. (In PROLOG, comma sequences of this kind indicate a simultaneous fulfillment requirement for *each* of the subgoal predicates).

Having transcribed the *a priori* percept/motor categories into a format suitable for logical predication, it becomes possible to specify the constraint rules for the shape-sorter environment in these terms. Any exhaustive description of such protocols must include logical constraints equivalent to the physical notions of object non-coincidence, object persistence under translation, object instability in the absence of supporting structures, and so on. While these notions are innate, or at least intuitive to human agents, they must be derived by our simulated agent via first-order logical induction in terms of the *a priori* percept categories applied to the outcomes of exploratory actions.

The simplified rules governing the protocols of the shape-sorter environment, which may be considered to act as a very generalized supervisor to the simulated open-ended learning agent, are hence rendered as the series of PROLOG clauses:

$move(X1, Y1, Z1, X2, Y2, Z2) :-$
$position(A, X1, Y1, Z1), is\_shape(A, L), inc\_z(Z1, Z3),$
$not(position(\_, X1, Y1, Z3)), not(position(\_, X2, Y2, Z2)),$
$inc\_z(Z4, Z2), position(B, X2, Y2, Z4), not(hole\_shape\_match(L, B)).$

$$move(X1,Y1,Z1,X2,Y2,Z2) :-$$
$$position(A,X1,Y1,Z1), inc_z(Z1,Z3), is\_shape(A,L1),$$
$$not(position(\_,X1,Y1,Z3)), not(position(\_,X2,Y2,Z2), inc\_z(Z4,Z2),$$
$$position(B,X2,Y2,Z4), is\_hole(B,L2), hole\_shape\_match(L1,L2),$$
$$orientation(L1,O1), orientation(L2,O2), not(O1 == O2).$$

$$move(X1,Y1,Z1,X2,Y2,Z2) :-$$
$$position(A,X1,Y1,Z1), is\_shape(A,L1), inc_z(Z1,Z3),$$
$$not(position(\_,X1,Y1,Z3)), position(B,X2,Y2,Z2),$$
$$inc\_z(Z2,Z3), not(position(\_,X2,Y2,Z3)), is\_hole(B,L2),$$
$$hole\_shape\_match(A,B), orientation(L1,O1), orientation(L2,O2),$$
$$O1 == O2.$$

(For the sake of clarity in the following arguments an additional iteration variable, T, that exists within the *move* and *position* predicates in order to delineate separate temporal stages [ie *move*$(T,X1,Y1,Z1,X2,Y2,Z2)$ and *position*$(T,A,X1,Y1,Z1)$] is omitted throughout).

Multiple clauses in PROLOG represent *alternative possibilities* for the satisfaction of the logical constraints. Clause one in the above rule-sequence hence represents the possibility of placing shapes onto support surfaces that are *not* holes matching the shape's morphology. Clause two represents the possibility of placing shapes onto holes that *do* match the shape's morphology, but which have differing orientations. Clause three represents the possibility of placing shapes *into* holes that both match the shapes' morphology *and* have identical orientations.

This latter rule represents the classical 'solution-state' of the shape-sorter puzzle. Thus, although we will not explicitly hardwire any goal-seeking characteristics into the agent, we shall later see that active testing of generalized rule inferences relating to the placing of shapes on top of surfaces (such as the first sparse random sampling of the environment is likely to give rise to) can result in agent behavior that is superficially similar to that implied by explicit goal-seeking. We might also speculate that this observation applies more generally; the uniqueness (in the sense of being an exception to the general rule) of the solution-state in a typical protocol-based puzzle environment would render its saliency for an active-learning agent much greater than would be the case for a passively-learning logical agent employing inductive methods.

3 Approach to First-Order Logical Induction

The first task of the cognitive bootstrapping agent, prior to any perceptual updating, is to infer the generalized rules governing the achievability of the exploratory actions under consideration. That is, the agent must attempt to derive the above clause structure (acting in the capacity of *supervisor* within the simulated environment) from specific instances of those rules' application. This is the task of *Inductive Logic Programming* [33].

The inductive logical approach that is most readily applicable to our environment is Muggleton's *PROGOL* [34]. PROGOL proceeds in a top-down manner, constructing the *most specific clause* (see below) covering the first positive exploratory example from a series of predicate *mode declarations*. Mode declarations define the type and number of a predicate's arguments, as well as its variable input/output structure; they also determine the number of permissible instantiations of the predicate variables. Since clauses in PROLOG are defined by head *and* body predicates, both types of mode declaration are required in PROGOL. A typical body mode declaration is hence of the form:

$: -\ \ modeb(n, p(+A, +B, -C, -D))$, with n the number of permissible instantiations, p the predicate functor, $[A, B]$ the set of input variables and $[C, D]$ the set of output variables.

The *most specific clause* is hence the clause with the most exhaustive predication consistent with the mode declarations. As such, it defines a lower bound of a *theta-subsumption lattice* which has as its upper bound the empty clause (one clause theta-subsumes another if and only if there exist a set of variable instantiations [ie a substitution] such that all atoms in the former clause will constitute a subset of the atoms in the latter clause). The lattice itself is navigated in a general-to-specific fashion via heuristically-guided $A^\star$ searching. The heuristic in question is the clause compressivity with respect to the remaining positive examples insofar it is consistent with the negative examples. The most effective of these is then selected as *background knowledge*, and all positive examples that are rendered redundant by it are removed from the total set of examples, after which the process begins again with the first of the remaining positive examples. PROGOL thus arrives at its estimate of the most compressive clause, or set of clauses, consistent with the example data.

The output of the attempted PROGOL induction of the above rule-set from the set of exploratory actions is hence the 'objective' model of legitimate environmental actions. We now turn to the question of how the corresponding 'subjective' model of perception appropriate to this objective model is to be derived and usefully employed within the puzzle environment.

4 Active Learning Via Cognitive Bootstrapping Utilizing First-Order Logical Induction

We have defined cognitive bootstrapping as the simultaneous inference of optimal object and percept models in such a way as to avoid problems of underdetermination, with a consistent *empirical* criterion for both object and percept model selection sustained throughout the learning process. We have indicated that perception-action learning provides a natural framework for this kind of updating, requiring that the perceptive capabilities of cognitive agents are determined by their active capabilities. Formalizing this notion as a condition of *bijectivity* between perceptual transitions and action states, we shall now set out to implement an iterative learning system that alternates between the two complementary phases of object-model inference and percept-remapping. *Perceptually-motivated exploration* is then the intermediary linking the two phases of cognitive bootstrapping.

The first of these phases is hence the attempted first-order logical inference, via PROGOL, of the clauses that determine the behavior of the shape-sorter from the set of cumulative exploratory actions labeled as legitimate or illegitimate by the supervising PROLOG clause-set. In the second phase, the remapping of perceptual variables deriving from this inference will directly suggest, in consequence of those variables' generalized nature, a novel set of exploratory action possibilities that can be employed to test the objective model. More specifically, because the perceptual updating carried-out in the second phase of cognitive bootstrapping is designed to give optimal representativity to the objective hypothesis derived in the first phase of cognitive bootstrapping, the novel action possibilities suggested by the remapped percepts are precisely those that are consistent with the model (in consequence of the bijectivity condition). The compressivity criteria for the acceptance of percept hypotheses ensures that these consistent action possibilities are accessed in far more efficient manner than they were in the *a priori* percept-motor space. The perceptual remapping thus, in effect, *re-parameterizes* the model of legitimate action transitions in the most compact manner possible.

Hence, we see that in having described a cognitive agent with generalizing logical induction capabilities under the condition of perception-action bijectivity, we have implicitly described an *active* cognitive learning system. In general terms, an agent that employs active learning is one that utilizes environmental hypotheses to *motivate* exploration in order that a more rapid convergence on the correct model can be achieved (cf eg [3, 32]). A passively learning agent, in contrast, derives it environmental hypotheses from randomized exploration. We might therefore expect the cognitive bootstrapping agent to be inherently more efficient at environmental model determination than a purely passive agent: we shall set out to test this hypothesis in section 5. It should be clear, however, that our cognitive model goes beyond that of the typical active exploratory agent in simultaneously seeking an optimal remapping of the *percept* domain, in effect, actively learning both the perceptual and objective environment at the same time. We therefore now turn to the method by which this perceptual remapping is achieved.

4.1 Remapping the Perceptual Variables

Imposing the condition of *bijectivity* between action states and perceptual beginning/end state pairings implies that there must exist a mapping $(P_{\text{initial}} \times P_{\text{final}}) \rightarrow A$ such that the bijectivity condition holds: ie, $\{P\}_{\text{initial}} \times \{P\}_{\text{final}} \Leftrightarrow \{A\}$, where $\{P\}$ is the complete set of perceptions, and $\{A\}$ the complete set of achievable actions. Hence, if, for the sake of example, we were to set about imposing this condition on the *a priori* perceptual space, we would proceed in following way. The default actions in the absence of any experimental data would have to be assumed to be those of the *a priori* motor space, $move(X1, Y1, Z1, X2, Y2, Z2)$. However, the possibility of specifying *any* individual (lower-case) proposition, $move(x1, y1, z1, x2, y2, z2)$, is also an assertion of the existence of an exhaustive set of singleton maps:

$$\{(x1, y1, z1)\} \rightarrow \{(x2, y2, z2)\} \quad \forall\ x1, y1, z1, x2, y2, z2 \tag{2}$$

Hence, comparing with the above perception-action bijectivity condition, and noting that the $(x1, y1, z1)$ range over the *same* space as the $(x2, y2, z2)$ we see that the default percepts $\{P\}$ are the set of *positions*, $\{(X, Y, Z)\}$ (ie co-instantiations of the X, Y and Z ordinates). In other words, given the *a priori* perceptual predicate structures available to the agent, such as $shape$, $orientation$ and so on, it is, under the condition of bijectivity, only the predicate structure $position$ that properly constitutes a percept in the absence of experimental constraint data. (This reflects the fact that a notion such as *shape* does not yet have any action-determined perceptual significance).

While this result is trivial (and to a certain extent tautological) for the default motor-space, it is indicative of the approach to perceptual updating adopted when exploration gives rise to results that break the *assumed* equivalence between the *a priori* motor-space and the set of possible actions.

Now, for the sake of demonstration, suppose that randomized sampling (ie exploration) of the *a priori* action domain $(\{(X, Y, Z)\} \times \{(X, Y, Z)\})$ has given rise to a sufficient number of legitimate actions (say, 5) for logical inference to be enacted. (This would imply around a two orders of magnitude greater number of illegitimate samples - see equation 1). Further suppose that the application of PROGOL to the cumulative exploratory data has given rise to the inference of the partially accurate legitimacy rule:

$move(X1, Y1, Z1, X2, Y2, Z2) :-$
$position(A, X1, Y1, Z1), inc\_z(Z3, Z2), position(B, X2, Y2, Z3),$
$inc\_z(Z4, Z1), not(position(A1, X1, Y1, Z4)),$
$not(position(A2, X2, Y2, Z2)).$

(which would correspond to a constraint-rule stating that only unimpeded objects may be placed on top of other objects.)

It is hence apparent that, in attempting to find a rule that both *generalizes* and *legitimizes* the training set, it has become necessary for the inference system to introduce a set of new variables $\{A, B, Z3, Z4\}$ beyond the existing six variables $\{X1, Y1, Z1, X2, Y2, Z2\}$ required to specify the *a priori* motor-space. It may be recalled that the body mode declarations specify the input and output structure of the variables in the individual atomic propositions constituting the inferred clause. Consequently, the atomic propositions can be represented as the nodes of a directed acyclic graph constituting the full clause (acyclic since we explicitly forbid recursion in the inferred clause).

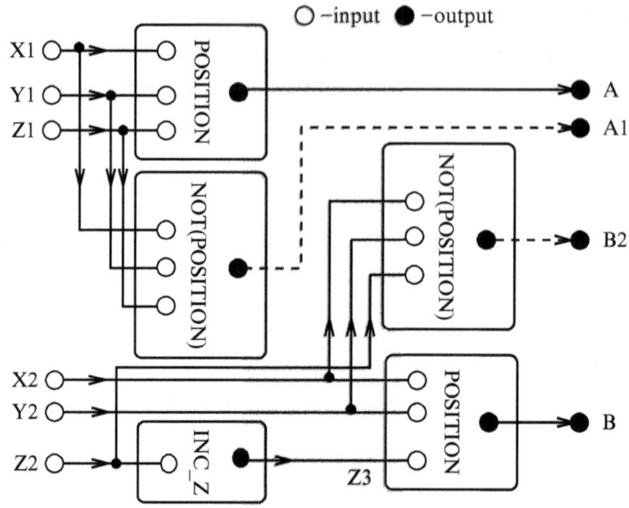

Fig. 1 Example schematic of clause structure (see footnote 10)

Hence, if we render the predicate input/output structure of this rule visually (as in figure 1), it is apparent that the six initial input variables in the *a priori* percept space (ie, the three variables specifying the initial percepts $(X1, Y1, Z1)$ and the three variables specifying the final percepts $(X2, Y2, Z2)$) are mapped onto only two output variables[9].

Consequently, if it were possible to define predicates in such a way as to be able to unambiguously *reverse* this variable mapping, it would be possible to define a clause structure in which the *input* variables A and B uniquely define the *output* variables $[X1, Y1, Z1, X2, Y2, Z2]$, such that the initial and final percept states that link together legitimate actions within the inferred model are preserved. (That is, so that the maps $\{(X1, Y1, Z1)\} \rightarrow \{A\}$ and $\{(X2, Y2, Z2)\} \rightarrow \{B\}$ are bijective). It would, in so far as it is permissible to regard variable instantiations as *ordinates*, hence be possible to 're-parameterize' the original six-dimensional percept space as a *two*-dimensional percept space characterizing permissible actions in the inferred rule. In intuitive terms, this new percept space consists of, respectively the set of *objects*, $\{A\}$, and the set of *surfaces*, $\{B\}$ (A being mapped onto any unimpeded object existing at location $(x1, y1, z1)$, and B being mapped onto any unimpeded objects existing *below* the intended movement location, $(x2, y2, z2)$). Thus we see that the compressed percept space represents a *higher level* of perception than that of the *a priori* space, in the sense that it expresses relational, action-relevant abstract concepts that were not present in the default percept space. Randomized exploration in

[9] Note that the predicates of the form $not(\ldots)$ do not generate output variables that can be meaningfully employed to address the percept space, and are hence indicated by dotted lines in figure 1 to denote their removal from the remapping process.

this higher level domain then consists in the placing of random objects onto random surfaces, rather than simply the movement of the gripping arm between random locations.

In order to achieve this remapping, it is necessary to strictly define predicates as bijective *functions* between input and output variables, rather than merely *constraints* upon them (both forms of predication are possible within PROLOG). This means that agent predication must be configured such that input variables relate to output variables via *single* instantiations[10]. Thus, when, for instance, describing the functional relation that exists between entity labels and three-dimensional position vectors, $position(A, B, C, D)$, we have adopted a predicate form that employs A as an output variable (acting over object-labels) in such a way that it is *always* uniquely specified by the input variable triple, (B, C, D), ranging over combinations of the X,Y and Z ordinates. This approach is thus in explicit contrast to the logically equivalent, but possibly more intuitive, description of positional occupation in terms of entity predicates with forms of the type: $hole(X, Y, Z)$, $shape(X, Y, Z)$, etc (which would be either *true* or *false* as appropriate for the given positional input variables). By rejecting this predicate form we are hence here demanding that predicates act as functions between their internal variables, rather than acting as characteristic (or indicator) functions from the internal variables to the Boolean set $\{true, false\}$.

Significantly, all 'essence'-like predication (of the type: $has\_quality\_1(X)$, $has\_quality\_2(X)$, etc) that acts over a common variable (or set of variables), X, can be specified in such a form by re-envisioning the various functor names $(has\_quality\_1, has\_quality\_2, \ldots)$ as *specific classes of sub-variable* ranging over the instances to which they apply. That is, we convert the n characteristic functions:

$$has\_quality\_n(X) \rightarrow \{true, false\} \tag{3}$$

to a set of disjoint subfunctions of the bijective master function:

$$\begin{aligned} f : X \rightarrow \{ \; & L_{\text{1st object with quality 1}}, \\ & L_{\text{2nd object with quality 1}}, \ldots, \\ & L_{\text{1st object with quality 2}}, \\ & L_{\text{2nd object with quality 2}}, \ldots\}, \end{aligned}$$

such that $(has\_quality\_n(X, L) \rightarrow true)$ only when the two conditions that: (X has the quality n) and ($L = f(X)$) are fulfilled.

In doing so, we obtain a novel set of predicates, $has\_quality\_n(X, L)$, with the required property of invertability, and which may be added to the mode-declarations of the PROGOL code in the form:

[10] The only permitted exception to this rule being predicates with mode declarations that specify only input variables (ie, those having the form $modeb(1, p([+variable\ type])))$, such as $hole\_shape\_match(A, B)$. In these cases, the predicate can be regarded as a variable-less terminating node of the directed acyclic clause structure.

$: -\ modeb(1, has\_quality\_n(+X, -L))$. The revised predicate thus becomes an explicit *subgoal* for which a PROLOG interpreter must find, proof-theoretically, an instance of the output variable(s) satisfying the predicate's logical condition in order to return a value of *true*. (Rather than simply returning the predicate's truth value as would be the case for a characteristic function).

It is hence apparent, for essence-like predication, that the magnitude of the set of output possibilities will always be less than that of the input set. More generally, however, *any* reasonable cognitive predicate will have a predominately constraining effect on the set of input perceptual variables, since we expect the initial, *generic*, sensorimotor space to be limited by the *specific* situation in which the agent to which it belongs is placed. We therefore expect that such predicates will typically map larger sets of input variables to smaller sets of output variables. There are thus two independent mechanisms that act to compress the percept space when we seek to instantiate, in reverse, the inferred clause's input variables by the output variables.

Hence, in having defined a more compact *but equally expressive* percept space by insisting on the reversibility of the clause structure, it becomes possible to explore both the validity of this perceptual hypothesis *as well* as the environmental hypothesis represented by the inferred clause. It does this via *perceptually-determined* exploration. Here, random instantiations of initial and final state pairings in the modified percept space correspond to *action propositions* via the condition of percept-action bijectivity. Moreover, because the modified percept space compactly represents only the action possibilities inherent in the inferred action-legitimacy hypothesis, these action proposition serve as potential *tests* of the inferred rule. This is then the basis for the *active* component of the perception-action learning of cognitive bootstrapping agents discussed in the next section.

It should be noted that in the inferred clause that we have selected to illustrate the mechanism of perceptual remapping, there is a clear distinction between initial and final percepts in consequence of the existence of the individual mappings $\{(X1, Y1, Z1)\} \rightarrow \{A\}$ and $\{(X2, Y2, Z2)\} \rightarrow \{B\}$. All reasonably accurate inferences will fall into this category (reflecting the *a priori* independence of $[X1, Y1, Z1]$ and $[X2, Y2, Z2]$, and the temporal symmetry of permissible actions within the shape-sorter environment). However, this is not necessarily the case, and it may thus appear that a strategy needs to be adopted to deal with rule inferences that spuriously attribute a relationship between the variable sets $[X1, Y1, Z1]$ and $[X2, Y2, Z2]$ (thereby attributing an action-independent relationship between initial and final percepts). This is most easily accomplished by ensuring that *all* of the variables within the inferred clause's I/O structure ($X1, Y1, Z1, X2, Y2, Z2, A, B, Z3$ and $Z4$) respect this distinction, rejecting outright any inferences that do not. However, it is equally possible simply to overlook the issue: this independence of input and output perceptual variables is only true of certain perception-action environments; other environments, for instance, where irreversible actions are possible, may require final perceptual states to depend upon initial percept states. Provided that all of the variables governing the *a priori* action space are uniquely instantiated (which is guaranteed by the clause reversibility condition given above), it does not operationally matter if the remapped percept space temporarily relates initial and final perceptual states. (Since the repeated random instantiations of exploratory actions will eventually reject this supposition if it proves to be unwarranted). Thus, we expect the final convergent model of legitimate actions within the shape-sorter environment to respect the independence of initial and final percept states, irrespective of what happens during the learning process.

4.2 Algorithmic Approach To Perceptual Remapping

When constructing a general clause to cover a set of specific examples, PROGOL will invoke new variables with novel labels as they are required in order to conform with the predicate mode declarations. Consequently, the perceptual remapping procedure recounted in the previous subsection can be described via the *inclusion relationship* that exists between the differing sets of variables $\{A_1, A_2, \dots A_n\}$ contained within the individual predicates, $pred_m(A_1, A_2, \dots A_{n_m})$. Specifically, if we exclude the possibility of set self-inclusion, and impose an additional directed edge between the input and output sets of variables within each of the predicates, then the inclusion map constitutes a *directed acyclic graph* in which the *sink* vertices are the *a priori* variables and the *source* vertices are the remapped perceptual variables. Determining which vertices are source vertices, and hence establishing the set of remapped perceptual variables for a given action-rule inference, is consequently a trivial matter of ascending the inclusion hierarchy.

However, it may be the case that a number of the remapped perceptual variables so derived range over *identical* domains. When this occurs for predicates contained within *different* clauses in the inferred rule-set, there exists a redundancy between them by virtue of the clauses' mutual *independence*. Hence, for maximal compaction of the perceptual variables, we can map the redundant variables onto each other. This amalgamation is accomplished by the set union operation conducted over variable *type*. (Variable types are designated in the body mode declarations; for instance, the *entity* and *angle* in the declaration '*modeb*$(n, orientation(+entity, -angle))$).

Hence, suppose that we obtain a set of α clauses denoted $\{C^1, \dots, C^\alpha\}$, such that each C contains the set of remapped perceptual variables $\{A_1^1, A_2^1, \dots, A_{na}^1\}$. We are thus presented with a series of potentially surjective mappings between the sets of remapped perceptual variables contained within the clauses and the set of variable types, $\{T_1, T_2 \dots T_n\}$;

$$\begin{aligned}
C^1 &: \{A_1^1, A_2^1, \dots, A_{n_1}^1\} \rightarrow \{T_1, T_2 \dots T_n\},\\
C^2 &: \{A_1^2, A_2^2, \dots, A_{n_2}^2\} \rightarrow \{T_1, T_2 \dots T_n\},\\
&\dots \rightarrow \dots\\
C^\alpha &: \{A_1^\alpha, A_2^\alpha, \dots, A_{n_3}^\alpha\} \rightarrow \{T_1, T_2 \dots T_n\}, \text{ etc}
\end{aligned} \tag{4}$$

We consequently propose to construct the *minimal* superset:

$$\mathcal{M} = \bigcup_{y=1}^{n} \left\{ M : M = \max_{x} \left| \left\{ \bigcup_{t} [A_t^x : C^x(A_t^x) = T_y] \right\} \right| \right\} \quad (5)$$

with the capability of collectively addressing each of the independent percept spaces (ie such that $C^n \subset \mathcal{M}, \forall n$).

Should the superset so formed be less compact than the set of *a priori* percept variables (that is, have a larger cardinality), then it is rejected over the original perceptual variables. In general, however, this is not the case, and the new space is substantially more compact. Note, that the method will not, in general, achieve *maximal* compaction, given that for fairly complex rule inferences only some of the constants within any variable type will be applicable. Individual samples in the compact space are hence required to be tested according to the inferred rule before actual, embodied exploration; however, the process is very much accelerated by the reduction in sample-space dimensionality.

As a alternative to equation 5, it is possible to treat the clauses sequentially by sampling in their individual perceptual spaces over alternating exploratory cycles. However, we adopt the above method in order to generate as coherent a higher-level perceptual domain as possible.

Thus, in summary, we see that in compactly remapping the *a priori* percept space on the basis of the inferred rule of action legitimacy, we define a space consisting of high-level, action-relevant concepts capable of mediating between the agent's motor possibilities and the physical restrictions of the environment. We have, in other words, defined a set of *affordances*.

The key logical constraint on PROGOL for achieving this, which is applicable in any non-recursive logical environment in which perceptual variable instantiations are linked via action-variable instantiations, is thus that of *functional predicate bijectivity*. Thus, if a head mode declaration (representing an *a priori* action) links one instantiation of perceptual variables to another;

$:- modeh(1, action(+[perceptual\ variables]_1, +[perceptual\ variables]_2))$,

we can always ensure perceptual compressibility in individual clauses by insisting that body mode declarations are of the form;

$:- modeb(n, p(+[inputvariablelist], -[outputvariablelist]))$, such that $n = 1$.

This can always be achieved (see earlier) by such means as writing entity-quality designations in the form ($has\_quality\_n(X, L) \rightarrow true$), rather than as characteristic functions for individual qualities. Perceptual variables that are contained in multiple clauses can then be compressed by application of equation 5. Because only a fraction of the generic *a priori* action-space is employed, this enforced predicate reversibility always maps the *a priori* perceptual domain into a higher-level one via the directed acyclic graph structure implicit in logical clauses. (*Reversible* predicates [ie those with uniquely instantiated mappings between input and output variables] can always be treated as graph nodes).

Hence, generic application of the method requires only that we specify the set of predicates pertaining to the most basic perceptual categories: in particular positional label attributions (colors, shapes, textures, etc) and spatial adjacency relations. (More general predicates obtained by stochastic clustering at lower levels of the hierarchy can also be added in the same manner). The relational cognitive bootstrapping method then ensures that composite perceptual structures such as 'surface' (ie, a composite of positional occupancy, vertical adjacency, and positional vacancy) are learned as required: that is, only in so far as these are meaningful to the embodied agent in action terms.

4.3 The Different Phases of Active Exploration

In defining a perceptual space appropriate to those actions deemed permissible by the inferred shape-sorter protocols, the cognitive bootstrapping agent should find data capable of falsifying the action hypothesis far more quickly than would otherwise be the case. (By default PROGOL requires only one instance to falsify a hypothesis, though this threshold can be increased to achieve varying degrees of error-sensitivity). Hence, in advocating the above method for achieving empirically meaningful perceptual inference within an inductive first-order logic framework, we have defined an agent that *incidentally* performs active learning in both the perception and action domains. However, this is not to imply that this is the most efficient or complete approach to active learning possible within the context of cognitive bootstrapping. To begin to approach typical active learning performance improvements (that is, with convergence times of the order $O(log(1/error))$ as opposed to $O(1/error)$ for idealized active perceptron learning models [5]) it is necessary to extend the active learning framework.

The most immediate such requirement is that necessitated by the elimination of local minima effects, which can arise when a *subset* of the totality of clauses describing permissible actions has been accurately inferred. Clauses are independently satisfiable, and hence exploration carried-out by a cognitive agent in terms of a correctly inferred rule-*subset* could mean that the agent would never experience any percept capable of falsifying these rules. This would then give rise to an incorrect impression that the agent had converged on the final model of permissible actions (the only criteria for model accuracy available to the agent being those of the legitimacy of its exploratory actions). To counteract this, it is necessary to induce the agent to carry-out exploration outside of the inferred hypothesis. This can be accomplished most straightforwardly via uniform random sampling of the *a priori* action space during an additional *passive* learning phase. However, the relative number of moves required for each phase is likely to vary from scenario to scenario, depending on the extent to which correctly inferred clauses constrict the space of exploratory actions. Therefore, when, in the next section, we attempt to quantify the convergence advantage attributable to cognitive bootstrapping, we shall employ a variety of passive-to-active exploratory move ratios in order to heuristically assess the ideal strategy for perceptual bootstrapping in the shape-sorter domain. In having to decide upon a policy that optimizes exploration with respect to both the known and the unknown perceptual spaces, the cognitive bootstrapping method thus resembles a classical reinforcement learner [21], for which the analogous dilemma is the agent's attempts to maximize its environmental reward with respect to the explored and unexplored domains (although no form of cognitive inference is implied in the latter case).

As well as the addition of a passive phase, it is also possible, in principle, to further refine the active phase so as to involve the direct testing of action possibilities that have the capability to distinguish between competing object hypotheses. This is hence classical active learning in the manner of [39]. However, any such approach potentially complicates the interpretation of the learning agent in terms of perception-action theory, since competing hypotheses operate in *differing* perceptual domains, and hence exploratory actions undertaken in any one of these domains would refine the model of action legitimacy only with respect to that perceptual model. Any such system would thus not strictly meet the criteria of cognitive-bootstrapping; the simultaneous derivation of environmental *and* perceptual models appropriate to the agent's active capabilities.

Hence, of these possibilities, we shall employ only the passive learning modification to the cognitive bootstrapping agent described in the previous subsection. Even without these other modifications, the alternation between active and passive learning produces an exploratory method capable of ascending performance gradients during the active phase, and (after undergoing random perturbations in the *a priori* action space), a method capable of finding alternative, potentially more global, gradients to ascend during the passive phase. The approach adopted might thus be considered a primitive form of *simulated annealing* [23].

To provide a performance benchmark for this method, we shall also generate a *purely* passive learner in which PROGOL inference is applied to cumulative exploratory actions derived from random sampling of the *a priori* action space $(X1, Y1, Z1, X2, Y2, Z2)$. For both the bootstrapping and non-bootstrapping learners, we shall cumulatively apply logical inference in batches of 10 exploratory actions for every learning iteration. For the cognitive bootstrapping learner this means that we alternate respectively between the accumulation of sets of 10 and $n \times 10$ exploratory moves via active and passive exploration: n is hence an integer multiple controlling the ratio of active to passive exploration for the bootstrap learner.

To initiate both the bootstrapping and non-bootstrapping learners, we employ a batch of 200 randomly-sampled exploratory moves in order to arrive at a position from which active learning can commence. Hence, we attempt first-order logical inference only after sufficient exploratory moves have been accumulated to be reasonably certain (that is, approximately 90% certain) that rule inference has taken place (randomly-selecting 10% of the illegitimate rules for reasons of efficency). However, since it is not always possible to guarantee that this has occurred, we revert, in the absence of a generalized rule, to a passive exploration cycle. Accuracy figures in this case are nominally given by the legitimacy rate of the random exploratory samples of the *a priori* motor space (ie, on average $(|shapes|\times(|x|\times|y|-p))/(|x|\times|y|\times|z|)^2)$, reflecting the fact that active learning in the absence of generalization would presumably necessitate the re-enacting of positive examples. Under normal circumstances, however, accuracy is determined by the error rate of the inferred rule (ie, the percentage of actions incorrectly classified as either legitimate or illegitimate) calculated over the *entire* uniformly-sampled set of action possibilities within the *a priori* motor space.

5 Experimental Results

The result of the experiments for the passive and active learners with differing values of n ($n = 1$, 3 and 5, respectively) are given in figures 2, 3 and 4, representing the average over 10 experimental runs (with standard deviations indicated by the error-bars). It can be observed that both the passive and active runs start from an initially high percentile accuracy as a consequence of the fact that even a partially accurate rule-inference is sufficient to correctly eliminate the vast majority of the $(|x1| \times |y1| \times |z1|)^2$ proposable transitions within the largely redundant *a priori* space.

It is also evident in each of the three figures that the active learning procedure achieves convergence considerably faster than the passive learning procedure. If we define convergence time as being the number of iterations required in order for

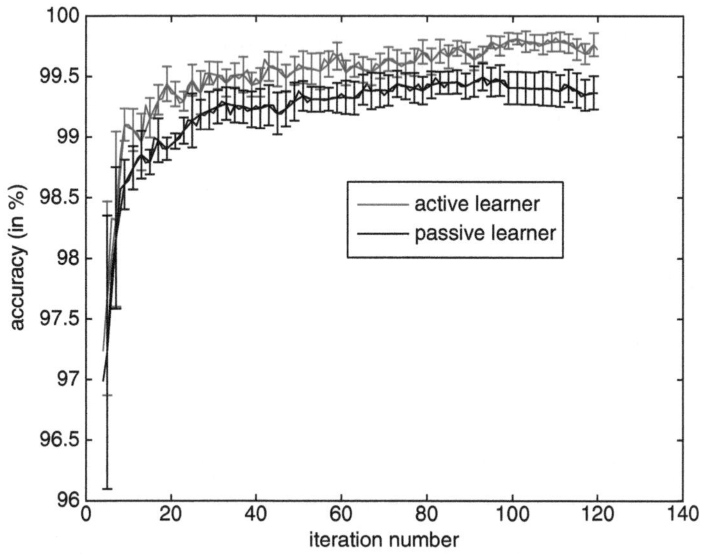

Fig. 2 Accuracy versus iteration number for the cognitive bootstrapping and passive learners (bootstrap learner active/passive ratio = 1)

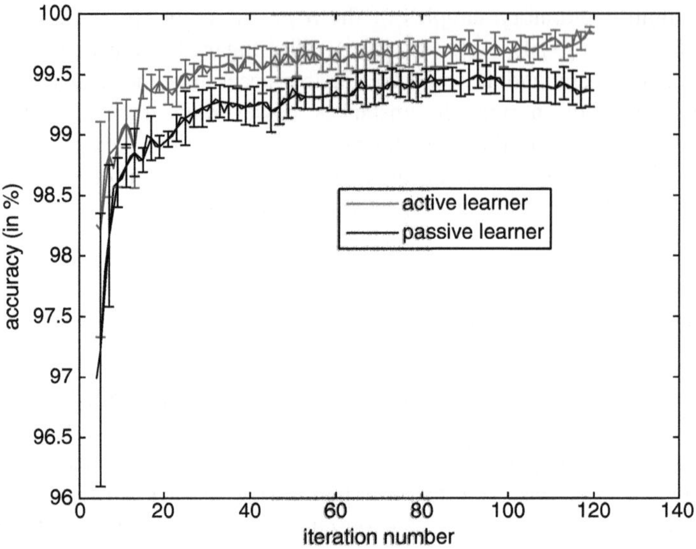

Fig. 3 Accuracy versus iteration number for the cognitive bootstrapping and passive learners (bootstrap learner active/passive ratio = 3)

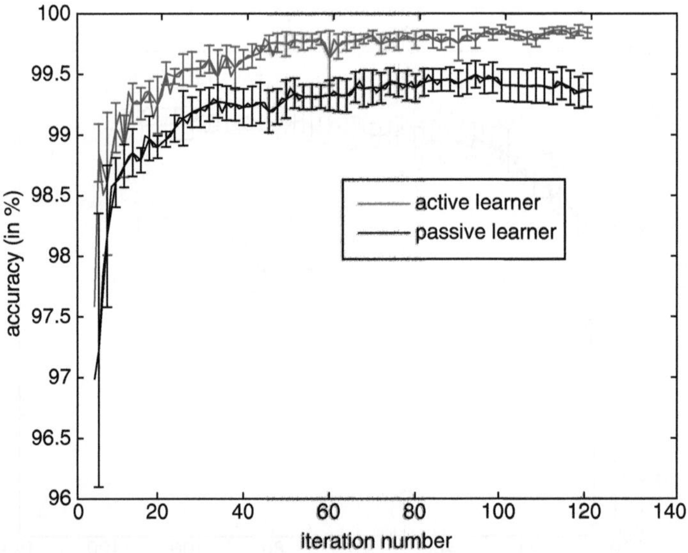

Fig. 4 Accuracy versus iteration number for the cognitive bootstrapping and passive learners (bootstrap learner active/passive ratio = 5)

Table 1 Iterations Required To Reach A Given Percentile Accuracy

Passive/Active Ratio	Accuracy Percentile		
	$x = 99\%$	$x = 98\%$	$x = 97\%$
$n = 1$	9.	6.	3.
$n = 3$	10.	4.	3.
$n = 5$	9.	5.	4.
$n = \infty$ (passive)	23.	7.	5.

Table 2 Bootstrap Learner Convergence Rates As A Multiple Of The Passive Learner Convergence Rate

Passive/Active Ratio	Accuracy Percentile		
	$x = 99\%$	$x = 98\%$	$x = 97\%$
$n = 1$	2.56	1.27	1.67
$n = 3$	2.3	1.75	1.67
$n = 5$	2.56	1.4	1.25

Table 3 Convergence Accuracy (Mean Accuracy After Having Achieved A Given Percentile Of The Maximum Accuracy Attained)

Passive/Active Ratio	Percentage of Maximum Attained		
	$x = 99.75\%$	$x = 99.50\%$	$x = 99.00\%$
$n = 1$	99.70 %	99.63 %	99.58 %
$n = 3$	99.71 %	99.65 %	99.61 %
$n = 5$	99.79 %	99.75 %	99.69 %
$n = \infty$ (passive)	99.38 %	99.35 %	99.28 %

performance to fall within a given accuracy percentile, x, then the average convergence times for the active and passive learners with respect to n and x are those indicated in Table 1 ($n = \infty$ defines a passive learner). In the most disparate case, when $n = 1$ or 5 and $x = 99.0$, this represents 9 iterations for the passive learner, and 23 iterations for the active learner; a 2.56-fold improvement in the convergence rate (see Table 2).

The respective absolute performance values on which the learners converge (defined as the average performance value after the systems have come within 1 percent of their maximum values) is 99.79 percent for the active learner when $n = 5$ and 99.38 percent for the passive learner. Values for other settings of n and x are given in Table 3. In all cases the absolute performance at convergence is higher for the active learner.

5.1 Alternative Experimental Domain

In order to assess the generalizability of these findings, we initiated a parallel set of tests in secondary experimental domain with differing action and motor possibilities, but (for the purposes of comparability) a similar perceptual predicate structure. We thus employ identical body mode declarations as before, but construct an alternative ground-truth clause, with a differing head mode declaration. The *a priori* action domain in this case is hence characterized by a simulated robot arm with planar movement potential along the X and Y axes only. However, it also has the capability to rotate any gripped entity to an arbitrary angle θ around the z axis. Again, there is assumed to be a 'gripping' gesture at the outset of the action, and a 'releasing' gesture at the end of the action. The *a priori* action command is hence *move(X1,Y1,*θ*1,X2,Y2,*θ*2,)* for the initial and final motor states $(X1, Y1, \theta 1)$ and $(X2, Y2, \theta 2)$, respectively.

The domain of application of this robot arm is an idealized 'jig-saw'-like environment in which any one of four puzzle pieces can be given a different position or orientation on the puzzle-board subject to one condition. This is that the piece is correctly aligned with any other puzzle pieces that happen to be adjacent to it (co-aligned objects can placed adjacently in either the X or Y directions).

The ground-truth rule protocols for this environment are hence specified as the simple four-clause sequence:

$$move(X1, Y1, \theta_1, X2, Y2, \theta_2) :-$$
$$free\_position(X2, Y2), position(L, X1, Y1), orientation(L, \theta_1,),$$
$$inc\_y(Y2, G), position(K, X2, G), orientation(K, P), theta_2 == P.$$

$$move(X1, Y1, \theta_1, X2, Y2, \theta_2) :-$$
$$free\_position(X2, Y2), position(L, X1, Y1), orientation(L, \theta_1,),$$
$$inc\_y(G, Y2), position(K, X2, G), orientation(K, P), theta_2 == P.$$

$$move(X1, Y1, \theta_1, X2, Y2, \theta_2) :-$$
$$free\_position(X2, Y2), position(L, X1, Y1), orientation(L, \theta_1,),$$
$$inc\_x(X2, G), position(K, G, Y2), orientation(K, P), theta_2 == P.$$

$$move(X1, Y1, \theta_1, X2, Y2, \theta_2) :-$$
$$free\_position(X2, Y2), position(L, X1, Y1), orientation(L, \theta_1,),$$
$$inc\_x(G, X2), position(K, G, Y2), orientation(K, P), theta_2 == P.$$

Predicates are defined as before, albeit in two-dimensional terms (ie *position (entity_label, x_ordinate, y_ordinate)*. X, Y and θ are quantized to 8,3 and 4 units, respectively.

The result of the experiments for the passive and active learners with differing values of n ($n = 1, 3$ and 5, respectively) are given in figures 5, 6 and 7, representing the average over 10 experimental runs (with standard deviations indicated by the error-bars).

It is again evident in each of the three figures that the active learning procedure achieves convergence considerably faster than the passive learning procedure. Convergence times (defined as above) are given in Table 4 for three sample-points. In the most disparate case, when $n = 3$ and $x = 99.5$, this represents 49 iterations for the passive learner, and 7 iterations for the active learner; a 7-fold improvement

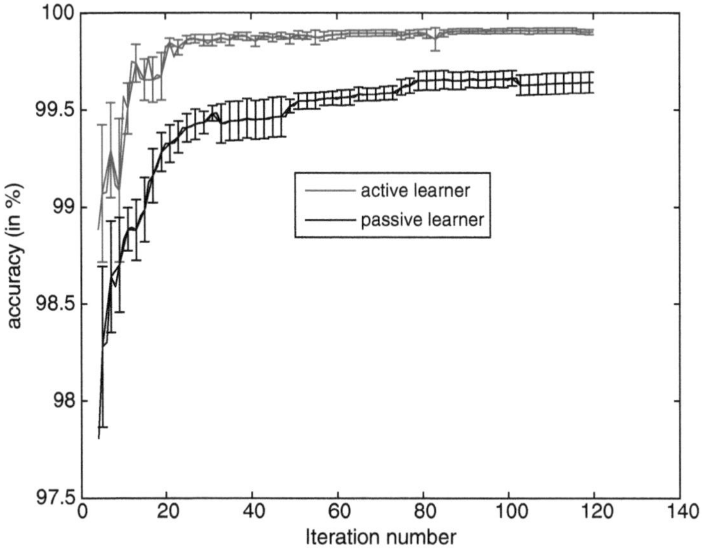

Fig. 5 Accuracy versus iteration number for the cognitive bootstrapping and passive learners (bootstrap learner active/passive ratio = 1). Note that the first five points are omitted for graph-scaling purposes

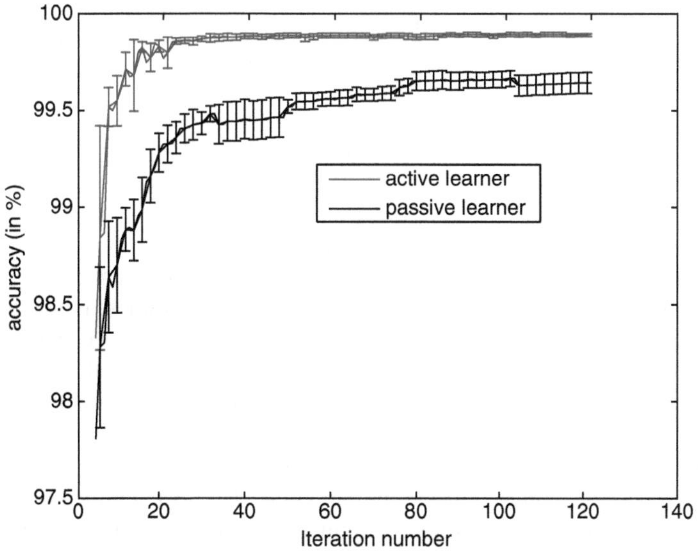

Fig. 6 Accuracy versus iteration number for the cognitive bootstrapping and passive learners (bootstrap learner active/passive ratio = 3). Note that the first five points are omitted for graph-scaling purposes

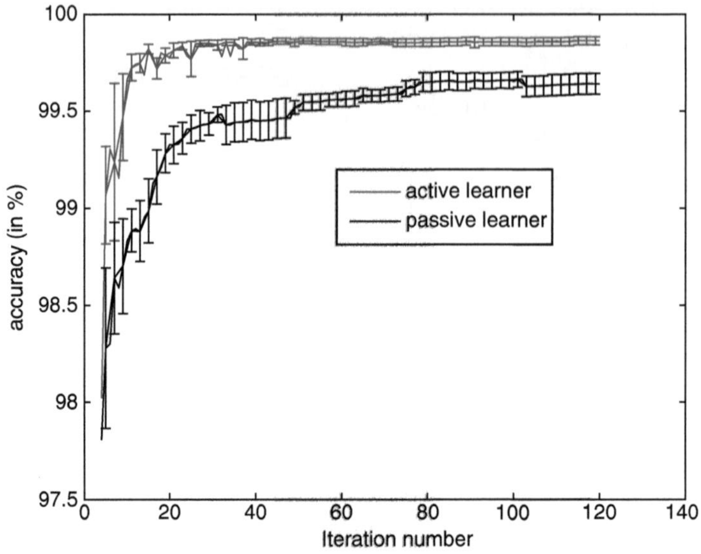

Fig. 7 Accuracy versus iteration number for the cognitive bootstrapping and passive learners (bootstrap learner active/passive ratio = 5). Note that the first five points are omitted for graph-scaling purposes

Table 4 Iterations Required To Reach A Given Percentile Accuracy

Passive/Active Ratio	Accuracy Percentile $x = 99.5\%$	$x = 98.5\%$	$x = 97.5\%$
$n = 1$	10.	4.	3.
$n = 3$	7.	5.	3.
$n = 5$	10.	5.	3.
$n = \infty$ (passive)	49.	7.	4.

in the convergence rate (see Table 5). The respective absolute performance values on which the learners converge (defined as the average performance value after the systems have come within 0.5 percent of their maximum values) is 99.86 percent for the active learner when $n = 3$ and 99.56 percent for the passive learner. Values for other settings of n and x are given in Table 6. In all cases the absolute performance at convergence is significantly higher for the active learner. However, it is the *rate* of convergence that we regard as being of principle significance: while the accuracy percentage gives a good indication of the progress of rule induction, it does not do so in a linear fashion. In particular, the most significant actions (in terms of the specificity of the protocols applicable to them) are those relating to the placing of accurately-aligned shapes into correspondingly shaped holes: however, these actions occupy only a very small subset of the total exploratory domain.

Results are hence substantially better in this domain than in the previous case.

Table 5 Bootstrap Learner Convergence Rates As A Multiple Of The Passive Learner Convergence Rate

Passive/Active Ratio	Accuracy Percentile		
	$x = 99.5\%$	$x = 98.5\%$	$x = 97.5\%$
$n = 1$	4.90	1.75	1.33
$n = 3$	7.00	1.40	1.33
$n = 5$	4.90	1.40	1.33

Table 6 Convergence Accuracy (Mean Accuracy After Having Achieved A Given Percentile Of The Maximum Accuracy Attained)

Passive/Active Ratio	Percentage of Maximum Attained		
	$x = 99.75\%$	$x = 99.50\%$	$x = 99.00\%$
$n = 1$	99.87 %	99.84 %	99.83 %
$n = 3$	99.86 %	99.86 %	99.85 %
$n = 5$	99.84 %	99.82 %	99.82 %
$n = \infty$ (passive)	99.56 %	99.51 %	99.48 %

6 Discussion and Conclusions

We have hence outlined a methodology for cognitive bootstrapping within a first-order logical environment that enables the simultaneous inference of optimized models of objects *and* percepts. The classical paradox associated with this type of simultaneous inference (namely, the potential meaninglessness of any empirical criterion for objective inference when the *interpretation* of empirical data is affected by perceptual updating, which is itself determined via observation) is overcome through the adoption of a *perception-action learning* mechanism. In the explicit coupling of perceptions to actions we thus aim to overcome the *underdetermination* (and hence the arbitrariness) of perceptual updating present in Quinean-like [37] models of perception, for which sensory updating can only ever be arbitrary with respect to sensory evidence. Another significant advantage that stems from this perception-action approach is the obviation of difficulties of *framing* [27], which occur when perceptual data is accumulated in a manner that is not related to the agent's active potential.

By formalizing the central notion of perception-action learning; *'action precedes perception'*, as the requirement for a condition of bijectivity between percept-transitions and individual actions existing within generalized models of action legitimacy, we hence provide a framework in which both perceptual and objective learning are made possible. More simply, cognitive bootstrapping seeks to create a space of action possibilities that are *always* perceptually realizable, and where redundant action possibilities are eliminated from perception.

As well as eliminating redundancy, the condition of bijectivity implies that perceptual updates must retain the expressiveness of the *a priori* space, permitting *every* legitimate action possibility to be perceived, so that the determination of action legitimacy constitutes the *environment model hypothesis*. The bijectivity condition hence allows us to eliminate difficulties typically associated with the *hermeneutic circle of interpretation* [11, 38], in which perceptual updates suggested by experimental observation can progressively lose information about the environment in trying to achieve the goal of *perceived* accuracy[11]. It is hence crucial that we retain the *a priori* perception-action space in order to *ground* the perceptual inferences in a hierarchical fashion.

Beyond this, the *explicit* linkage of perception to anticipated action possibilities in percept-action learning means that perceptual inferences are also action inferences and vice versa. Hence, the application of the condition of perception-action bijectivity to a *generalized* hypothesis-learner capable of compact representation of its input-space (such as PROGOL equipped with the reverse variable mapping stage) implies that *novel* (that is, previously unexperienced) action and percept possibilities can be addressed by the learning system.

Moreover, since such novel action possibilities are necessarily more constricted than those of the *a priori* space, they have the capacity, under random sampling, to find data capable of falsifying the object model hypothesis very much more rapidly than would be the case for random sampling of the *a priori* space. The cognitive bootstrapping mechanism we have described is hence also an *active learning* mechanism.

This investigation hence set out, firstly, to demonstrate that cognitive bootstrapping (simultaneous object and percept inference) can be accomplished self-consistently within a first order logical environment via the expedient of remapping the predicate variable structure. Secondly, it sought to demonstrate that such a system, as an active learner, is capable of achieving convergence on an objective model faster than a purely passive learner (ie, one that does not attempt perceptual updating) when randomly sampling the *a priori* space.

To this end, a PROLOG-based simulated shape-sorter environment was constructed, in which first-order logical inference was carried out via PROGOL, with predicate mode declarations so constructed as to allow for the reverse mapping of the perceptual variables. The typically more compact perceptual variable set arising from this reverse mapping can then be randomly instantiated in order to provide a set of proposed experimental actions in the *a priori* space for the cognitive bootstrapping agent to perform.

In carrying out this experimental instantiation of cognitive bootstrapping in section 5, we did indeed find that the simulated agent gave rise to significantly faster training (up to 2.56 times faster) than an equivalent, but non-bootstrapping agent by virtue of this implicit active learning potential. In a secondary test-environment this performance improvement was at the seven-fold level.

We should note that from a practical perspective, would be possible, in principle, to bypass cognitive bootstrapping and achieve active learning in the shape-sorter domain by sampling exploratory actions uniformly from the *a priori* six-dimensional space $(X1, Y1, Z1, X2, Y2, Z2)$, filtering them for consistency with the inferred rule. However, it is much more computationally efficient to find an appropriate parameterization of the legitimate action-space denoted by the inferred rule. Moreover, it would not, in general, be possible to guarantee that uniform sampling within an

[11] It is always possible to map percepts to smaller subsets such that the ability to discriminate 'difficult areas' is lost, and falsifying action possibilities are no longer perceived.

a priori space would be correlated with uniform sampling within an inferred percept space. This would apply if, for instance, we were to *chronologically* relax the instantiation uniqueness condition on predicate mode declarations, and permit reverse mapping to be *injective* over time. Hence, an action specified at the top level of the hierarchy might now be instantiated after *some number* of action sub-stages in the hierarchical perception-action level immediately beneath it (subgoals must hence always have perceivable consequences). This latter possibility is more representative of biological cognition; inferred object-motor categories such as affordances [13, 28] do not *immediately* constrain every motor-possibility at the muscular level, but rather apply progressive constraints hierarchically as different sub-perceptions and action possibilities become apparent. Hence progressive hierarchical levels serve to ground the high-level action in the *a priori* sensorimotor system existing at the muscular-neuronal level by specifying sensorimotor *sub-goals* at each stage of the hierarchy.

Beyond the exploration of this possibility, another objective of future research will be to eliminate the requirement of predicate mode declarations as they are currently specified. Hence, rather than having to assuming a given predicate form before inferring the clause variable structure (this being a limitation of PROGOL), we would instead infer *both* the predicate form and its variable structure. In this way, the method could be made sufficiently generic so as to enable the agent to be placed in a completely arbitrary environment with no prior assumptions beyond that of the *a priori* sensor-motor complex.

This possibility is very naturally encompassed by the cognitive bootstrapping framework via its ability to utilize exploratory findings in high-level perception-action domains to feedback *specification* information to the lower levels in such a way as to eliminate feature redundancy and feature under-specification. Hence, suppose that we have resolved to treat perceptions as the means of differentiating actions and, furthermore, regard the compressive capability of perceptual components as being an essential component of cognition (after [49]). It then becomes natural to amalgamate those percepts that have logically-identical action relations at the *a priori* level. (Thus, for instance, a 'chess-board'-pattered shape that was initially labeled via multiple percepts by the *a priori* image segmentation system, might instead be more logically addressed a single percept after exploratory manipulation has determined their connectedness). Equally, and conversely, it becomes natural for the cognitive bootstrapping mechanism to demand an expanded *a priori* percept domain when unable to disambiguate action outcomes (for instance, by refining *a priori* image segmentation parameters in order to generate a broader range of percept labels). Thus the *a priori* percept domain presented to logical induction need have no significant structure beyond that of a *label*, the action relations between percepts serving as the *sole* means of determining objective models of the external environment. It is then possible to autonomously derive a predicate of the type '$is\_hole(A)$' with exactly equivalent characteristics to that hitherto specified via background knowledge. This can be achieved by exploiting the fact that percepts or combinations of percepts capable of distinguishing holes from shapes (for instance color-labels) can be coupled to the action-derived meaning of the concept 'hole' (discovered by random exploration in the *a priori* motor-domain) by generalizing over specific instances in the usual manner. Groups of percepts that show this property consistently can hence then be collected and combined with the *a priori* motor variables in order to specify novel predicates for reuse via the first-order logical induction system. Such an approach may be thought of as the active selection of the feature-space underlying perception, as distinct from the active selection of the *operative subspace* underlying perception as is the case for the method we have outlined hitherto.

More broadly, the cognitive bootstrapping method we have adopted, that utilizes perception-action learning with bijectivity constraints in order to provide a self-consistent framework for perceptual updating, is potentially applicable to *any* learning methodology capable of generalization and compression. Thus, as well as logical quantification over variable instances, it is also possible to generalize over feature-space vectors via the usual mechanisms of statistical pattern recognition. Hence, cognitive bootstrapping can equally be applied to perceptual variables governing the *parameters* of a statistical distribution within the sensory domain. For instance, environmental classes inferred via unsupervized learning may be utilized to propose agent actions that result in active modification of the parameters governing the clustering behavior in a way that allows novel action-relevant object classes to be derived. Here again, the problem of framing is resolved by adaptively filtering-out perceived entities that have no bearing on the action capabilities of the agent, these entities being actively classed as noise via the statistical generalization mechanism. Such a statistical approach to active perceptual reparameterization would almost certainly need to be adopted for any real-world implementation of the cognitive bootstrapping agent: this is hence our next research objective.

Another issue of interest is in relation to supervised learning; a cognitive bootstrapping agent can learn not merely to imitate activity undertaken by the supervisor, but also to *perceive* the external environment in the same manner as the supervisor. In addition to the benefits associated with active learning, this alignment between the agent's and supervisor's perceptual spaces could, for instance, be used to bootstrap linguistic communication between the agents (in the manner of [41,42]). Linguistic communication, in distinction to other types of agent behavior, can trigger actions in response to sensory data referring to hypothetical and/or generalized situations. As such, the communicating agent needs to ensure that its internal environmental representation correlates with that of the communicating agent. In the majority of agent-based linguistic models this correlation is given *a priori*. However, a cognitive bootstrapping agent is theoretically capable of *learning* a linguistic framework from the actions of a trainer, so that linguistic models (such as those governing word reference) can be confirmed via a mixture of exploratory and communicative testing. Without this embodiment in the active domain, ultimately no definite assurance of a commonality of meaning between communicating agents is possible (refer, for instance, to Wittgenstein's *Philosophical Investigations* [48], or to Millikan's notion of biological cognition [30,31]). We hence, again, note the importance of the perception-action framework in overcoming the underdetermination of perceptual-updating in cognitively-autonomous agents: this constitutes the principle motivation for our work.

Acknowledgments The work presented here was supported by the the European Union, grant DIPLECS (FP 7 ICT project no. 215078)[12]. We also gratefully acknowledge funding from the UK Engineering and Physical Sciences Research Council (EPSRC) (grant EP/F069421/1)

References

1. Michael L. Anderson. Embodied cognition: A field guide. *Artif. Intell.*, 149(1):91–130, 2003.
2. R. Bowden, L. Ellis, J. Kittler, M. Shevchenko, and D. Windridge. Unsupervised symbol grounding and cognitive bootstrapping in cognitive vision. In *Proc. 13th Int. Conference on Image Analysis and Processing*, pages 27–36, September 2005.
3. David A. Cohn, Zoubin Ghahramani, and Michael I. Jordan. Active learning with statistical models. *J. Artif. Intell. Res. (JAIR)*, 4:129–145, 1996.
4. Cottingham, Stoothoff, Murdoch, and Kenney. *The Philosophical Writings of Descartes.* Cambridge University Press, Cambridge, UK, 1991.
5. Sanjoy Dasgupta, Adam Tauman Kalai, and Claire Monteleoni. Analysis of perceptron-based active learning. In Peter Auer and Ron Meir, editors, *COLT*, volume 3559 of *Lecture Notes in Computer Science*, pages 249–263. Springer, 2005.
6. J. Dewey. The reflex arc concept in psychology. *The Psychological Review*, (3):356–370, 1896.
7. Thomas G. Dietterich. Hierarchical reinforcement learning with the maxq value function decomposition. *J. Artif. Intell. Res. (JAIR)*, 13:227–303, 2000.
8. J.A. Fodor. *The Modularity of Mind.* MIT Press, Cambridge, MA, 1983.
9. S. Gallagher. *How the Body Shapes the Mind.* Oxford University Press, Oxford, 2005.
10. Ph. Gaussier. Towards a cognitive system algebra: A perception/action perspective. In *European Workshop on Learning Robots*, pages 88–100, Prague, Sept 2001.
11. A. Geir. *Explanation and Understanding: The Hermeneutic Arc.* PhD thesis, University of Oslo Department of Philosophy, 2001.
12. R. W. Jr Gibbs. *Embodiment and Cognitive Science.* Cambridge University Press, 2005. ISBN-10: 0521811740 — ISBN-13: 9780521811743.
13. J. J. Gibson. *The ecological approach to visual perception.* Houghton-Mifflin, Boston, 1979.
14. G. Granlund. Organization of architectures for cognitive vision systems. In *Proceedings of Workshop on Cognitive Vision*, Schloss Dagstuhl, Germany, 2003.
15. G. H. Granlund. An associative perception-action structure using a localized space variant information representation. In *Proceedings of Algebraic Frames for the Perception-Action Cycle (AFPAC)*, Kiel, Germany, September 2000.
16. G. H. Granlund and A. Moe. Unrestricted recognition of 3d objects for robotics using multi-level triplet invariants. *AI Magazine*, 25(2):51–67, 2004.
17. S. Harnad. The symbol grounding problem. *Physica D*, (42):335–346, 1990.
18. M. Heidegger. *Being and Time.* Blackwell, 1996.
19. Scott B. Huffman and John E. Laird. Flexibly instructable agents. *Journal of Artificial Intelligence Research*, 3:271–324, 1995.
20. E. Husserl. *Ideas Pertaining to a Pure Phenomenology and to a Phenomenological Philosophy – First Book: General Introduction to a Pure Phenomenology.* The Hague, 1982.
21. “Leslie Pack Kaelbling, Michael L. Littman, and Andrew P. Moore”. Reinforcement learning: A survey. *J. Artif. Intell. Res. (JAIR)*, 4:237–285, 1996.
22. Immanuel Kant. *Critique of Pure Reason.* Cambridge University Press, 1999. ISBN: 0521657296.

---

[12] However, this paper does not necessarily represent the opinion of the European Community, and the European Community is not responsible for any use which may be made of its contents.

23. S. Kirkpatrick, C. D. Gelatt, and M. P. Vecchi. Optimization by simulated annealing. *Science, Number 4598, 13 May 1983*, 220, 4598:671–680, 1983.
24. G. Lakoff and M Johnson. *Philosophy in the Flesh : The Embodied Mind and Its Challenge to Western Thought*. Harper Collins Publishers, 1999.
25. G. Lakoff and R. Núñez. *Where Mathematics Comes From: How the Embodied Mind Brings Mathematics into Being*. Basic Books, 2000.
26. D. Magee, C. J. Needham, P. Santos, A. G. Cohn, and D. C. Hogg. Autonomous learning for a cognitive agent using continuous models and inductive logic programming from audio-visual input. In *Proc. of the AAAI Workshop on Anchoring Symbols to Sensor Data*, 2004.
27. J. McCarthy and P.J. Hayes. Some philosophical problems from the standpoint of artificial intelligence. *Machine Intelligence*, (4):463–502, 1969.
28. J. McGrenere and W. Ho. Affordances: Clarifying and evolving a concept. In *Proceedings of Graphics Interface 2000*, pages 179–186, Montreal, Canada, 2000.
29. M. Merleau-Ponty. *Phenomenology of Perception*. New York: Humanities Press, 1962.
30. R. G. Millikan. *Language, Thought, and Other Biological Categories: New Foundations for Realism*. The MIT Press; Reprint edition, December 1987. ISBN: 0262631156.
31. R. G. Millikan. *White Queen Psychology and Other Essays for Alice*. The MIT Press; Reprint edition, March 1995. ISBN: 0262631628.
32. Andrew Moore. Variable resolution dynamic programming: Efficiently learning action maps in multivariate real-valued state-spaces. In L. Birnbaum and G. Collins, editors, *Machine Learning: Proceedings of the Eighth International Conference*, 340 Pine Street, 6th Fl., San Francisco, CA 94104, June 1991. Morgan Kaufmann.
33. S. Muggleton. *Inductive Logic Programming*. Academic Press, 1992.
34. Stephen. Muggleton. Inverse entailment and progol. *New Gen. Comput.*, 13:245–286, 1995.
35. A. Newell and H. Simon. The theory of human problem solving; reprinted in collins & smith (eds.). In *Readings in Cognitive Science, section 1.3.*, 1976.
36. K. Popper. *The Logic of Scientific Discovery. (translation of Logik der Forschung)*. Hutchinson, London, 1959.
37. W. V. O. Quine. *Ontological Relativity*. Columbia, 1977.
38. P. Ricoeur. *Hermeneutics and the Human Sciences: Essays on Language, Action and Interpretation*. Cambridge: Cambridge University Press, 1981.
39. H. S. Seung, Manfred Opper, and Haim Sompolinsky. Query by committee. In *Computational Learning Theory*, pages 287–294, 1992.
40. G . Sommer, E. Bayro-Corrochano, and Th. Blow. Geometric algebra as a framework for the perception-action cycle. In *Workshop on Theoretical Foundation of Computer Vision, Ed. F. Solina*, Wien, 1997. Springer Verlag.
41. L. Steels. The origins of syntax in visually grounded robotic agents. In M. Pollack, editor, *Proceedings of the 10th IJCAI, Nagoya*, pages 1632–1641. AAAI Press, Menlo-Park Ca., 1997.
42. L. Steels and F. Kaplan. Bootstrapping grounded word semantics. In Ted Briscoe, editor, *Linguistic Evolution through Language Acquisition: Formal and Computational Models*, chapter 3. Cambridge University Press, 2002.
43. R. Sun. Desiderata for cognitive architectures. *Philosophical Psychology*, 17(3), 2004.
44. R. Sun, T. Peterson, and E. Merrill. A hybrid architecture for situated learning of reactive sequential decision making. *Applied Intelligence*, 11:109–127, 1999.
45. Bas. van Fraassen. *The Scientific Image*. Oxford: Oxford University Press, 1980. ISBN 0-19-824427-4.
46. D Windridge and J Kittler. Cognitive bootstrapping: A survey. Technical report, CVSSP, University of Surrey, UK, 2005.
47. D Windridge and J Kittler. Epistemic constraints on autonomous symbolic representation in natural and artificial agents. *Applications of Computational Intelligence in Biology: Current Trends and Open Problems, Studies in Computational Intelligence (SCI)*, 122, 2008.
48. L. Wittgenstein. *Philosophical investigations : the German text with a revised English translation by Ludwig Wittgenstein*. Oxford : Blackwell, 2001. ISBN 0631231277.
49. J G Wolff. Cognitive development as optimisation. In L Bolc, editor, *Computational Models of Learning*, pages 161–205, Heidelberg, 1987. Springer-Verlag.

Detection of Auditory Cortex Activity by fMRI Using a Dependent Component Analysis

Carlos A. Estombelo-Montesco, Marcio Jr. Sturzbecher, Allan K.D. Barros, and Draulio B. de Araujo

Abstract Functional MRI (fMRI) data often have low signal-to-noise-ratio (SNR) and are contaminated by strong interference from other physiological sources. A promising tool for extracting signals, even under low SNR conditions, is blind source separation (BSS), or independent component analysis (ICA). BSS is based on the assumption that the detected signals are a mixture of a number of independent source signals that are linearly combined via an unknown mixing matrix. BSS seeks to determine the mixing matrix to recover the source signals based on principles of statistical independence. In most cases, extraction of all sources is unnecessary; instead, a priori information can be applied to extract only the signal of interest. Herein we propose an algorithm based on a variation of ICA, called Dependent Component Analysis (DCA), where the signal of interest is extracted using a time delay obtained from an autocorrelation analysis. We applied such method to inspect functional Magnetic Resonance Imaging (fMRI) data, aiming to find the hemodynamic response that follows neuronal activation from an auditory stimulation, in human subjects. The method localized a significant signal modulation in cortical regions corresponding to the primary auditory cortex. The results obtained by DCA were also compared to those of the General Linear Model (GLM), which is the most widely used method to analyze fMRI datasets.

Keywords Dependent Component Analysis · Independent Component Analysis · Mixture of signals · Recover the source signals · Signal of interest · fMRI · GLM · ICA

M. Jr. Sturzbecher and D.B. de Araujo
Department of Physics and Mathematics, FFCLRP, University of Sao Paulo, Ribeirao Preto, SP, Brazil
e-mail: marcio@biomag.usp.br; draulio@usp.br

C.A. Estombelo-Montesco (✉)
DCOMP/UFS Depto. de Computaçao da Universidade Federal de Sergipe, Cidade universitaria Prof., Jose Aloisio de Campos, Jardim Rosa Elze, CEP 49100-000, São Cristóvão - SE
e-mail: estombelo@gmail.com

A.K.D. Barros
Department of Electrical Engineering, Federal University of Maranhao, Sao Luis, Maranhao, Brazil
e-mail: allan@ufma.br

A. Hussain et al. (eds.), *Brain Inspired Cognitive Systems 2008*, Advances in Experimental Medicine and Biology 657, DOI 10.1007/978-0-387-79100-5_7,

1 Introduction

Functional neuroimaging tools are becoming more widely used in the framework of cognitive neuroscience and cognitive psychology. These studies are generally focused at measuring specific aspects of brain function looking towards understanding the relationship between activity in certain brain areas and specific mental functions. A common method in functional neuroimaging is the functional Magnetic Resonance Imaging (fMRI) that can measure localized changes in cerebral blood flow related to neural activity [1]. The principal goal of fMRI is mapping and characterizing, non-invasively, certain aspects of the human brain function. Currently, there are many processes involved in manipulating the image contrast that follows neuronal activation, most of which are based on Blood Oxygenation Level Dependent (BOLD) mechanism that rely on hemoglobin oxidative state changes. More specifically, the signal changes in BOLD fMRI are determined by the paramagnetic properties of deoxyhemoglobin (deoxy-Hb), as it enhances the spin phase dispersion, and thereby affects the transverse relaxation (T2) and especially the non-refocused transverse relaxation ($T2^*$). This effect increases with deoxy-Hb concentration and is particularly pronounced if the deoxy-Hb is compartmentalized, e.g. in red blood cells and within blood vessels.

In more details, when brain cells are active, they consume oxygen carried by hemoglobin in red blood cell from local capillaries. The local response to this oxygen utilization is an increase in blood flow to regions of increased neural activity. This leads to local changes in the relative concentration of Hb and deoxy-Hb and changes in local cerebral blood volume in addition to this change in local cerebral blood flow. The oxyhemoglobin is diamagnetic and deoxyhemoglobin is paramagnetic and this difference makes that each one has different magnetic susceptibility. When there is a change into the magnetic susceptibility in a local region of the brain, the characteristic relaxations time of the tissue in response to an appropriate sequence of radio frequency pulse also changes [1]. As a consequence there is a change into the image contrast (detected by MRI scanner) called as BOLD (Blood Oxygen Level Dependent) effect [1].

In short, during an increase in neuronal activation, there is an increase in local cerebral blood flow, but only a small proportion of the oxygen is used. There is therefore a net increase in the tissue concentration of oxyhemoglobin and a net reduction in the tissue concentration of the paramagnetic deoxyhemoglobin in the local capillary bed, and draining venules, leading to an increase in signal intensity on $T2^*$-weighted images.

Unfortunately, BOLD-fMRI often have low signal-to-noise-ratio (SNR) (about 2%–4% with 1.5T magnetic field strength) and are contaminated by strong interference from other physiological sources that complicates data analysis.

Due to the low SNR, whole brain volume images are usually acquired many times during the task performance. The protocols commonly used compare many periods in which the subjects are performing a specific task with respect to control or rest periods. When the task is alternated with rest in one-by-one symmetric pattern, the protocol is named block paradigm. This is the most frequent method used in fMRI, especially when fMRI is applied into the clinical practice.

Generally the methods applied to fMRI data can be divided into two categories [2]: hypothesis-driven analyses and the data-driven analyses. The first one is when we test a time course of a *voxel* to see if it is active, therefore we are really testing how well that time course matches the idealized waveform that an active voxel should exhibit. Nevertheless, the assumptions of hypothesis-driven analyses may not always be valid, specifically under complicated experimental conditions, or under pathological conditions that affect neurovascular coupling [3]. Moreover, most fMRI analyses methods are based on the assumption of linear behavior, that is, responses to long-duration stimuli can be predicted from responses to shorter duration stimuli. This theoretical consideration supports the majority of the classic methods, which require a model for the hemodynamic response in order to predict the fMRI signal, such as the general linear model (GLM) [4], the cross-correlation [5]. A few emerging methods do not rely on such assumption, as those based on information theory [6–8], and the analysis of variance (ANOVA) [9].

Another possibility is to use data-driven approaches, in which the structure of the data is explored in order to obtain and localize patterns related to task activations, also known as exploratory analyses. One such analysis, known as blind source separation (BSS), or independent component analysis (ICA), is designed to extract signals of interest, even under low SNR conditions. ICA is based on the assumption that the detected signals are a mixture of a small number of independent source signals (called components) that are linearly combined via an unknown mixing matrix. ICA seeks to determine the mixing matrix and to recover the source signals (components) based on principles of statistical independence. Observe that an ICA analysis can be conducted completely blind with respect to the experimental task or hypotheses. ICA has been successfully applied to fMRI analysis in order to find independently distributed spatial patterns that depict source processes in the data [10, 11].

One of the main challenges lies in deciding how many components to isolate from a given data set. This is even more critical in fMRI time series, where the number of time series is enormous. For instance, an experiment where the EPI matrix consists of 64 × 64 voxels will have, in principle, 4096 possible independent signals. One straight forward strategy to reduce such number is to compute the ICA only over cortical regions, where the neuronal activity of interest is mostly prominent [12]. Another possibility is to use *a priori* information of the signal of interest, as its temporal structure. Herein we aim to apply one of such approaches, already successfully implemented in different biomedical signals, as fetal magnetocardiography [13] and magnetogastrography [14].

2 Objectives

Herein we propose an algorithm based on a variation of ICA, called Dependent Component Analysis – DCA, where the signal of interest is extracted using a weighted matrix based on time-delay obtained from an autocorrelation analysis of fMRI time series obtained from auditory stimulation.

3 Methods

Subjects: A total of 10 non-symptomatic subjects participated in an auditory fMRI experiment. Subjects were comfortably positioned in the scanner and foam padding packed on the sides of the head was used to reduce head motion during the experiment.

Auditory Task: The stimulus was delivered by a MRI compatible headphone, in a block design, with six blocks of rest (27.5 sec each), interleaved with five blocks of activity (27.5 sec each). During the task, the volunteers passively listened to a complex story with standard narrative structure. Immediately after the exam they had to report the content of the story.

Recordings: BOLD fMRI data were acquired on a SIEMENS 1.5T scanner (Magnetom Vision, Erlangen, Germany). Measurements were performed in a single session, containing 64 brain volumes of 16 slices each, using an EPI-BOLD sequence. The image parameters were TE = 60 ms, TR = 4.6 sec, 128×128 matrix, FOV = 220mm, slice thickness = 5 mm, flip angle = 90°. Furthermore, for superposition of the statistical maps onto a high resolution image, we used 154 sagital cross-sections of 1 mm^3, covering both hemispheres of the brain, obtained with a T1 gradient echo sequence, SPGR (TR = 9.7 ms; TE = 4 ms; flip angle = 12°; matrix = 256×256; FOV = 256 mm; slice thickness = 1 mm).

Data Analysis Pre-processing: The fMRI data were pre-processed with a slice scan time correction, motion correction, high-pass filtered (0.01 Hz). Statistical analysis was based on a novel method called Dependent Component analysis (DCA) [15]. The strategy proposed is based on previous works [13, 14] of data-driven analysis, which successfully extracted the component of interest even in cases of low SNR.

Dependent Component Analysis: One of the most common models of blind source separation (BSS) assumes that the signal is a linear mixture of independent random sources. While BSS can be applied regardless of the temporal structure of the signal, the presence of temporal structure can aid in isolating the component of interest. DCA is a method based on multivariate analysis that uses *a priori* delay based on the temporal characteristics of the auditory hemodynamic response function to be extracted. For such, we use a temporal component having a characteristic time delay, or frequency. The method for extraction and artifact removal is fully described in [13, 14] and a short description follows (Figure 1).

Consider n sources $\boldsymbol{S} = [s_1, s_2, \ldots, s_n]^T$ that are mixed into vector $\boldsymbol{X}$ through the following linear combination: $\boldsymbol{X} = \boldsymbol{AS}$, where $\boldsymbol{A}$ is an $n \times n$ invertible matrix.

Our goal here is to find the source of interest, s_i. In general, the number of independent components can be as large as the dimension of $\boldsymbol{X}$. As only a single source is to be extracted, the signal can be expressed as $\boldsymbol{y(k)} = \boldsymbol{w^T x(k)}$, where $\boldsymbol{w}$ is a weight vector for that single source. Defining the error as $\boldsymbol{e(k)} = \boldsymbol{y(k)} - \boldsymbol{y(k\text{-}p)}$ and minimizing the mean squared error $\boldsymbol{MSE(w)} = \boldsymbol{E[e^2]}$, we find:

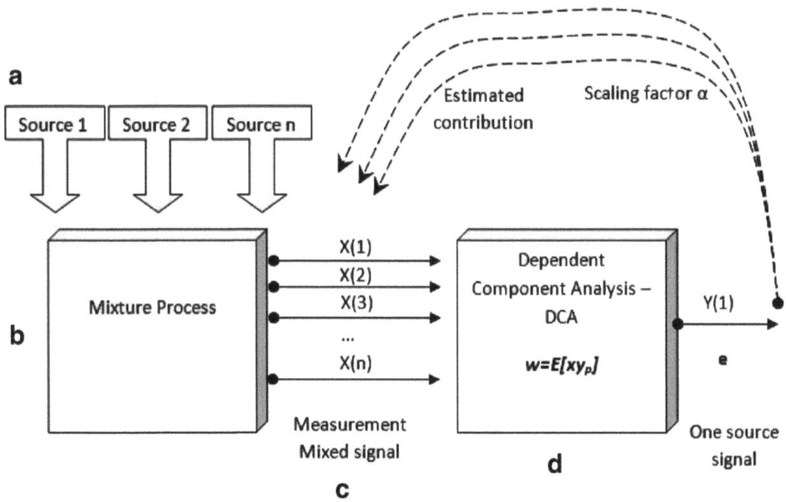

Fig. 1 DCA schematics. **a)** First each source (auditory cortex, heart, tissue, artifacts, etc) produces a signal, actually not seen directly. **b)** Then each source is mixed with other sources and it is represented by the block (linear) mixture process. **c)** When the signals are acquired actually we are measuring the signals from the mixture process obtaining one time series for each voxel. **d)** Then separation/extraction of the source of interest can be done and further evaluated through DCA process using w. e) The output contains a single time series of interest (the extracted component)

$$w = E[xy_p],$$

where $y_p = y(k\text{-}p)$. We use sequential signal extraction along with *a priori* information about the autocorrelation function.

One practical problem is how to estimate the optimal time delay, p. A simple solution is to calculate the autocorrelation function of the signals and find the feature, in our case a peak with appropriate time-lag, corresponding to the signal of interest. In order to accomplish this, we model the system using auto-regression [13].

Here we must consider that there are sixteen axial brain slices being measured continuously over the time. Each slice is consisted of an n × n matrix, sampled 64 times. Therefore each slice has a dimension of [n×n×nPoints], where nPoints = 64. This tri-dimensional matrix suffers a transformation, for our purposes, to the dimension [n^2 × nPoints], which corresponds to matrix X in our model.

It is important to note that all processing was carried out exclusively in voxels within the brain, and all others were removed using a mask, which included gray matter, white matter and liquor. Also, for comparison purposes, another method was applied, called the General Linear Model (GLM), which is often used in the fMRI framework.

General Linear Model: This approach considers the linearity of the hemodynamic response, and the fMRI data is represented in a linear model [4]. The model is described as follow: the observed data is equal to a weighted combination of several

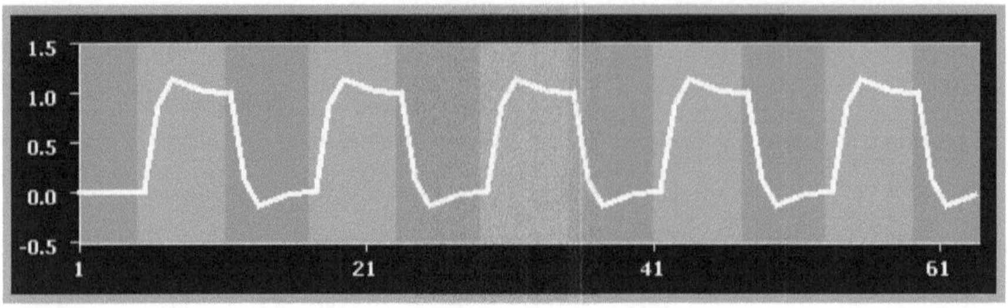

Fig. 2 Predictor function used for the hemodynamic response function. It is based on the modification of the box-car function. The green bars represent periods when the task was being performed, and in gray are the baseline periods. The white line is the specific model which resembles a hemodynamic response. X scale is the number the points sampled

model factors (hypothesis) plus an error. The parameters weights are the relative contribution of each factor to the overall observed data.

The solution of this linear model equation involves the knowledge of the measurement experimental data and the predictor function (model factor, external waveform, a priori hypothesis) generally based on evidences from another techniques. Figure 2 represents a typical predictor function used into the GLM analysis. The green bars represent periods when the task was being performed, and in gray are the baseline periods. The white line is the specific model which resembles a hemodynamic response.

The estimation of a specific set measured of time points, $\boldsymbol{X}$, is based on a model, $\boldsymbol{G}$, by adjusting a specified set of weight factors, $\boldsymbol{\beta}$, aiming serves to minimize the error term, $\boldsymbol{\varepsilon}$, as follows:

$$\boldsymbol{X} = \boldsymbol{G} \times \boldsymbol{\beta} + \boldsymbol{\varepsilon},$$

$\boldsymbol{X}$ is the original data; $\boldsymbol{G}$ is the design matrix that specifies how the model factor changes over timer; $\boldsymbol{\beta}$ represents the experimental parameters to be estimated; and the error $\boldsymbol{\varepsilon}$ is to be minimized.

4 Results

After processing the fMRI data with DCA the one component extracted is shown in Figure 3, which is characterized by a strong correlation with the experimental protocol used. All components extracted, from each slice, is shown in Figure 3.

As can be observed from Figure 3, each component extracted has a periodic and quasi-periodic characteristic. A closer inspection of the component behavior can be estimated by the average of all components extracted (Figure 4).

With the source of interest extracted with DCA, one can generate a statistical map, as to locate where in the brain such component was most prominent. Again those pixels that stand out a specified statistical threshold are shown in color scale.

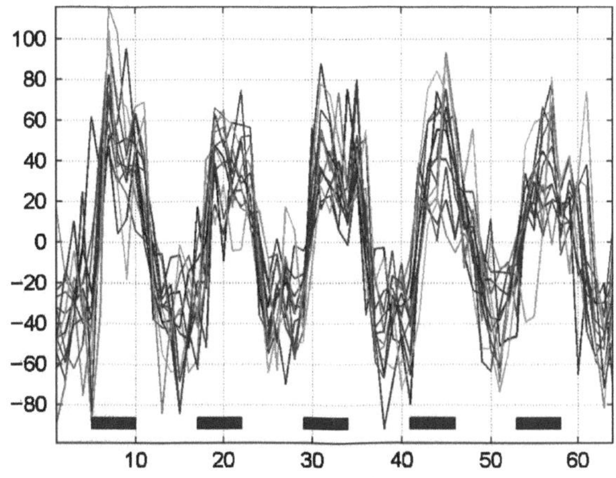

Fig. 3 Source components extracted by the DCA method. One component is extracted for each slice. The number of slices adquired are sixteen. Red block correspond to the auditory stimulus. X scale is the number the points sampled

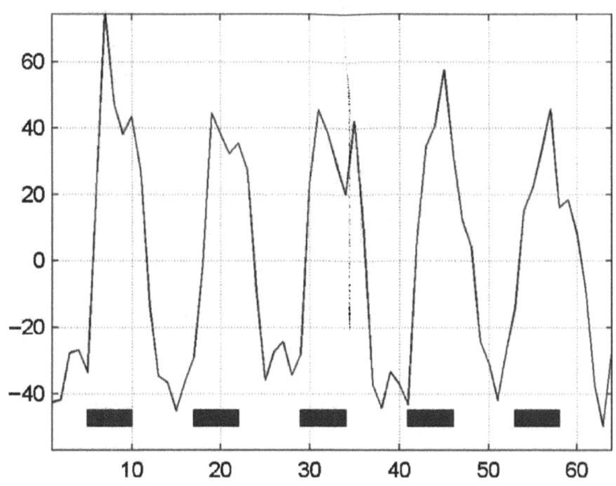

Fig. 4 Averaging of the sources components extracted (solid line) from Figure 3. Red block correspond to auditory stimuli periods. For a quasi-periodic signal, DCA identifies the signal component (solid line) based on the (inner) time delay determined from the temporal characteristics of the auditory hemodynamic response in fMRI measurement. X scale is the number the points sampled

The final co-registration onto a high resolution anatomical image, from a representative subject, is shown in Figure 5.

The results show areas of activity corresponding to the superior temporal gyrus, consistent with the primary auditory cortex.

For comparison purposes, GLM maps were also obtained from the same auditory protocol (Figure 6). Figure 6 shows the output statistical map, onto a high resolution image, after GLM computation. The results shown in Figure 6 agree with well established functional maps of the auditory cortex.

Although statistical threshold was the same for both analysis ($p < 0.05$), visual inspection of the DCA maps reveal a more limited area of brain modulation. In contrast, preliminary GLM results shown a wider area, more spread with respect to the DCA maps.

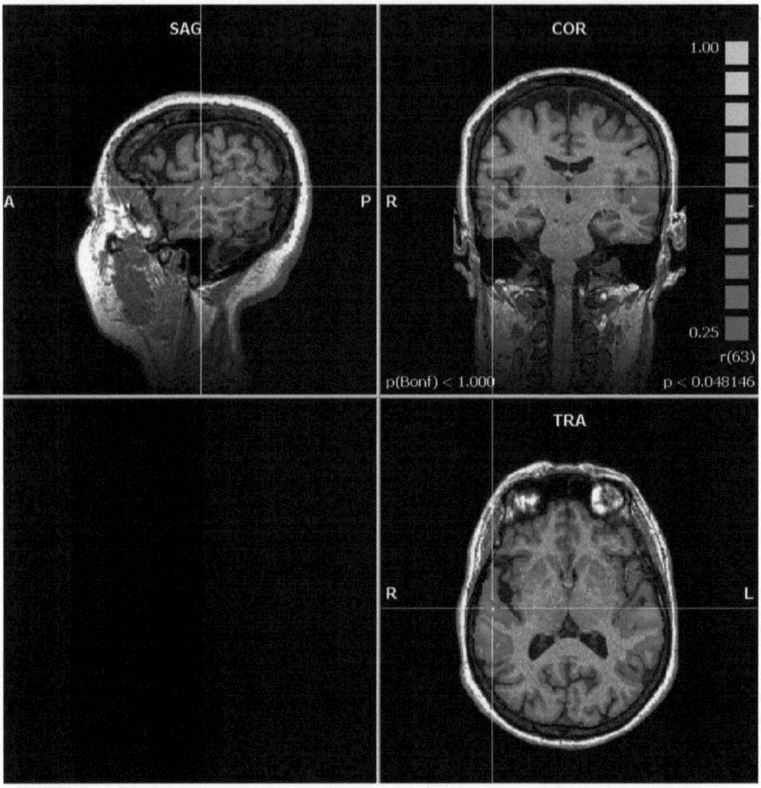

Fig. 5 DCA Result with a given characteristic in this case this characteristic is based on the time delay of the fMRI data. The areas correspond to the superior temporal gyrus, consistent with the primary auditory cortex

5 Discussion and Conclusion

The results of this study support the idea that our approach is efficient in evaluating the autocorrelation structure of the components, and extracting the component of interest. Therefore, it tends to be more general than those methods that attempt to correlate the dataset with a specific model [4, 5]. In every subject, the areas found by DCA are in agreement to the auditory cortex activity observed in the literature [16, 17].

Although based on a qualitative analysis, it was observed that the statistical maps generate from DCA are more restrictive and better related to the localization of the auditory cortex foci. For comparison purposes we maintained the same statistical threshold ($p < 0.05$) for both methods.

Some differences between the methods are important to understand. First in GLM it is necessary to have a model, which is to be used in the design matrix, also called a

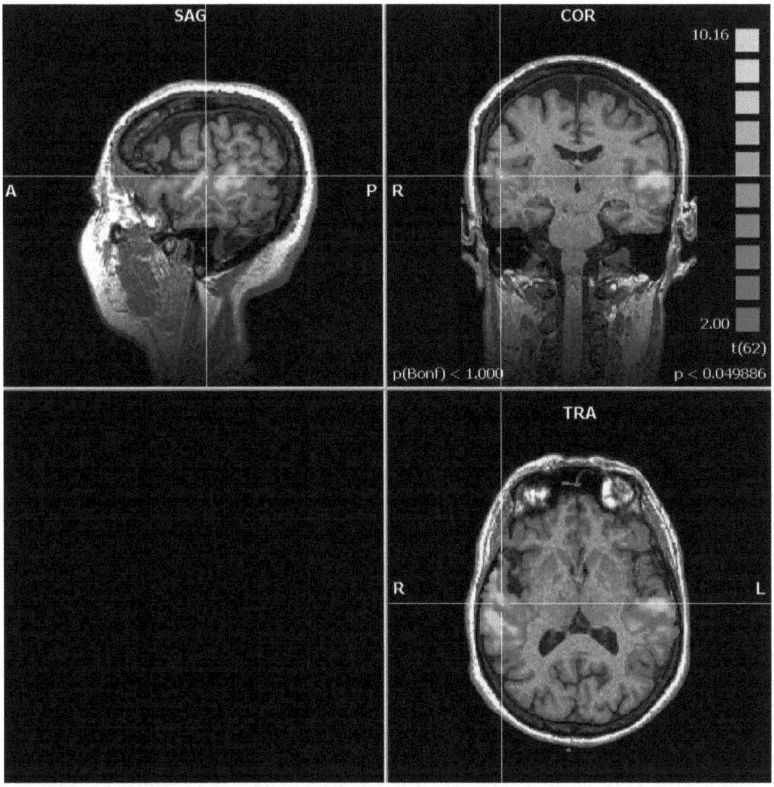

Fig. 6 GLM result for a representative subject submitted to an auditory stimulus. The main brain areas modulated and identified by the GLM are consistent with structures involved in auditory processing

response predictor. The predictor function generally comes from hypothetical models that should fit the real hemodynamic response function, to be calculated through combination of weights.

On another hand, DCA doesn't need an external reference. Instead, DCA considers a kind of internal information a priori based on the fMRI data itself. This a priori information can be deduced from the multivariate auto-regressive modeling of the signals and the calculation the poles of a transfer function which models the whole system. From this pole it can be estimated the optimal timed delay to be used in DCA method.

Therefore, one of the main concerns in classical methods, for fMRI analysis, is the assumption that the hemodynamic response can be standardized [18]. The DCA analysis makes no assumptions about the shape and timing of the hemodynamic response, whatsoever. Regardless, the only assumption made was that the protocol modulated brain activity with a specific frequency. This allows the extraction and detection of specific components of interest.

Besides, most previous ICA methods applied to fMRI were designed to extract all components embedded in the data. Here, instead, we can recover only one specific source of interest from the measured signals. Furthermore, as DCA extract one component it avoids the permutation problem widely knows in ICA [19].

Acknowledgments This work was supported by FAPESP (Grant No. 05/03225-7), CNPq and CAPES, and the CINAPCE Program (FAPESP: 05-56447-7). The authors also wish to thanks Prof. Antonio Carlos dos Santos for supporting data acquisition and Sandra Moroti for the technical support.

References

1. Logothetis NK (2002) The neural basis of the blood-oxygen-level-dependent functional magnetic resonance imaging signal. Philosophical Transactions of the Royal Society of London Series B-Biological Sciences 357: 1003–1037.
2. Schad LR (2002) Functional magnetic resonance imaging (fMRI). Part 2: Data analysis and applications. Radiologe 42: 756–764.
3. Andrade KC, Pontes-Neto OM, Leite JP, Santos AC, Baffa O, et al. (2006) Quantitative aspects of brain perfusion dynamic induced by bold fMRI. Arquivos De Neuro-Psiquiatria 64: 895–898.
4. Friston KJ, Frith CD, Frackowiak RSJ, Turner R (1995) Characterizing Dynamic Brain Responses with Fmri - a Multivariate Approach. Neuroimage 2: 166–172.
5. Bandettini PA, Jesmanowicz A, Wong EC, Hyde JS (1993) Processing Strategies for Time-Course Data Sets in Functional Mri of the Human Brain. Magnetic Resonance in Medicine 30: 161–173.
6. de Araujo DB, Tedeschi W, Santos AC, Elias J, Neves UPC, et al. (2003) Shannon entropy applied to the analysis of event-related fMRI time series. Neuroimage 20: 311–317.
7. Tedeschi W, Muller HP, de Araujo DB, Santos AC, Neves UPC, et al. (2004) Generalized mutual information fMRI analysis: a study of the Tsallis q parameter. Physica a-Statistical Mechanics and Its Applications 344: 705–711.
8. Tedeschi W, Muller HP, de Araujo DB, Santos AC, Neves UPC, et al. (2005) Generalized mutual information tests applied to fMRI analysis. Physica a-Statistical Mechanics and Its Applications 352: 629–644.
9. Clare S, Humberstone M, Hykin J, Blumhardt LD, Bowtell R, et al. (1999) Detecting activations in event-related fMRI using analysis of variance. Magnetic Resonance in Medicine 42: 1117–1122.
10. Calhoun VD, Adali T, Stevens MC, Kiehl KA, Pekar JJ (2005) Semi-blind ICA of fMRI: a method for utilizing hypothesis-derived time courses in a spatial ICA analysis. Neuroimage 25: 527–538.
11. McKeown MJ, Makeig S, Brown GG, Jung TP, Kindermann SS, et al. (1998) Analysis of fMRI data by blind separation into independent spatial components. Human Brain Mapping 6: 160–188.
12. Formisano E, Esposito F, Di Salle F, Goebel R (2004) Cortex-based independent component analysis of fMRI time series. Magnetic Resonance Imaging 22: 1493–1504.
13. de Araujo DB, Barros AK, Estombelo-Montesco C, Zhao H, da Silva ACR, et al. (2005) Fetal source extraction from magnetocardiographic recordings by dependent component analysis. Physics in Medicine and Biology 50: 4457–4464.
14. Estombelo-Montesco CA, de Araujo DB, Silva ACR, Moraes ER, Barros AK, et al. (2007) Dependent component analysis for the magnetogastrographic detection of human electrical response activity. Physiological Measurement 28: 1029–1044.

15. Barros AK, Cichocki A (2001) Extraction of specific signals with temporal structure. Neural Computation 13: 1995–2003.
16. Amaro E, Williams SCR, Shergill SS, Fu CHY, MacSweeney M, et al. (2002) Acoustic noise and functional magnetic resonance imaging: Current strategies and future prospects. Journal of Magnetic Resonance Imaging 16: 497–510.
17. Yoo SS, O'Leary HM, Dickey CC, Wei XC, Guttmann CRG, et al. (2005) Functional asymmetry in human primary auditory cortex: Identified from longitudinal fMRI study. Neuroscience Letters 383: 1–6.
18. Vazquez AL, Noll DC (1998) Nonlinear aspects of the BOLD response in functional MRI. Neuroimage 7: 108–118.
19. Suzuki K, Kiryu T, Nakada T (2002) Fast and precise independent component analysis for high field fMRI time series tailored using prior information on spatiotemporal structure. Human Brain Mapping 15: 54–66.

Brain–Computer Interface Using Wavelet Transformation and Naïve Bayes Classifier

Thiago Bassani and Julio Cesar Nievola

Abstract The main purpose of this work is to establish an exploratory approach using electroencephalographic (EEG) signal, analyzing the patterns in the time-frequency plane. This work also aims to optimize the EEG signal analysis through the improvement of classifiers and, eventually, of the BCI performance. In this paper a novel exploratory approach for data mining of EEG signal based on continuous wavelet transformation (CWT) and wavelet coherence (WC) statistical analysis is introduced and applied. The CWT allows the representation of time-frequency patterns of the signal's information content by WC qualiatative analysis. Results suggest that the proposed methodology is capable of identifying regions in time-frequency spectrum during the specified task of BCI. Furthermore, an example of a region is identified, and the patterns are classified using a Naïve Bayes Classifier (NBC). This innovative characteristic of the process justifies the feasibility of the proposed approach to other data mining applications. It can open new physiologic researches in this field and on non stationary time series analysis.

Keywords Pattern analysis · Classification · Signal processing · Brain computer interface

---

This work was supported in part by CAPES (Coordenadoria de Aperfeiçoamento de pessoal de Nível Superior) and PUCPR (Pontifícia Universidade Católica do Paraná).

T. Bassani
Instituto Nacional de Inovação em Diagnósticos para a Saúde Pública - fisica.ufpr.br/INIDSP
CPGEI–Programa de Pós-Graduação em Engenharia Elétrica e Informática Industrial
www.cpgei.cefetpr.br
and
UTFPR - Universidade Tecnológica Federal do Paraná
www.utfpr.edu.br, www.ppgia.pucpr.br/~tbassani
e-mail: thiago.bassani@gmail.com

J.C. Nievola (✉)
is with PUCPR at PPGIa (Programa de Pós-Graduação em Informática). He is a Full Professor at PUCPR and the Team Leader of the Knowledge Discovery and Machine Learning Research Group
e-mail: nievola@ppgia.pucpr.br

A. Hussain et al. (eds.), *Brain Inspired Cognitive Systems 2008*, Advances in Experimental Medicine and Biology 657, DOI 10.1007/978-0-387-79100-5_8,

1 Introduction

From the first moments in this world, a human initiates a communication exchange, using different interfaces, such as the cry vocalization of a newborn infant, after a typical childbirth. Communication is the basis of human development, it allows us to interact with others, express ideas, desires, feelings etc. However, for individuals with the locked-in syndrome the communication is a difficult challenge, as for subjects with amyotrophic lateral sclerosis (ALS) disease. Patients with ALS lose the autonomy by a progressive neurodegenerative disease that causes the loss of control over voluntary muscles. In these cases the only interface remaining for communication is their brains activity, and the Brain-Computer Interface (BCI) propose to allow users with motor disabilities to communicate, improving the life quality of the locked-in patients.

The BCI uses electrophysiological measurements of brain activity to enable communication with external devices, such as computers and prosthesis [1]. Generally, electrical signals represent these brain states, and these may be organized into large datasets. In this field the patterns analysis is an essential step to understand the brain states features present in these datasets. The state-of-the-art describes various algorithms to identify these patterns. Nonetheless, comparison of these algorithms is a difficult task given the diversity of BCI systems for different aspects such as target application, neuromechanism used, the amount of data tested, number of subjects, experimental paradigms, and other factors. The BCI represents a novel interdisciplinary knowledge field; the many available challenges for researchers provide a step toward the current state-of-the-art. In this way, a strategic contribution for this field will be a useful and understandable pattern extraction method for knowledge discovery in databases (KDD).

The KDD becomes then an important subject for academia and industry. This is a process of extraction of a novel, useful and understandable pattern from a collection of data. For the extraction of novel information the continuous wavelet transformation (CWT) is presented as a powerful technique. Although wavelet transformations have attracted many attentions in the Data Mining community, there has been no defined exploratory approach where the patterns are visualized graphically, and the most prominent ones are then studied individually, for instance with a classifier method. In this paper we introduce and apply an exploratory approach for data mining based on non stationary electroencephalographic (EEG) signal.

This work intends to design and apply a framework to analyze time-frequency patterns illustrated on CWT and WC qualitative analysis, and classifies those patterns through a quantitative measurement extracted from the CWT to analyse the electroencephalographic (EEG) signal pattern. The objective of such framework is to introduce and apply an exploratory approach for Data Mining based on EEG signal provided from users with motor disabilities. This research intends also to improve the results of classification and eventual BCI performance. A vital feature of BCI system is the capability to distinguish between the attended and ignored events with speed and accuracy. These characteristics differentiate artificial pattern recognition systems applied on BCI. For that reason, the aim of this research is to improve the bit rate measurement, as a speed parameter, with an acceptable accuracy. Therefore, the accuracy is not the only characteristic of the classifier analyzed. The speed of communication is another desirable characteristic for BCI systems.

This work is organized as follows: Section 2 extends the introduction and presents the state-of-art; Section 3 introduces methodology of the proposed framework. Results are given in Section 4 with specifications and parameters used. Finally, Section 5 presents a conclusion of the results, and Sect. 6 discuss the final issues of the paper.

2 State-of-Art

The current work is based on EEG data set acquired and organized by Hoffmann et al [1]. They described a BCI based on a six-choice P300-based system. The P300 is a positive deflection in the human EEG, appearing approximately 300 ms after the presentation of rare or surprising task-relevant stimuli to all subjects [2]. Four disabled and four able-bodied subjects tested the system on four different sessions, with six individual runs, one for each possible choice of the system. Hoffmann et al made the data set available for downloading on the website of the École Polytechnique Fédérale de Lausanne (EPFL) in BCI group (http://bci.epfl.ch/p300)[1].

Hoffmann et al captured the activity of the brain using the EEG scalp electrodes using the 10-20 standards, a widely used noninvasive technique [1]. It acquired a brain electrical signal that has differences of potential in the range (0 - 100μV) [2]. With such a small amplitude the contamination of EEG data at many points during the recording process is common. Therefore, an artifact removal and filtering procedure must be applied before the analysis of EEG signals. Even so, the signal must be filtered to avoid contaminations, and also focus in a particular oscillation frequency. Such oscillation relies on the neurophysiologic observations, that large populations of neurons in the respective cortex are sending in rhythmical synchrony when a subject is not engaged with one of his limbs, i.e. movements, tactile senses, or just mental introspection [3].

The recognition of dissimilarities between a target and non target EEG response by the presented stimuli is a pre-requirement for a reliable P300 Speller system. Such dissimilarities could be investigated through the EEG coherence measurement, as a well-established tool to analyze the linear relationship between two signals [4] [5]. The classical Fourier analysis requires stationary feature within each window analyzed, which is not found in brains dynamical signal of EEG. A more appropriate approach suggested by the authors of this paper is the CWT. This method could analyze fractal structure in time series that contain non stationary power at different frequencies [6], so the dissimilarities between target and non target EEG signal are recognized.

What distinguishes the Discrete Wavelet Transform (DWT) and the CWT is the scale shift domain, which are in $\mathbf{Z}^+$ and in $\mathbf{R}^+$, respectively, although both transformations allow working with discrete sampled time series, like the EEG signals. These characteristics are valid with different wavelet functions: the use of orthogonal basis implies the use of DWT, while a non orthogonal wavelet function can be used with either the DWT or CWT [7]. For analysis purposes the orthogonal CWT is better suited because its redundancy allows good legibility of signal information content [8], in contrast to the DWT, which doesn't permit that analysis. The CWT also uses complex wavelet basis functions that capture the amplitude and the phase information from the signal [9].

Reviewing the literature, few works related to the present study were found. Lachaux et al in their work [10] studied and applied single-trial brain signals using WC using CWT. They show the statistical properties of the WC method and compare with Fourier coherence in this particular time series analysis. They also present a qualitative approach to compare two non stationary neural signals, although for BCI system a quantitative measurement is still needed for classification purposes.

Sakkalis et al [11] applied a method to analyze schizophrenic brain activity to test a known hypothesis of disconnection and working at memory deficits. They used the WC with graph theory measurements to evaluate distance functional connectivity in complex neural networks. The results presented are networks used to distinguish healthy from schizophrenic disturbances connections. The WC method used showed to be capable of revealing novel patterns in neural signals.

Bostanov et al [12] applied a method based on CWT and Student's t-statistic, named *t-CWT* method, for single-trial event-related potentials (ERP) signal of BCI Competition 2003. They applied a simple pattern recognition system called the *linear discriminate analysis* (LDA). This approach introduces a quantitative measurement based on the CWT method for pattern classification. However, for further studies in BCI systems an exploratory approach gives useful information about the underlying process, and allows working with a specific pattern.

The performance of a pattern recognition system depends on the features and the classification algorithm employed. This work chooses a simple classification system, a Naïve Bayes Classifier (NBC) to enforce the power of the pre-processing method. For EEG studies, NBC can be used as a simple and reliable classifying method [13].

3 Methodology

The classification framework developed in this study has five steps based on traditional KDD processes, as described in figure 1. The first step is the data cleaning process. This process attempts exclusion of discontinuities of the signal, the application

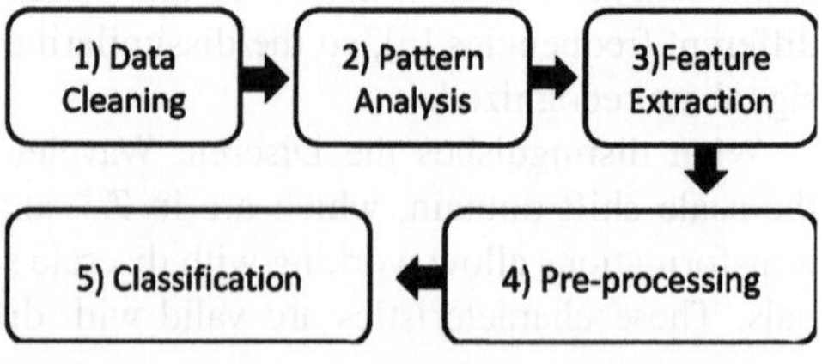

Fig. 1 An overview of the proposed framework

of a filter process, and the signal division on blocks. The subsequent step analyzes the blocks of trials using the CWT and WC statistical processes. The objective of this analysis is to identify visually significant patterns that show dissimilarities, with confidence levels higher than 95% .In the third step, these patterns are represented by the scale-averaged wavelet power, which extracts the frequency features in a time vector. In step four, this vector is down-sampled to reduce its dimensionality, and normalized between 0 and 1 using all samples in the data set. The fifth step uses the pre-processed vectors for cross-validation training of a NBC classification algorithm. The outcome of this step is the performance measurements that might change a specific preceding step or restart the entire process.

3.1 Signal Dataset

The signals used in this work were acquired, digitalized and make available by Hoffmann et al on EPFL in Switzerland [1]. The data set contains data from four disabled and four able-bodied subjects. According to them, the disabled subjects were all wheelchair-bounded, however they had different communication and limb muscle control abilities as described by them.

Each one of the eight subjects of the dataset was instructed to face a laptop screen, with six images as shown in fig. 2. The images exemplify an application scenario, where the user could control other devices like a television or a radio using the BCI. To evoke the P300 response each image flashes in a random sequence. Each subject was submitted to four sessions, with six runs (one sequence of flashes). Each run the subject was asked to count the number of flashed in specified images. The first flash comes 400 ms after the beginning of the EEG recording, and lasts for 100 ms, and during 300 ms none of the images was flashed. The EEG used was recorded at 2048 Hz sampling rate from 32 electrodes placed at the standard positions of the 10-20 international system. The Biosemi Active Two ® was used to amplify and digitalize the EEG signals, exemplified on fig. 3.

The stimuli are spaced by 400 ms for each other, and the maximum time that a stimulus is repeated is 4.400 ms, and the minimum is 400 ms. The stimulus flashes in a random sequence divided on blocks of 2 seconds of length. Using those parameters is easy to build an algorithm to identify each random sequence, here called blocks. These blocks with the six stimuli, or flashes, represent each image from the screen. The target is the image which the subjects is asked to count, and the non target is the other ones. The first stimulus begins at 400 ms, or at sample 820 with a 2048 Hz sample rate. The last stimulus finishes around 60 s. Summarizing, the run contains around 150 stimuli, with 25 target stimuli.

3.2 Finite Impulse Response Filter

Hoffmann et al applied a *6th* order forward-backward Butterworth bandpass filter, with 1 Hz to 12 Hz of cutoff frequencies [1]. Even so, this work used also a finite

Fig. 2 The six images flashed used for evoking the P300. Source: [1]

impulse response (FIR) digital filter to select the bands analyzed: delta (2-4 Hz); theta (4-8 Hz); alpha (8-12 Hz); beta (12-30 Hz); delta and theta band (0-8 Hz); a delta to beta (0-30 Hz); and theta and alpha (4-12 Hz). Fig. 4 shows the magnitude response of the filter 4 to 12 Hz.

The filter used is a FIR with the equiripple method, a high-order filter with a linear phase and are stable. This filter is close to the ideal with about 10dB/Hz slope in the transition bands, which frequency response is finite and rectangular. The Butterworth Filter also used is an infinite impulse response (IIR); in contrast

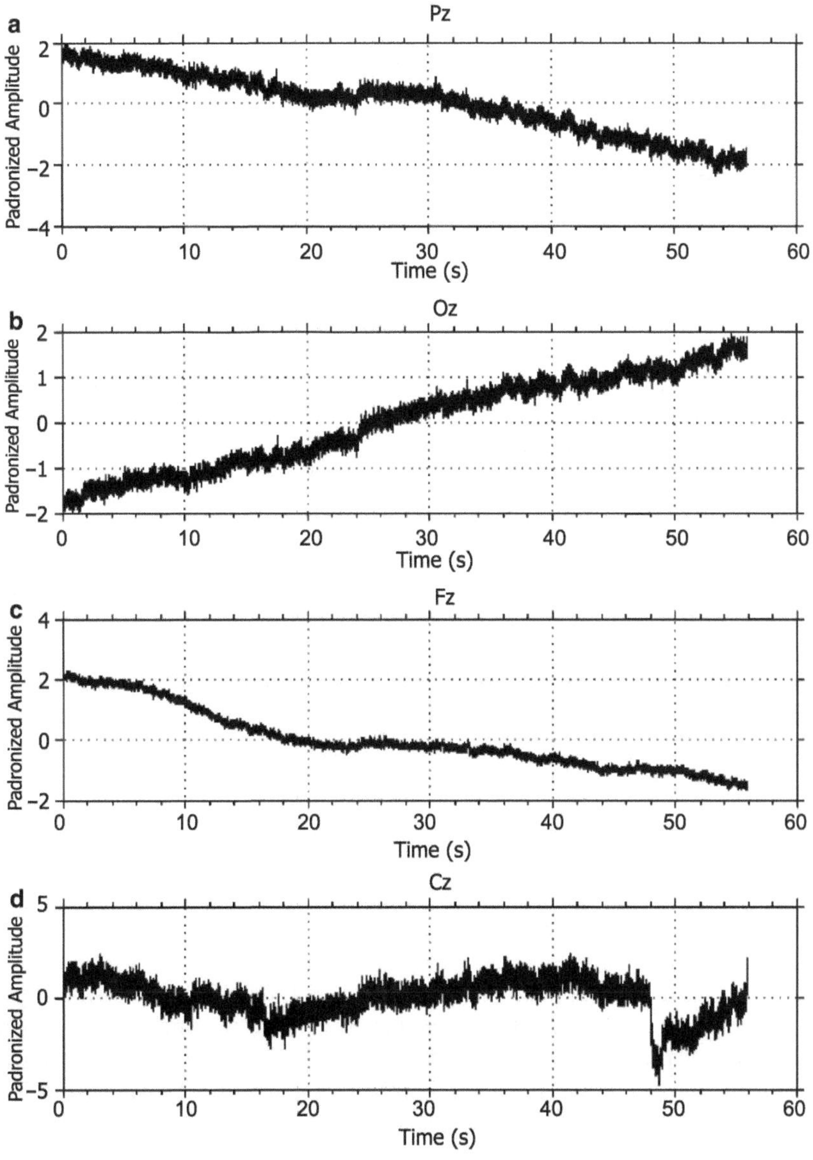

Fig. 3 A example of the padronized EEG signal during an entire run, showing the main electrodes Pz, Cz, Fz and Cz

to the FIR filter all poles are not located at the origin, and is therefore not always stable. Whereas in other cases the IIR filters could be preferred over the FIR filter since the IIR filters could achieve sometimes a sharper transition region the FIR filter with the same order.

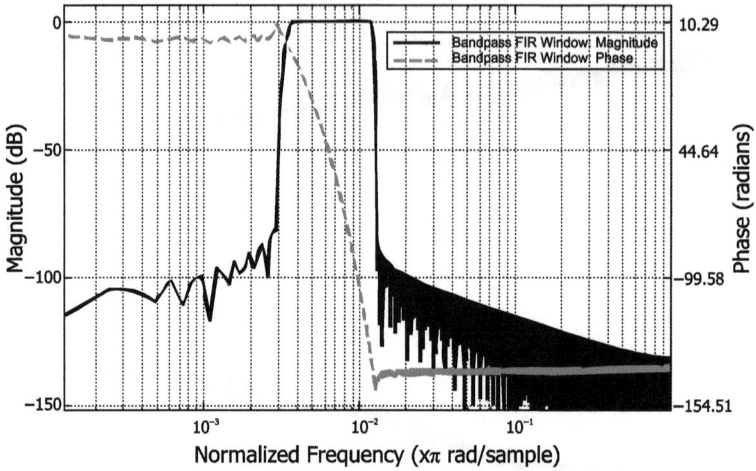

Fig. 4 The Gain, and magnitude, of the band pass FIR filter (4-12 Hz)

3.3 Continuous Wavelet Transformation

Each signal block is analyzed using a CWT, represented by $W_n(s)$ as defined on equation 1. The CWT is a convolution of x_n with a scaled and translated version of the wavelet function ψ, defined in the next section.

$$W_n(s) = \sum_{n'=0}^{N-1} x_{n'} \psi^* \left[\frac{(n' - n)\,\delta_t}{s} \right] \tag{1}$$

The asterisk(*) indicates the complex conjugated of ψ, and s is the wavelet scale [9]. Here, x_n represents a time series spacing by δ_t and the vector $n' = 0, \ldots, N-1$, where N is the number of points in the time series. The CWT is calculated varying the wavelet scale s and translating along the time index n. A graph representation showing both the amplitude of any feature versus the scale and how this amplitude varies with time can be constructed.

3.3.1 Cone of Influence

The time series studied of EEG signal have discontinuities at the beginning and at the end. For Fourier Transformations these edges signify a problem, since it assumes that the time series is cyclic. Nonetheless, the EEG signals are considered in this work a non cyclic time series, and the edges will affect the analysis. This edge effect in CWT is called cone of influence (COI), and it is composed by two edges effects; each one forms a cone of influence through the scale and frequency.

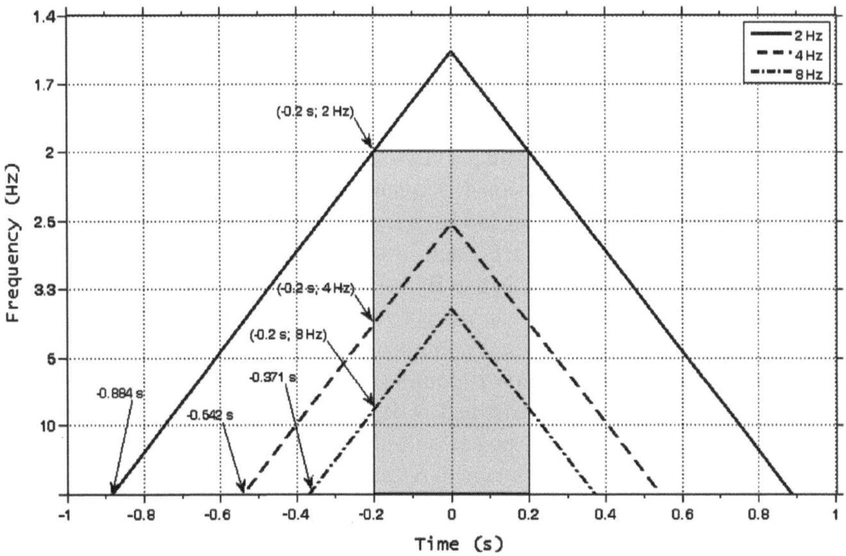

Fig. 5 The Cone of Influence region in the time-frequency plane for frequency bands of 2, 4 and 8 Hz

Exemplifying, for higher scale periods (lower frequencies) fewer are the reliable results of the analysis, since cone decreases until the influence of the beginning intersect the influence of the end.

The EEG signal study shows some specific frequency bands with many interesting background studies: as the alpha, beta, theta and delta band [14]. The CWT presents as a powerful technique to analyze those frequency bands. However, in order to produce reliable results this technique requires the correct use of the COI. An example of this effect is shown at fig. 5: the wavelet spectrum analysis center at 0 second, from -0.2 to 0.2 seconds, require different windows size depending on which frequency band analyzed: 2, 4 and 8Hz. More specifically exemplified, to analyze frequencies of 2 Hz, painted on gray, a time window from -0.884 to 0.884 seconds is required.

3.3.2 Morlet Wavelet Function

In order to apply the CWT technique the choice of mother wavelet function is an important issue. It could be orthogonal or non orthogonal, based on **C** or **R** domain, and many other fundamental requirements depending on the kind of features one wants to extract from the signal. To apply an acceptable function one must look at its reproducing kernel, which characterizes its space, scale and angular selectivity. For this work a complex-valued wavelet Morlet function was chosen; it is the most commonly used, and is indicated in eq. 2.

$$\psi(t) = \pi^{-1/4} e^{i w_0 t} e^{-t^2/2} \tag{2}$$

where $\psi(t)$ is the wavelet function that depends on a non dimensional time parameter t, and i denotes the imaginary unit, $(-1)^{1/2}$. This wavelet function forms two exponential functions modulating a Gaussian envelope of unit width, where the parameter ω_0 is the non dimensional frequency parameter, here taken to be 6 to satisfy the admissibility condition and have a zero average [7].

In spite of that, the method presented is generally applicable to other wavelet functions, for instance the Mexican hat. By using the Morlet instead of the Mexican hat, the wavelet transformation can extract features that are better located in the frequency domain, e.g. phase-locked gamma oscillations [12].

As the Morlet wavelet function is complex, so is the CWT, $W_n(s)$ defined on eq. 1. Hence, the power spectrum defined as $|W_n(s)|^2$ is commonly used to represent this wavelet transformation. This power is used on WC and for representing a scale-averaged wavelet power, and is shown on figures 6 and 7. The outer elliptical region

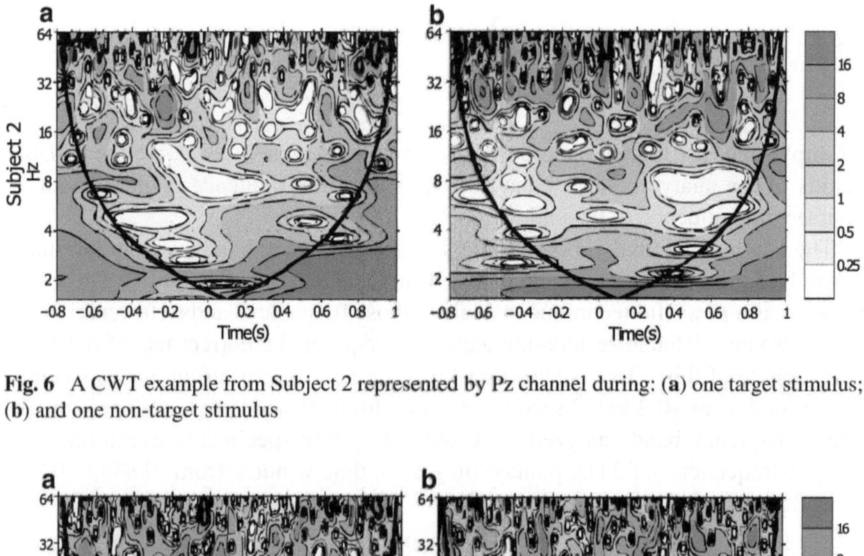

Fig. 6 A CWT example from Subject 2 represented by Pz channel during: (**a**) one target stimulus; (**b**) and one non-target stimulus

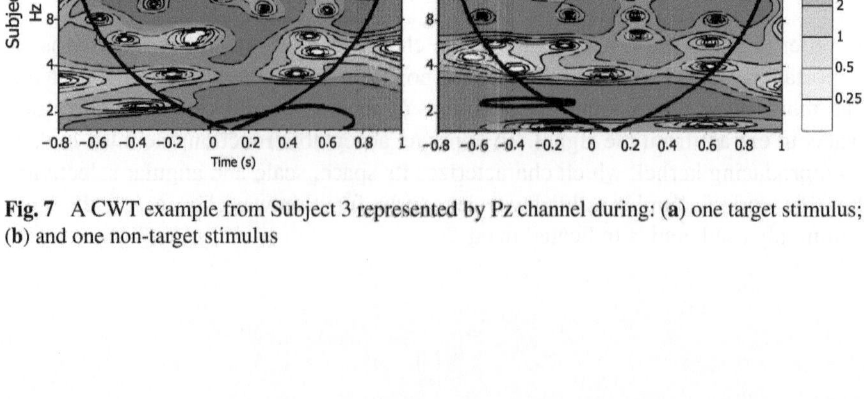

Fig. 7 A CWT example from Subject 3 represented by Pz channel during: (**a**) one target stimulus; (**b**) and one non-target stimulus

at the edges of the second graph with wide contour indicates the cone of influence (COI) in which errors may be apparent due to the transformation of a finite-length series EEG signal. The Monte Carlo estimation of the significance level requires the order of 10.000 surrogated data set pairs. The thick contour designates the 5% significance.

3.4 Wavelet Coherence

From the CWTs, average target W_n^x and non target W_n^y EEG response WC to analyze the similarities and synchronicity between the signals can be constructed. This could be illustrated as local correlation between both CWTs. This measurement is defined as the square of the cross-spectrum, defined on eq. 3, and normalized by the individual power spectra. This gives a quantity between 0 and 1, and measures the cross-correlation from two time-series as a function of frequency, expressed on eq. 4, where S is a smoothing operator. This operator smoothes the time and then smoothes the scale axis in both CWTs applied. The design of the smooth operator was based on Grinsted et al [15].

$$W_n^{xy}(s) = W_n^x(s)\, W_n^y(s)^*\,, \tag{3}$$

$$R_n^2(s) = \frac{\left|S\left(s^{-1} W_n^{xy}(s)\right)\right|^2}{S\left(s^{-1}\left|W_n^x(s)\right|^2\right) S\left(s^{-1}\left|W_n^y(s)\right|^2\right)}, \tag{4}$$

where the W_n^{xy} is the cross-spectrum,between W_n^x and W_n^y , the CWTs of target and non target EEG response. The statistical significance level of the wavelet coherence is estimated using Monte Carlo methods. A large ensemble of surrogate data set pairs was generated with the first order autoregressive coefficients $AR(1)$ for each calculated WC.

$$PC_n(s) = tan^{-1} \frac{\Im\left\{W_n^{xy}(s)\right\}}{\Re\left\{W_n^{xy}(s)\right\}} \tag{5}$$

The phase difference is calculated using the complex phase angle. $PC_n(s)$ is the phase-coherence defined on eq. 5, over regions with higher than 5% statistical significance that is outside the COI to quantify the phase relationship.

3.5 Scale-Averaged Wavelet Power

The scale-averaged wavelet power (SAWP) is used to represent a selected range of scales, here defined as non dimensional frequencies s, defined on eq. 6.

$$\bar{W}_n^2 = \frac{\delta_j \delta_t}{C_\delta} \sum_{j=j_1}^{j_2} \frac{\left|W_n\left(s_j\right)\right|^2}{s_j}, \tag{6}$$

where W_n^2 is the weighted sum of the wavelet power spectrum over scales s_1 to s_2, i.e. representing an average fluctuation power non dimensional $j_1 = 1Hz$ and $j_2 = 4Hz$ of wavelet scale. The symbol δ_t is the time series spacing, the δ_j is the scale spacing, j is the scale series from $j_1, (j_1 + \delta_j), \ldots$ to j_2. The parameter C_δ is the reconstruction factor (the Morlet function uses it), which is empirically set as equal to 0.776.

3.6 Naïve Bayes Classifier

The SAWP measurement is a time-series extracted from the brains signal, organized on vectors which represent the user intention. The Naïve Bayes Classifier (NBC) learns the user intentions from a set of training vectors. The NBC is characterized by two main advantages: the simplicity of its structure, and the speed of the learning algorithm it employs.

The probabilistic approaches make strong assumptions about how data is generated, and posit a probabilistic model about these assumptions. The NBC is the simplest of these models, it assumes all attributes of the example as independent of each other. While this assumption is intuitively false, in most real-world tasks the model often performs very well [13].

The NBC is a probabilistic classifier. This method simply classifies, for example, the vector $\boldsymbol{x}$ in the class c_k if it has the highest probability $P\left(c_k|\boldsymbol{x}\right)$, where k is the number of classes. Following the Bayes theorem, and the assumption of the independence between the features of the vector $\boldsymbol{x}$, this probability could be calculated using eq. 7.

$$P\left(c_k|\boldsymbol{x}\right) = P\left(c_k\right) \times \frac{\prod_{j=1}^{d} P\left(\boldsymbol{x}_j|c_k\right)}{P\left(\boldsymbol{x}\right)} \tag{7}$$

4 Experimental Results

4.1 Pattern Analysis

A vital feature of BCI system is the capability to distinguish between the attended and ignored events with speed and accuracy. These characteristics differentiate artificial pattern recognition systems applied on BCI. This research develops a pattern recognition framework based on CWT and FIR filter feature extraction method. This framework intends to investigate the patterns recognized by those methods. Additionally, the framework review the EEG events with a pattern analysis method based on CWT.

An essential issue of pattern analysis is to comprehend, and understand the resulting patterns of the entire process. The CWT allows the illustration of patterns of each stimulus, and assists the staff and the user to comprehend the natural meaning of EEG patterns.

For each pair of target and non target stimuli the wavelet coherence was calculated. Like on CWT, the outer elliptical region at the edges with wide contour indicates the Cone of Influence. The Monte Carlo estimation was also used for the significance level, and it requires the order of 10.000 surrogated data set pairs. The thick contour designates the 5% significance level in figures 6 and 7. The number of scales per octave should be high enough to capture the rectangle shape of the scale smoothing operator while minimizing computing time. Empirical tests were run with $2, 5, 10, 15$ and 20 scales per octave; the satisfactory computational costs obtained 5 scales per octave.

The electrodes, or channels, used for this study were the Fz, Cz, Pz and Oz, respectively from the frontal, central, parietal and occipital region. The objective of this work is to classify the patterns through each run. Each subject has four sections divided on six runs. Summarizing, the data set has a total of 192 runs analyzed. In order to study the EEG response for CWT the filtered target and non target stimuli waveform were compared using the WC. A total of 768 figures was created to analyze a common pattern across all WCs, representing a sample of the database, such as the figure 8.

The CWT illustrates the patterns of each five averaged blocks. On Figure 6 one could visualize a common example of subject 2 pattern both target and non target time-frequency spectrum. There, no significant region is presented by the frequency between 2-30 Hz, and the figure 6-A shows a slow climb after the stimulus start approximately at 4 Hz. Figure 6-B shows a slow decrease of the wavelet power over that region. Another example of pattern is on frequencies higher than 8 Hz, shown on figure 6. It shows a low average power on 6-A, in contrast to the higher average power on 6-B. These two patterns could be visualized also on figure 7-A, where the WC shows few significant patterns, most of it higher than 8 Hz.

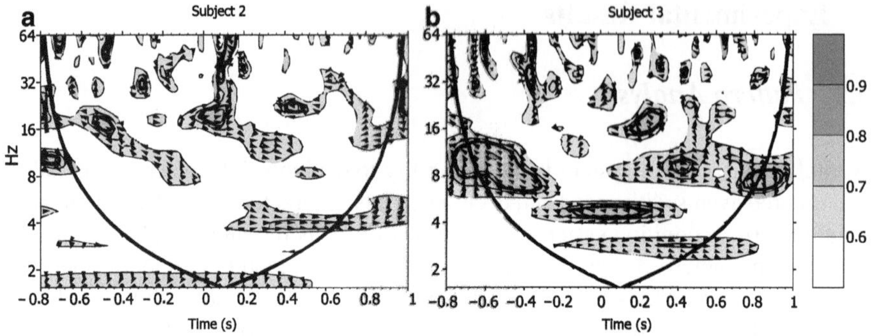

Fig. 8 A WC example from (**a**) Subject 2 represented by Pz channel between a target and non-target stimulus; (**b**) Subject 3 represented by Oz channel between a target and non-target stimulus

4.2 *Feature Extration*

The feature extraction procedure to achieve those capabilities selects pre-defined frequency bands representations. The bands analyzed were: delta (2-4 Hz), theta (4-8 Hz), alpha (8-12 Hz) and beta (12-30 Hz). Additionally, the oscillations of the delta and theta band (0-8 Hz), a delta to beta (0-30 Hz), and the band of 4-12 Hz applied by Hoffmann et al on a Butterworth filter [1]. Those frequencies were used by three feature extraction methods: the filter, the CWT and the combination of the filter and the CWT. The first one uses a FIR filter process and also the Butterworth filter. On this experiment the CWT were not applied, and the FIR filter process is tested with each analyzed band, e.g. the filter selects a particular frequency band delta (2-4 Hz) setting the cutoff frequency to 2 and 4 Hz, shown on 4. Additionally, the data set was tested without the filter process, to mark a base line for both methods. The second set of experiments tests the performance of the CWT without the filter process. A scale-averaged wavelet power uses the EEG trials to process selecting each frequency band from the CWT. And the third set of experiments combines the filter process first with the scale-averaged wavelet power. The filter selects a specific frequency band and the CWT is applied on this band also.

4.3 *Classification Validation*

The general performance was obtained through k-fold cross-validation using 10x10-folds method [16]. This technique divides the data randomly into ten parts, each part is held out in turn and the learning scheme is trained on the remaining nine. The procedure is repeated ten times and the average for the ten parts is calculated, for cross-validation training and validation procedure.

4.4 Performance Metrics

The decision made by the classifier is organized on a structure known as a confusion matrix or contingency table. The confusion matrix has four categories: True Positives (TP), False Positives (FP), True Negatives (TN) and False Negatives (FN). Given the confusion matrix, we can define: the sensitivity, specificity and accuracy. The Sensitivity and Specificity measurement defined on eq. 8 and eq. 9 are appropriated metrics for the classifier performance. The Sensitivity represents the percentage of target stimulus which is classified as a target. The specificity represents the proportion of non target stimulus which is classified as non target. This two metrics is both important; nonetheless, the specificity is more essential for this BCI application, because the false positive values are the most unwanted event in the system.

$$Sensitivity = TP/\,(TP + FN) \tag{8}$$

$$Specificity = TN/\,(TN + FP) \tag{9}$$

Each run contains 150 stimuli, with 25 target stimuli and 125 non target stimuli. This unbalanced proportion of samples in class targets with 16% and non target with 83% of stimulus difficult the training process and could generate a tendentious response of the classification process. If the classifier responds always a non target class, its accuracy, defined by eq. 10 will be 83%. An acceptable solution should analyze all stimuli as a different class, avoiding tendentious responses. This work considers only the accuracy of the target class, even if its accuracy is lower than the non target. This decision not only avoids the tendentious response to the system, but also allow to review the system performance with reliable metrics.

$$Accuracy = (TP + TN)/(TP + TN + FP + FN) \tag{10}$$

The speed of communication is an important characteristic of a BCI system. This feature depends on interstimulus interval, the number of different stimuli, the classification accuracy, and the control flow algorithm. The bit rate is a theoretical measurement that simulates all these factors as a performance metric for the speed of communication, and it is used to compare BCI systems. The bit rate b in bits/min can be computed according to the following eq. 11 [14].

$$b(N, p, t) = \left(log_2(N) + p(log_2) + (1 - p) log_2 \left(\frac{1 - p}{N - 1} \right) \right) \frac{60}{t}, \tag{11}$$

where the variable N denotes the number of different commands a user can send, which is six for this approach. Furthermore, p denotes the probability that a command is correctly recognized by the system. The t is the time in seconds that is needed to send one command. This work studies each stimuli individually, the stimuli time window is one second, and it begins 0.2 seconds before its occurrence.

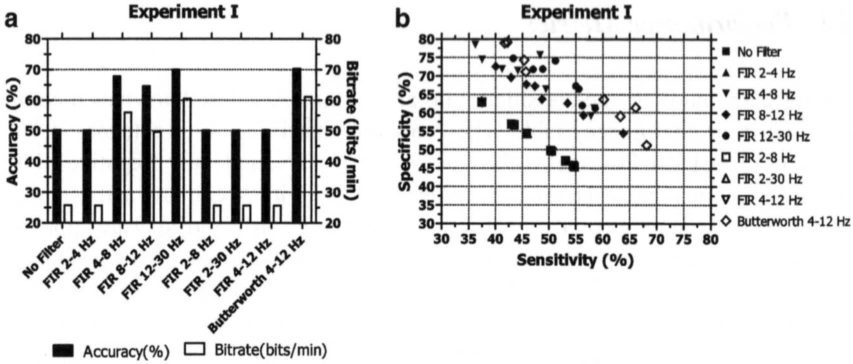

Fig. 9 Experiment I results represented by: (**a**) the averaged accuracy of all subjects; (**b**) the specificity and sensitivity of each subject

4.5 Experiment I

The first set of experiments aims at obtaining the classification performance of the FIR filter process and also the Butterworth filter. In this experiment the CWT were not applied, and the FIR filter process is tested with each analyzed band, e.g. the filter selects a particular frequency band delta (2-4 Hz) setting the cutoff frequency to 2 and 4 Hz. Additionally, the data set was tested without the filter process, to mark a base line in the research. The filtered signal measurement is used as an input signal for the NBC algorithm directly. As a result, the data set is optimally classified into two classes, targets and non targets following a validation process, and an average accuracy metric is then calculated; the results can be visualized on fig. 9.

4.6 Experiment II

The second set of experiments tests the performance of the CWT without the filter process. The wavelet transformation uses the scale-average wavelet power measurement to select each analyzed frequency band setting the parameters s1 and s2 in equation 6. The SAWP measurement is then applied on NBC algorithm, and the results of this classification are shown on fig. 10.

4.7 Experiment III

The third set of experiments combines the target filter process with the CWT. The filter selects a specific frequency band and the CWT is applied also on this band,

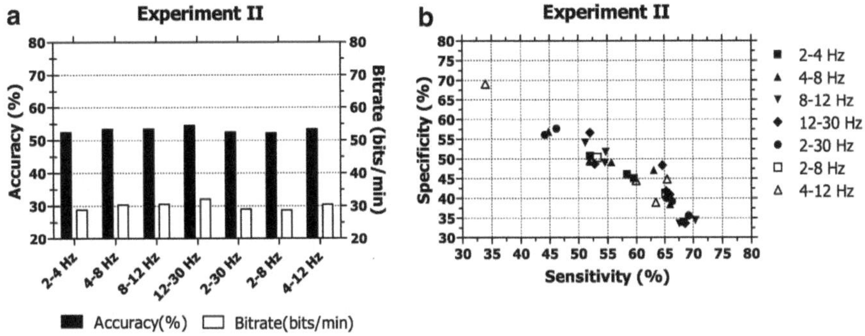

Fig. 10 Experiment II results represented by: (**a**) the averaged accuracy of all subjects; (**b**) the specificity and sensitivity of each subject

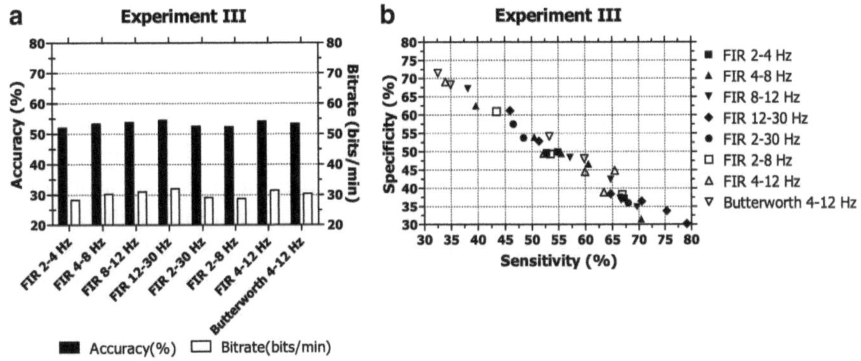

Fig. 11 Experiment III results represented by: (**a**) the averaged accuracy of all subjects; (**b**) the specificity and sensitivity of each subject

generating a vector with the SAWP applied on the NBC, the results are presented on fig. 11.

5 Conclusion

The results on figure 9 could be described as two groups. The first group is represented by all tests which performed under 55% of accuracy and under 25 bits/min, which includes the tests without filter, and with the filter FIR 2-4 Hz, FIR 2-8 Hz, FIR 2-30 Hz, and FIR 4-12 Hz. Those methods do not present reliable results, due a low accuracy result.

The second group is composed by the tests which achieve a higher accuracy and bit rate performance, such as FIR 4-8 Hz, FIR 8-12 Hz, FIR 12-30 Hz and Butterworth 4-12 Hz. The highest accuracy was 70,34%, and the highest bit rate was 61,15 bits/min, both from the Butterworth filter. The FIR 12-30 Hz achieves also a high accuracy of 70,04% and bit rate 60,51 bit/min. Furthermore, the average specificity of this test (68,74%) was higher than the Butterworth test (67,38%). This difference is significant because the Butterworth has a higher false positive proportion; it means that this method has a tendency to classify more non target stimulus as being a target stimulus than FIR 12-30 Hz. And the difference of specificity between these tests is higher than the difference between their accuracies. The FIR 12-30 Hz is reasonable method to obtain more reliable results, then the Butterworth filter.

The experiment II and III present a lower accuracy compared with experiment I. The accuracy results for these experiments did not achieve a significant performance. The bitrate is approximately 29 bits/min and the averaged accuracy is approximately 52%.

Hoffmann et al. achieved an accuracy as high as 100%, and achieved 29 bits/min [1]. Although experiment I had a lower accuracy, 70,04%, their bit rate performance is higher, 60,51 bit/min, compared to the state-of-art. This means that the lower accuracy performance obtained in this work doesn't represent a system that is performing worst or better then the state-of-art.

The increase on the bit rate measurement occurs due the reduction of the time window analyzed, with only one second of length. This time window enables 60 characters per minute. A subject without disabilities types on a computer an average of 95 characters per minute, while composing a text [17]. Therefore, in this case an increase communication speed of characters per minute causes a decrease of the accuracy. The chosen solution of FIR 12-30 Hz could represent a reliable solution with an averaged accuracy of 70,04% and 60,51 bit/min, enabling a communication of 60 characters per minute.

6 Disscussion

The main purpose of this work was to develop an exploratory approach for EEG signal, in which the patterns could be studied on the time-frequency plane. This innovative characteristic of the technique justifies the feasibility of the proposed approach on other data mining applications. This approach allows the study of not only the most prominent pattern, and at the same time it allows the visualization and classification of other time-frequency windows. Furthermore, it can also open new physiologic researches in this field, and researches on different non stationary time series analysis.

The algorithmic approach sketches the idea of using statistically-based feature vectors in the time-scale CWT domain in order to select the most relevant time-frequency segments able to show the most prominent task changes out of the background signal. Results suggest that the proposed methodology is also able of identifying regions WC spectrum during the specified task. Moreover, in the identified regions, patterns could be used by a classification algorithm, as the NBC, to translate the EEG-signal to control commands. Further studies are necessary to determine the extent and possible causes of the patterns recognized.

Acknowledgements The authors acknowledge Ph.D. Elisangela F. Manffra, Ph.D. Luiz R. Aguiar, and M.Sc. Guilherme Nogueira for the fruitful discussions. Also a special thanks for the Laboratory of Rehabilitation Engineering (LER) Research Group at PUCPR.

References

1. U. Hoffmann, J. Vesin, T. Ebrahimi, K. Diserens, Journal of Neuroscience Methods **167**(1), 115 (2008)
2. S. Sutton, M. Braren, J. Zubin, E.R. John, Science **150**, 11871188 (1965)
3. S. Lemm, B. Blankertz, G. Curio, K.R. Muller, IEEE Transactions on Biomedical Engineering **52**, 1541 (2005)
4. C.J. Stam, B.W. van Dijk, Physica D: Nonlinear Phenomena **163**, 236 (2002)
5. S. Micheloyannis, V. Sakkalis, M. Vourkas, C.J. Stam, P.G. Simos, Neuroscience Letters **373**, 212 (2005)
6. H. Kantz, T. Schreiber, *Nonlinear time series analysis* (Cambridge University Press, 1997)
7. M. Farge, Annu. Rev. Fluid Mech. **24**, 395 (1992)
8. G. McKhann, D. Drachman, M. Folstein, R. Katzman, D. Price, E.M. Stadlan, Neurology **34**(7), 939 (1984)
9. C. Torrence, G.P. Compo, Bulletin of the American Meteorological Society **79**(1), 61 (1998)
10. J.P. Lachaux, A. Lutz, D. Rudrauf, D. Cosmelli, M.L.V. Quyen, J. Martinerie, F. Varela, Neurophysiol Clin **32**(3), 157 (2002)
11. V. Sakkalis, T. Oikonomou, E. Pachou, I. Tollis, S. Micheloyannis, M. Zervakis, in *28th Annual International Conference of the IEEE Engineering in Medicine and Biology Society* (2006)
12. V. Bostanov, IEEE Transactions on Biomedical Engineering **51**(6), 1057 (2004)
13. C.T. Lin, K.L. Lin, L.W. Ko, S.F. Liang, B.C. Kuo, I.F. Chung, EURASIP Journal on Advances in Signal Processing **2008**, 10 (2008)
14. J. Wolpaw, N. Birbaumer, W. Heetderks, D. McFarland, P. Peckham, G. Schalk, E. Donchin, IEEE Transactions on Rehabilitation Engineering **8**(2), 164 (2000)
15. A. Grinsted, J.C. Moore, S. Jevrejeva, Nonlinear Processes in Geophysics **11**, 561566 (2004)
16. T. Dietterich, Approximate statistical tests for comparing supervised classification learning algorithms. Research report, Computer Science Dept., Oregon State University (1997)
17. W.O. Galitz, *The Essential Guide to User Interface Design: An Introduction to GUI to User Interface Design*, 3rd edn. (John Wiley and Sons, 2007)

Neuromorphic Systems: Past, Present and Future

Leslie S. Smith

Abstract Neuromorphic systems are implementations in silicon of elements of neural systems. The idea of electronic implementation is not new, but modern microelectronics has provided opportunities for producing systems for both sensing and neural modelling that can be mass produced straightforwardly. We review the the history of neuromorphic systems, and discuss the range of neuromorphic systems that have been developed. We discuss recent ideas for overcoming some of the problems, particularly providing effective adaptive synapses in large numbers.

Keywords Neuromorphic systems · Electronic cochlea · Electronic retina · Electronic neural systems

1 Introduction

Neuromorphic systems are electronic implementations of neural systems. Such implementations may take place at a number of different levels. For example, one may model sensory or sensorimotor systems, or one may model specific neural systems at many different levels, ranging through (at least) whole brain, brain region, cortical column, mid-brain or brainstem nucleus, neural microcircuits, single neurons, structural parts of neurons (dendrites, axons, soma), patches of membrane, down to ion channels encased in the neural bilipid membrane. Some go further, suggesting modelling quantum effects at synapses and in the dendrite [Ham07] [Hir91], but there do not appear to be electronic (as opposed to software) implementations of these models.

Different technologies have been used for these implementations at different times, reflecting the prevailing electronic technologies, and current systems are (of course) implemented in either analogue or digital very large scale implementation (aVLSI or dVLSI). According to Wikipedia (http://en.wikipedia.org/wiki/Neuromorphic) the term was coined by Carver Mead (see below), in the late 1980's,

L.S. Smith (✉)
Department of Computing Science and Mathematics, University of Stirling,
Stirling FK9 4LA, U.K.
e-mail: l.s.smith@cs.stir.ac.uk

A. Hussain et al. (eds.), *Brain Inspired Cognitive Systems 2008*, Advances in Experimental Medicine and Biology 657, DOI 10.1007/978-0-387-79100-5_9,

but the ideas have roots that go back well before this. The idea of an equivalent circuit for a neuron goes back at least to 1907 ([Lap07]), where a neuron is modelled by a resistor and a capacitor. Since then, various technologies have been used to model both neurons and sensory surfaces, and these are discussed in sections 2 to 4. There are particular issues that arise in these different types of models, such as matching the speed of the implementation to the environmental changes of interest, or finding ways to store large numbers of possibly adaptive synaptic weights. More recently, a number of new methods of using VLSI technology have been proposed for implementation, as well as some novel techniques for storing adaptable weights, and these are reviewed in section 5.

A different form of electronic implementation of neural systems is to directly implement a formal model of a neuron, or of a neural system, for example the McCulloch-Pitts neuron [MP43], or the weightless systems used by Aleksander and Stonham [AS79]. These have been the basis for silicon implementations: in the McCulloch-Pitts case, one example (of many) is the Intel chip [Cor90], and in the weightless case, the work of Aleksander et al simply using memory technology [ATB84], and Austin using more specialised silicon implementations [AK98]. We do not discuss these or related work further in this chapter, since they are not strictly "neuromorphic": nonetheless, these remain interesting approaches.

2 Early Forms of Neuromorphic Systems: Systems Built from Discrete Components

The 20th century saw huge leaps in the capabilities of electronic components. Although these were used initially primarily for communication purposes, and later for digital calculation and computation, a small number of researchers also saw electronic technology as a mechanism for modelling neural systems. Indeed, Hodgkin and Huxley's work on the dynamics of active cells [HH52] is often presented both as a set of equations and as an electrical equivalent circuit. In the 1960's for example, FitzHugh used analog computers to model single neurons, gaining detailed insights into their operation from these implemented models [Fit66] (see http://www.scholarpedia.org/article/FitzHugh-Nagumo_model). Others followed with simple electronic models of neural processes [Lew68][JH69],[Roy72] [Bro79]. On the larger scale, Runge [RUV68] built a fairly detailed model of the pigeon retina using 145 Cadmium Sulphide sensors and 381 neural analogue circuits taking up 50 circuit boards. Early electrical models of the cochlea are discussed in [KS59], including a transmission line of 178 sections, with each section comprising two inductors and four capacitors. Although this type of work did permit the building of electronic models, and their testing (and comparison with real neural and sensory systems), they were not intended for direct incorporation into equipment, but were purely research models of particular systems.

The approach of using discrete components is limited by the complexity and expense of building such systems. It is not practical to build large numbers of such systems, and as a result, they were useful for research purposes, attempting to improve understanding of the original system by re-creating it in hardware, but not for more general applications.

3 Modern Neuromorphic Systems

In 1989, Carver Mead published his second seminal book (his first was [MC80]), "Analog VLSI and Neural Systems" [Mea89]. This book brought together ideas on the ways in which analogue field-effect transistor based circuits, particularly those operating in the subthreshold domain (where currents change exponentially with gate voltage change), were similar operationally to neural membranes. The book describes at length how circuits which emulate elements of neural systems may be constructed. It also contains some highly influential system examples, drawn from the auditory and visual domain. Even now, most current works in neuromorphic systems reference some of the content of this book. It laid down a path for those who wanted to implement models of neural systems directly in VLSI hardware. This has the major advantage of being easy to replicate, so that a successful design can be built relatively cheaply in bulk and possibly become a low-cost system component. However, unlike systems built from discrete components, it is not possible to alter the circuitry once it has been built, nor even to monitor internal voltages unless that voltage is made available off-chip. Further, there is generally a long delay between submission of a design to a chip foundry and receiving testable chips back. These problems make developing such systems much more difficult.

Nonetheless, researchers in a number of groups have worked to develop circuits for a number of different neural types of applications, based partly on the ideas and circuits in Mead's book. Quite a number of novel circuits developed by the analogue designers in the neuromorphic systems community are discussed in [LKI+02]. Some of these are essentially related to sensory systems, whilst others are more related to the actual underlying neurons themselves. In addition there is a body of work on emulating aspects of the cortex itself, whether modelling the components of the cortex or attempting to build complete models of small parts of cortex [MG07], as in the Blue Brain project (http://bluebrain.epfl.ch/): these are primarily software simulations, and are outside the scope of this review. Below, we present a brief review of the current range of neuromorphic systems: a full-scale review is beyond the scope of this article.

3.1 Neuromorphic Systems for Sensory Processing

Implementations of neural systems are generally only considered to be neuromorphic if they work in real time. Real time operation is an absolute requirement for sensory systems, and many neuromorphic systems have been targeted specifically at sensory systems. Below, we review particularly visual, auditory, olfactory and tactile neuromorphic systems. There has recently also been interest in proprioceptary sensors as well [WZPF06]. So far as the author is aware, there have not been any neuromorphic implementation of systems for taste! The other characteristic of neuromorphic implementations is that they operate in parallel, and this has particular advantages for sensory systems as discussed in section 3.1.4.

3.1.1 Neuromorphic Systems for Vision

The earliest neuromorphic visual system appears to be that of Runge, discussed earlier. In terms of VLSI based designs, one of the earliest is described in Mead's book, chapter 15, which was a version of [MM88]. This paper discusses a "silicon retina" which implements both the transduction (basic photoreceptor circuit which has a near-logarithmic response), and a horizontal resistive layer which models the outer plexiform layer of the retina. The original retina had 48 by 48 pixels. The system produces an output which is the difference between the centre intensity and a weighted average of the the surrounding intensities, which is quite unlike what happens on a digital camera. The overall effect is that the response to a static edge is a spatial derivative, rather like what happens in a real retina. In addition, the response to a featureless surface is essentially zero independent of the brightness.

Many other papers have developed the ideas within this paper. Better photoreceptors are proposed and analysed in detail in [DM96]. These are used to develop a more effective contrast-sensitive retina in [AMSB95]. By performing additional processing at the silicon retina, one can make the system able to model specific visual capabilities. For example, a time to collision detector[Ind98], and a model of the fly elementary motion detector has been built [HK98], and more recently a depth from motion system [YMW$^+$06]. One particularly interesting recent advance has been a system which detects intensity changes in an overall brightness independent way, and transmits these serially using AER, thus enabling sensing of rapidly altering visual scenes without the huge data rates implied by the need to process whole frames [LPD08].

3.1.2 Neuromorphic Systems for Audition

There has been considerable interest in cochlear models since the work of Helmholtz and von Bekesy on the nature of the transduction of the pressure wave in the cochlea. Unlike the case in vision, pressure wave to electrical signal transduction is external to the device, and uses a (traditional) microphone. Early electrical models of the cochlea [KS59] discussed earlier were large and unwieldy. Building neuromorphic auditory subsystems that might actually be used to provide auditory capabilities for whole systems really had to wait for VLSI capabilities of the type discussed by Mead. Some of the earliest integrated neuromorphic systems are discussed in Mead's book. Lyon and Mead [LM88] describe the implementation of a sequence of filters used to build what they called a silicon cochlea: implementation techniques have been codified in [FA95] . This work has been extended to be more biologically realistic in [LM89a], and applied to auditory localization [LM89c], pitch perception [LM89b], voiced speech [LAJ93b] and speech analysis [LW97], [LAJ93a]. The real cochlea is active, and attempts have been made to implement this in [FvSV97], and much more recently in[HJvST08]. This is particularly important because of the very wide dynamic range of the biological cochlea, and the way in which the selectivity alters with changing sound pressure level.

It is well known that early auditory processing is not simply a matter of transduction, and researchers have considered the processing that occurs in the auditory brainstem as well. Considerable work has been done on silicon modelling of the cochlear nucleus, the first port of call of the axons of the auditory nerve [SFV96]. Some of the neurons in the brainstem nuclei respond to what are clearly features, for example amplitude modulation [SV97] or onsets [GHS02].

3.1.3 Other Sensory Neuromorphic Systems

Neuromorphic systems have been designed for other sensory modalities as well. There has been considerable interest in olfaction: the electronic nose is a device that could have considerable application in various industries (such as brewing and perfumery). In this sensory domain, the sensors detect electrical changes due to the odorant molecules [SG92]. It is possible to integrate the sensors on to the actual CMOS chip [Pea97]. Although the idea is relatively straightforward (altering the electrical properties of an insulating (polymer) layer as a result of the arrival of air-borne molecules), there are difficulties both in delivery, and, because of the chemical nature of the sensing, due to issues of drift and poisoning. A spike-based implementation is described in [KHT$^+$05].

Tactile sensing systems transduce pressure or motion into electrical signals. Neuromorphic motion based systems based on models of rodent vibrissae have been developed: earlier versions do not generally have the vibrissae directly incorporated on to the CMOS VLSI system, but use the outputs from these sensors directly, and are intended for robot based applications [PNG$^+$06]. More recently, both the sensor and the electronics have been integrated [AHW$^+$07]. Skin-like sensor arrays have been developed as well: these are of interest as sensors for grippers, and more generally, as general tactile sensors. A capitative technique is used in [Roy06], to produce an array of 59 sensing units which he calls *taxels*, and a polycrystalline silicon technique is used in [VAV$^+$06].

3.1.4 Parallelism in Sensory Neuromorphic Systems: Greedy Processing

Hardware implementations, particularly fully-implemented ones, in the sense used by Hecht-Nielsen [HN90] (page 267), have the advantage of permitting true parallelism, unlike software implementations, or even systems which are only partially implemented (where, for example, a digital floating point multiplier is shared between a number of synapses). As as result, they can perform numerous different transformations on incoming (parallel) sensory signals simultaneously. This processing can proceed all the time, whether it is needed or not. This "greedy" processing means that values are available immediately should they be required. Such processing appears to be the case in animal systems, where (for example) in the brainstem auditory nuclei, all the auditory nerve signals appear to be continuously processed to produce a representation which arrives at the mid-brain inferior colliculus. The visual domain is a little different, since the foveal section of the retina can only examine a small visual angle at a time: there, the parallelism seems to occur in the early stages of cortical visual processing (particularly V1).

In neuromorphic systems, this greediness takes the form of, for example, continuously processing signals from all the pixels, to produce event based signals which may then be sent down an AER bus (as in [LPD08]). In the auditory domain, the numerous band-passed signals are simultaneously processed to search for onsets or for amplitude modulation, mirroring processing processing the the auditory brainstem [SV97]. Greedy processing is useful for computing features that are invariant under expected alterations in the sensory processing environment (such as overall illumination changes in vision, or level variation in audition). In addition, using greedy processing should mean that a higher-level system which required particular information about some part of an image, or some part of the spectrum would be able to retrieve this immediately without having to wait for it to be computed.

3.2 *Neuromorphic Models of Neural Circuitry, Neurons and Membranes*

In [Mea89] Mead draws an analogy between ionic currents passing across cell membranes through ion channels and electron currents through field effect transistors operating in subthreshold mode. He also provides a description of an implementation of neural axons (chapter 12 of [Mea89]). Since that time there has been considerable interest in neuromorphic models of neurons, implemented at a range of different levels, from simple single compartment models of neurons through to patches of cell membrane which could be assembled into multi-compartment neuron models. We delay discussion of neuromorphic models of synapses to section 5. At a rather higher level, there has been work on modelling neural circuits at a range of different levels: mostly this work has been in software, but there is growing interest in hardware based implementations.

3.2.1 Single Compartment Neuromorphic Models

Single compartment neural models treat the neuron's state variable (sometimes called activation, and sometimes described as the depolarisation of the neuron, depending on the level of neural realism intended) as a single value: they thus ignore the complexity of the dendrites, and consider all the conductances as lumped together. These models generally generate a spike (which may be a realistic neural spike, or simple an event characterised purely by time of occurrence) when this activation crosses some positive threshold from below. Such models are useful (and indeed, commonplace) in simulations, since both integrate-and-fire neurons and spike response models (see sections 4.1 and 4.2 of [GK02]) are of this form. Even such straightforward model neurons can have different degrees of faithfulness to reality: for example, they may (or may not) implement a refractory period (time after firing when the neuron cannot fire), relative refractory period (time after firing when it is possible, but more difficult to make the neuron fire), and may have all the conductances from the membrane gathered into a single conductance (or "leak"), or may consider a number of conductances independently.

One of the earliest neuromorphic single compartment models was [WMT96] in which a quite detailed leaky integrate and fire (LIF) neuron was implemented: it also exhibited other characteristics of real neurons, such as facilitation, accommodation and post-inhibitory rebound. A more recent model built in silicon exhibits the spiking behaviour of a number of classes of cortical neurons [WD08]. Others have been more interested in using LIF neurons as system components, aiming to use their computational properties rather than model real neurons: these are used by [SV97] as part of his amplitude modulation detecting system, and by [GHS02] as part of a sound analysis system. In these cases, the neuron implemented was a much simpler form of the leaky integrate and fire neuron. The primary advantage of these simpler models is that they still display useful computational capabilities (such as synchronization and co-incidence detection), but require less circuitry to implement. As a result it becomes possible to implement larger numbers of them on a single chip.

3.2.2 Neuromorphic Models of Elements of Neurons

Single compartment models entirely ignore the dendrites of neurons, yet these are frequently large and complex structures. There has been interest in neuromorphic implementations of dendrites as the timings of the different inputs to these, and the way in which they are combined can provide powerful computational capabilities even before signals reach the soma of the the neuron. In the models developed in [Eli93][NE96] straightforward linear summation can occur, but interaction between nearby synapses can permit discrimination between different pulse intervals and patterns, as well as detecting correlations between spike trains. A different aspect of dendrites is their ability to transmit signals integrated from synapses forward (towards the soma) at the same time as transmitting action potentials backwards, and this has been demonstrated in a neuromorphic circuit in [RD01]. This capability is important for determining when synaptic characteristics should be altered. At the other end of the neuron, [MHDM95] describe a neuromorphic implementation of an axon, permitting low power transmission of a pulse at slow speeds: this could be useful for matching the speed of silicon with that of events in the real world.

At the level of the interaction of the ion channels on the membrane itself, the first major model demonstrating both the sodium and potassium conductances was implemented in [MD91]: this model demonstrated that spike generation in a neuron-like way could be emulated electronically using subthreshold aVLSI circuitry. This work has been extended and improved in [RDM98][RD00]. A different approach which uses novel semiconductor fabrication techniques is adopted by [NLBA04]: this approach uses properties of he semiconductors more directly, rather than using circuitry. In [SNT$^+$08] they extend this approach and use a tunnel diode to regenerate electronic signals. The effect is like that of neural axonic conduction, but at a rather higher speed. Logically, both of these approaches would allow a hardware implementation of a multi-compartment neuron to be built up from these patches of membrane.

3.2.3 Modelling Neural Circuitry

As well as modelling neurons or parts of neurons, there is also interest in neuromorphic implementations of neural circuitry. There are two different motives for this: firstly engineers would like to be able to use circuits of neurons to achieve particular processing functions, and secondly, modellers would like to be able to model particular arrangements of neurons that they find from neuroanatomy. An example of the first of these is a model of a *winner takes all* (WTA) network. These essentially choose one (or possibly more than one) of a set of interconnected neurons with different inputs, and select the one with the greatest activation (the winner). This is a useful capability, and models of these have been around for some time [LRMM89][PAS97]. Improving these by making them smaller, or by adjusting the timescale of the inputs is still an area of research [MG07].

Developing models of small volumes of cortex in silicon is an avowed aim of a number of groups (for example the *Brains in silicon* group at Stanford University, as well as Rodney Douglas at the Zurich Institute for Neuroinformatics (personal communication)), and the subject of at least one PhD thesis [Mer06]. As matters stand, however, this are is still dominated by huge software simulations [DLJ$^+$08].

4 Implementation Technologies for Neuromorphic Systems

Emulation of neural systems can be implemented in many different ways, ranging from software on standard digital computers, through to application specific integrated circuits (ASICs) implemented either in digital or analogue technology, possibly with additional technologies piggybacked on to implement elements that are difficult in CMOS. In general, to be called neuromorphic, there has to be some element of direct hardware implementation: otherwise one simply has a software model. (There is some discussion about whether an implementation based on efficient software on a multiple core processor might yet be called neuromorphic: we will not enter this discussion!) The other factor that is required in neuromorphic implementations is real-time operation, as discussed in section 3.1.

In Mead's original work, the implementation technology was sub-threshold analogue VLSI: however, many systems which are taken to be neuromorphic use digital technologies. There remain many choices still about the form of the implementation. One possibility is to use field programmable devices, whether field-programmable digital arrays, or field-programmable analogue arrays. These have the major advantages of being much easier and faster to configure, and avoiding the long delays inherent in the fabrication of ASICs. In addition, they may be reprogrammed unlike ASICs. However, the circuit density achievable is much less than that of ASICs, and the power consumption is much higher: this can be a particular problem for small or mobile devices. Further, where one is integrating sensors as well, field-programmable devices are not an option: one needs to choose an ASIC based implementation. It is important to note that the designers of neuromorphic ASICs generally have to use the technologies and foundries developed for digital chips, since this market is far bigger, and it is not currently possible to develop manufacturing technologies for this small niche market.

Even ASICs do not solve all the problems. In particular issues relating to interconnectivity and adaptiveness present problems. The planarity of CMOS devices places severe limits on interconnection possibilities. Further, most CMOS technologies are intended to produce components which are stable and always behave the same way, rather than components whose characteristics can evolve. There are also issues of ensuring that the speeds within the system (for example integration times) actually fit with events in the world whose timescales are often of the order of hundreds of milliseconds, rather than the nano- to micro-second range more usually encountered in the ASIC domain. Recently, a number of groups have developed novel technologies to address the issues of interconnectivity and adaptivity, and these are discussed in section 5.

4.1 Signal Coding

An important issue for implementation technologies is the nature of the signal representation both on-chip, and for transfer between chips. Essentially there are three overall possibilities: analogue, digital, and pulse-based signals. Analogue representations essentially imply analogue implementation, which has the advantages of implicit real-time operation and perhaps lower power. Further, signals in the world are normally analogue, simplifying interfacing. However, transistor variation across chip can make reliable circuit design very difficult, and the standard digital fabrication processes may not be as reliable for analogue as for digital designs. Digital designs, on the other hand, offer the usual digital advantages: noise immunity, ease of manufacture, high density, and effective use of the standard manufacturing processes. However real-time operation needs to be ensured in the design, and the digital signals require to be interfaced to the analogue world: in addition, some operations which can be small and simple in analogue implementation, such as multiplication, can require relatively large amounts of chip real estate to implement digitally.

One compromise is to use pulse or spike based representations. It is well known that spikes are used for communication between most neurons in animal brains, although the precise nature of the representations used in these is still under debate. Spikes provide a mechanism for asynchronous communication between electronic circuits: the actual spike is binary (it is an all or nothing event), but the precise time of the spike is essentially analogue. Between the elements on a single chip, spikes may be sent directly, but the planar nature of chips means that being able to send spikes between any pair of circuits on chip may not always be possible. One solution to this is to use internal routing as in [MHH$^+$08]. Equally clearly, the small number of connections coming off a chip means that simple spike coding cannot in general, be used for communication between arbitrary circuits on different chips. One technique for overcoming this limitation is to use a bus. For such a bus to work effectively, it needs to be standardised. The address-event bus[Boa00] (see http://www.pcmp.caltech.edu/aer/) is the current standard in this area. Of course, such a bus implies a maximal rate at which spiking events may be transferred, and a maximal precision to the timing of transmitted events: however, digital transmission busses do have a very high (and known) bandwidth so that one can calculate whether this is likely to be a problem at the design stage.

5 Adaptivity and Interconnectivity for Neuromorphic Systems

Two characteristics of neural and neuromorphic systems that set them aside from traditional computer systems are adaptivity and a high degree of interconnection between the elements. The adaptive elements are normally at the interconnections between the neural circuit elements, since they are generally modelled on synapses whose plasticity is an important element in learning in real neural systems. If there are n neurons there are normally $O(n^2)$ interconnecting synapses, so that it is important that the circuitry modelling the synapse is small and low-power. (There are other possibilities: if synapse implementation circuitry is sufficiently fast, it may be shared in time between a number of emulated synapses (partially implemented in the terminology of [HN90]). This does, however, make the circuitry more complex, but may be appropriate in digital implementations where synapses can include large multipliers.)

Synapses need to provide some specific capabilities: they need to be able to transmit a signal from the pre-synapse neuron (however this signal is coded), whilst modulating this signal, providing some particular alteration (in voltage or current, again depending on the nature of the implementation) at the post-synaptic neuron. The size of this alteration will need to be adjustable, if the synapse is adaptive. The earliest adaptive elements for the first generation of hardware neural networks included novel devices such as the memistor [Wid60], an electrochemical adaptive resistive element (named for being a memory resistor). By modern standards, these are large and slow. However, the production of reliable adaptive devices which are small enough to be able to be deployed in large numbers remains difficult. One possibility is to use a memory word, implemented digitally. The value coded in the word defines the alteration at the post-synaptic neuron. Though practical for truly digital implementations, this requires both digital to analog conversion and a technique for altering the value in analogue implementations. Another possibility is to use floating gate devices: this is the same technology used to create flash memory devices, and has been shown to be highly reliable for creating digital memories. It has been used for synapses [DPMM96][HFD02], using electron injection and electron injection to increase and decrease the voltage on the floating gate. However, these have not been used in large numbers. In addition to requiring a technique for long-term storage of synaptic effectiveness, synapses also change their effects on a shorter time scale (called synaptic facilitation and depression). A simple short-term adaptive synapse permitting both facilitation and depression is described in [CBD$^+$03], and a more sophisticated implementation including NMDA voltage-gated channels is described in [BI07].

Recently, there have been some new technologies which have shown promise as new implementation techniques for synapses. These are based on nanowire cross-bars whose junctions are built from metal oxides with hysterectic resistances: tin oxides and zirconium oxides are often used[SPS07]. These devices are actually memristors [Bus08][SSSW08], "a class of passive two-terminal circuit elements that maintain a functional relationship between the time integrals of current and voltage" (Wikipedia), which makes them able to be resistors with memory. These are built from a technology known as CMOL (CMOS and molecular electronics), and are two terminal devices which can be made very small. If these can be implemented appropriately, they offer for the first time a technology which could provide the appropriate number of synapses which cold be adaptive. However, this is still very much a matter of research.

6 Looking Forward: Where Are Neuromorphic Systems Headed?

There remains considerable interest in auditory and visual neuromorphic systems as technologies for eventually producing synthetic sensing systems with the same types of capabilities as biological auditory and visual systems. The rapid response of the neuromorphic camera in[LPD08] without the use of large-scale frame technology represent a real step forward. However, neuromorphic auditory systems have yet to prove themselves capable: this may be because they have yet to properly incorporate the brainstem processing in more than very simple way on to the filtering technology.

Another area of progress is likely to be in the integration of different types of sensors on to CMOS systems. Light sensors have been around for a long time, and polymer based sensors are in use in olfactory neuromorphic systems. In addition, micro-electromechanical systems (MEMS) microphones have now been developed (e.g. the Infineon SMM310, or the Akustika AKU1126), and these seem to be good candidates for integration directly on to CMOS substrates. Different types of sensors for olfactory and other senses may be based on ion sensitive FETs (ISFETs)[HAC04] and chemical sensitive FETs (chemFETs) [HN02]: these new technologies are based directly on FETs and so are clearly integratabtle on to CMOS systems, although dealing the the nearness of liquids presents some novel problems for such electrical equipment. There is also work ongoing in the development of proprioceptary sensors [WZPF06]. Being able to directly incorporate the transduction on to the CMOS system both reduces complexity and component count, and permits processing to be applied directly to the signal allowing the chip to produce usable outputs directly.

6.1 When Will Neuromorphic Systems Come Out of the Lab?

It is now almost 20 years since Mead's book [MC80] was published. At this point neuromorphic systems have had some applications, in robotics, and in some sensors, as well as in a rather interesting system for training neurophysiologists in how cells in the visual system respond (see http://www.ini.ethz.ch/~tobi/friend/chip/index.php). They have also been applied to toys: some of the Wowwee toys use neuromprphic hardware (http://www.wowwee.com/. They have yet to really catch on even in the autonomous robotics area. Why is this? So far they have been very much a low budget interest area for researchers, built using technologies developed for other purposes. However, there are signs that this is changing. Recently, DARPA announced they SyNAPSE initiative (http://www.darpa.mil/baa/BAA08-28.html), and this may lead to rather larger sums being available for development.

Part of the problem has been the simple capabilities of standard processor technologies: it has become quite practical to place full-scale computer systems on even quite small autonomous robots. However, it remains very difficult to perform the sorts of sophisticated processing that truly autonomous robots require to be useful in the relatively unpredictable real world (as opposed to on top of a table in an experimenter's laboratory). For real applications the capabilities of truly parallel neuromorphic systems (for example in dealing with varying light levels and real acoustic situations) may become more important.

The other likely application area is in interfacing computer-based systems to both users and the environment. Currently, most computers still use only the keyboard and mouse for input, and a screen and loudspeaker for output. There are improvements, for example touch screen and multi-touch screens which are being introduced. Yet the hardware of the user interface still conforms to the "make the user adjust to the machine" paradigm that disappeared from the software for the user interface many years ago. Further, if one is building systems that interact directly with their environment (without human mediation), then the system must sense its environment, and make sense of the sensory data directly. This implies both richness and complexity of the sensory interface (as well as real-time operation) and sophisticated processing that can cope with variation in this sensory data, and extract the important (invariant) information that is required for behaviour in the environment. Hardware solutions, as well as integrated sensors appear appropriate for his area, and it may be that this is where neuromorphic systems will finally make their mark.

Acknowledgements This work started off as a talk at BICS 2008, and was revised through being given as a seminar at Stirling University and the University of Surrey, and in the light of the referees' comments.

References

[AHW+07] P. Argyrakis, A. Hamilton, B. Webb, Y. Zhang, T. Gonos, and R. Cheung. Fabrication and characterization of a wind sensor for integration with neuron circuit. *Microelectronic Engineering*, 84(1749-1753), 2007.

[AK98] J. Austin and J. Kennedy. PRESENCE, a hardware implementation of bunary neural networks. In M. Boden L. Niklasson and T. Ziemke, editors, *ICANN98: Proceedings of the 8'th international conference on artificial neural networks*, volume 1, pages 469–474, 1998.

[AMSB95] A.G. Andreou, R.C. Meitzler, K. Strohben, and K.A. Boahen. Analog VLSI neuromorphic image acquisition and pre-processing systems. *Neural Networks*, 8(7-8):1323–1347, 1995.

[AR88] J.A. Anderson and E. Rosenfeld, editors. *Neurocomputing: Foundations of Research.* MIT Press, Cambridge, 1988.

[AS79] I. Aleksander and T.J. Stonham. Guide to pattern recognition using random access memories. *IEE Proceedings: Computers and Digital Techniques*, 40:2–29, 1979.

[ATB84] I. Aleksander, W.V. Thomas, and P.A. Bowden. WISARD a radical step forward in image recognition. *Sensor Review*, 4(3):120–124, 1984.

[BI07] C. Bartolozzi and G. Indiveri. Synaptic dynamics in analog VLSI. *Neural Computation*, 19:2581–2603, 2007.

[Boa00] K.A. Boahen. Point-to-point connectivity between neuromorphic chips using address-events. *IEEE Transactions on Curcuits and Systems II*, 47(5):416–434, 2000.

[Bro79] W.H. Brockman. A simple electronic neuron model incorporating both active and passive responses. *IEEE Transactions on Biomedical Engineering*, BME-26:635–639, 1979.

[Bus08] S. Bush. HP nano device implements memristor. *Electronics Weekly*, May 2008.

[CBD+03] E. Chicca, D. Badoni, V. Dante, M. D'Andreagiovanni, G. Salina, L. Carota, S. Fusi, and P. Del Giudice. A vlsi recurrent network of integrate-and-fire neurons connected by plastic synapses with long term memory. *IEEE Transactions on Neural Networks*, 14(5):1409–1416, 2003.

[Cor90] Intel Corporation. 80170NN electrically trainable analog neural network. *Datasheet*, 1990.

[DLJ+08] M. Djurfeldt, M. Lundqvist, C. Johansson, M. Rehn, O. Ekeberg, and A. Lansner. Brain-scale simulation of the neocortex on the IBM Blue Gene/L supercomputer. *IBM Journal of Research and Development*, 52(1/2), 2008.

[DM96] T. Delbruck and C.A. Mead. Analog VLSI transduction. Technical Report CNS Memo 30, California Institute of Technology Computation and neural Systems Program, Pasadena California, USA, April 1996.

[DPMM96] C. Diorio, P. Hasler, B.A. Minch, and C.A. Mead. A single-transistor silicon synapse. *IEEE Transactions on Electron Devices*, 43(11):1982–1980, 1996.

[Eli93] J.G. Elias. Artificial dendritic trees. *Neural Computation*, 5(4):648–664, 1993.

[FA95] P. Furth and A.G. Andreou. A design framework for low power analog filter banks. *IEEE Transactions on Circuits and Systems*, 42(11):966–971, November 1995.

[Fit66] R. Fitzhugh. An electronic model of the nerve membrane for demonstration purposes. *Journal of applied physiology*, 21:305–308, 1966.

[FvSV97] E. Fragniére, A. van Schaik, and E.A. Vittoz. Design of an analogue VLSI model of an active cochlea. *Analog Integrated Circuits and Signal Processing*, 12:19–35, 1997.

[GHS02] M. Glover, A. Hamilton, and L.S. Smith. Analogue VLSI leaky integrated-and-fire neurons and their use in a sound analysis system. *Analog Integrated Circuits and Signal Processing*, 30(2):91–100, 2002.

[GK02] W. Gerstner and W. Kistler. *Spiking Neural Models*. Cambridge, 2002.

[HAC04] P.A. Hammond, D. Ali, and D.R.S. Cumming. Design of a single-chip pH sensor using a conventional 0.6-μm CMOS process. *IEEE Sensors Journal*, 4(6):706–712, 2004.

[Ham07] S.R. Hameroff. The brain is both a neurocomputer and a quantum computer. *Cognitive Science*, 31:1033–1045, 2007.

[HFD02] D. Hsu, M. Figueroa, and C. Diorio. Competitive learning with floating-gate circuits. *IEEE Transactions on Neural Networks*, 13:732–744, 2002.

[HH52] A.L. Hodgkin and A.F. Huxley. Currents carried by sodium and potassium ions through the membrane of the giant squid axon of loligo. *Journal of Physiology*, 116:449–472, 1952.

[Hir91] N. Hirokawa. Molecular architecture and dynamics of the neuronal cytoskeleton. In R.D. Burgoyne, editor, *The neuronal cytoskeleton*. Wiley-Liss, 1991.

[HJvST08] T.J. Hamilton, C. Jin, A. van Schaik, and J. Tapson. An active 2-d silicon cochlea. *IEEE Transactions on biomedical circuits and systems*, 2(1):30–43, 2008.

[HK98] R.R. Harrison and C. Koch. An analog VLSI model of the fly elementary motion detector. In M.I. Jordan, J. Kearns, and S.A. Solla, editors, *Adcanves in Neural Information Processing Systems 10*, pages 880–886. MIT Press, 1998.

[HN90] R. Hecht-Nielsen. *Neurocomputing*. Addison-Wesley, 1990.

[HN02] A.M. Hodge and R.W. Newcomb. Evaluation of the VLSI adaptation of the chemfet, a biosensor forfluid analysis. In *IEEE International Symposium on Circuits and Systems, ISCAS 2002.*, volume 2, pages II580–II583, 2002.

[Ind98] G. Indiveri. Analogue VLSI model of locust DCMD neuron response for computation of object aproach. In L.S. Smith and A. Hamilton, editors, *Neuromorphic Systems: engineering silicon from neurobiology*, pages 47–60, 1998.

[JH69] R.H. Johnson and G.R. Hanna. Membrane model: a single transistoranalog of excitable membrane. *Journal of Theoretical Biology*, 22:401–411, 1969.

[KHT+05] T.J. Koickal, A. Hamilton, S.L. Tan, J.A. Covington, J.W. Gardner, and T.C. Pearce. Analog VLSI circuit implementation. of an adaptive neuromorphic olfaction chip. In *IEEE International Symposium on Circuits and Systems (ISCAS) IEEE International Symposium on Circuits and Systems (ISCAS),Kobe, JAPAN*, volume 54, pages 60–73, 2005.

[KS59] W.J. Karplus and W.W. Soroka. *Analog Methods: Computation and Simulation*. McGraw-Hill, 1959.

[LAJ93a] W. Liu, A.G. Andreou, and M.H. Goldstein Jr. Analog cochlear model for multiresolution speech analysis. In *Advances in Neural Information Processing Sytems 5*, pages 666–673, 1993.

[LAJ93b] W. Liu, A.G. Andreou, and M.H. Goldstein Jr. Voiced speech representation by an analog silicon model of the auditory periphery. *IEEE Trans. Neural Networks*, 3(3):477–487, 1993.

[Lap07] L. Lapique. Sur l'excitation electrique des nerfs. *Journal of Physiology, Paris*, pages 620–635, 1907.

[Lew68] E.R. Lewis. An electronic model of the neuroelectric point process. *Kybernetik*, 5:30–46, 1968.

[LKI+02] S-C. Liu, J. Kramer, G. Indiveri, T. Delbruck, and R. Douglas. *Analog VLSI: Circuits and Principles*. MIT Press, 2002.

[LM88] R.F. Lyon and C. Mead. An analog electronic cochlea. *IEEE Transactions on Acoustics, Speech and Signal Processing*, 36(7):1119–1134, 1988.

[LM89a] J. Lazzaro and C. Mead. Circuit models of sensory transduction in the cochlea. In *Analog VLSI implementations of neural networks*, pages 85–101. Kluwer, 1989.

[LM89b] J. Lazzaro and C. Mead. Silicon modeling of pitch perception. *Proceedings of the National Academy of Sciences of the United States*, 86(23):9597–9601, 1989.

[LM89c] J. Lazzaro and C.A. Mead. A silicon model of auditory localization. *Neural Computation*, 1(1):47–57, 1989.

[LPD08] P. Lichtsteiner, C. Posch, and T. Delbruck. A 128 x 128 120db 15 μs latency asynchronous temporal contrast vision detector. *IEEE Journal of Solid-state Circuits*, 43(2):566–576, 2008.

[LRMM89] J. Lazzaro, S. Ryckebush, M.A. Mahowald, and C.A. Mead. Winner-take-all networks of O(n) complexity. In D. Touretzky, editor, *Advances in Neural Inforpation Processing Systems*, pages 703–711, 1989.

[LW97] J. Lazzaro and J. Wawrzynek. Speech recognition experiments with silicon auditory models. *Analog Integrated Circuits and Signal Processing*, 13:37–51, 1997.

[MC80] C. Mead and L.A. Conway. *Introduction to VLSI Systems*. Addison-Wesley, 1980.

[MD91] M. Mahowald and R. Douglas. A silicon neuron. *Nature*, 354(6354):515–518, 1991.

[Mea89] C. Mead. *Analog VLSI and Neural Systems*. Addison-Wesley, 1989.

[Mer06] P. Merolla. *A Silicon Model of the Primary Visual Cortex: Representing Features Through Stochastic Variations*. PhD thesis, Department of Bioengineering, University of Philadelphia, PA, USA, 2006.

[MG07] C.M. Markan and P. Gupta. Neuromorphic building blocks for adaptable cortical feature maps. In *IFIP International conference on VLSI*, pages 7–12, 2007.

[MHDM95] B.A. Minch, P. Hasler, C. Diorio, and C. Mead. A silicon axon. In G. Tesauro, D.S. Touretzky, and T.K. Leen, editors, *Advances in Neural Information Processing Systems 7*. MIT Press, 1995.

[MHH+08] L. McDaid, J. Harkin, S. Hall, T. Dowrick, Y. Chen, and J. Marsland. EMBRACE: emulating biologically-inspired architectures on hardware. In *NN'08: Proceedings of the 9th WSEAS International Conference on Neural Networks*, pages 167–172, Stevens Point, Wisconsin, USA, 2008. World Scientific and Engineering Academy and Society (WSEAS).

[MM88] C. Mead and M.A. Mahowald. A silicon model of early visual processing. *Neural networks*, 1(1):91–97, 1988.

[MP43] W.S. McCulloch and W. Pitts. A logical calculus of ideas immanent in nervous activity. *Bulletin of Mathematical Biophysics*, 5, 1943. Reprinted in [AR88].

[NE96] D.P.M. Northmore and J.G. Elias. Spike train processing by a silicon neuromorph: The role of sublinear summation in dendrites. *Neural Computation*, 8(6):1245–1265, 1996.

[NLBA04] A. Nogaret, N.J. Lambert, S.J. Bending, and J. Austin. Artificial ion channels and spike computation in modulation-doped semiconductors. *Europhysics Letters*, 68(6):874–880, 2004.

[PAS97] P.O. Pouliquen, A.G. Andreou, and K. Strohben. Winner-takes-all associative memory. *Analog Integrated Circuits and Signal Processing*, 13(1-2):211–222, 1997.

[Pea97] T.C. Pearce. Computational parallels between the biological olfactory pathway and its analogue 'the electronic nose'. 2. sensor-based machine olfaction. *Biosystems*, 41(2):69–90, 1997.

[PNG+06] M. Pearson, M. Nibouche, I. Gilhespy, K. Gurney, C. Melhuish, B. Mitchison, and A.G. Pipe. A hardware based implementation of a tactile sensory system for neuromorphic signal processing applications. In *IEEE International Conference on Acoustics, Speech, and Signal Processing*, volume 4, May 2006.

[RD00] C. Rasche and R.J. Douglas. An improved silicon neuron. *Analog Integrated Circuits and Signal Processing*, 23(3):227–236, 2000.

[RD01] C. Rasche and R.J. Douglas. Forward- and backpropagation in a silicon dendrite. *IEEE Transactions on Neural Networks*, 12(2), 2001.

[RDM98] C. Rasche, R.J. Douglas, and M. Mahowald. Characterization of a silicon pyramidal neuron. In L.S. Smith and A. Hamilton, editors, *Neuromorphic Systems: Engineering Silicon from Neurobiology*. World Scientific, 1998.

[Roy72] G. Roy. A simple electronic analog of the squid axonmembrane: the neuroFET. *IEEE Transactions on Biomedical Engineering*, BME-18:60–63, 1972.

[Roy06] D. Roy. Design and developmental metrics of a 'skin-like' multi-input quasi-compliant robotic gripper sensor using tactile matrix. *Journal of Intelligent and Robotic Systems*, 46(4):305–337, 2006.

[RUV68] R.G. Runge, M. Uemura, and S.S. Viglione. Electronic synthesis of the avian retina. *IEEE Transactions on Biomedical Engineering*, BME-15:138–151, 1968.

[SFV96] Andr Van Schaik, Eric Fragniere, and Eric Vittoz. An analogue electronic model of ventral cochlear nucleus neurons. In *5th International Conference on Microelectronics for Neural Networks and Fuzzy Systems (MicroNeuro '96)*, 1996.

[SG92] H.V. Shurmer and J.W. Gardner. Odor discrimination with an electronic nose. *Sensors and Actuators B-Chemical*, 8(11):1–11, 1992.

[SNT+08] A. Samardak, A. Nogaret, S. Taylor, J. Austin, I. Farrer, and D.A. Ritchie. An analogue sum and threshold neuron based on the quantum tunnelling amplification of neural pulses. *New Journal of Physics*, 10, 2008.

[SPS07] G.S. Snider, P.J. Kuekes, and D.R. Stewart. Nanoscale lacth-array processing engines. US Patent number 7227993, June 2007.

[SSSW08] D.B. Strukov, G.S. Snider, D.R. Stewart, and S.R. Williams. The missing memristor found. *Nature*, 453:80–83, 2008.

[SV97] Andre Van Schaik and Eric Vittoz. A silicon model of amplitude modulation detection in the auditory brainstem. In *Advances in NIPS 9*, pages 741–747. MIT Press, 1997.

[VAV+06] G. Vasarhelyi, M. Adam, E. Vazsonyi, A. Kis, I. Barsony, and C. Ducso. Characterization of an integrable single-crystalline 3-d tactile sensor. *IEEE Sensors journal*, 6(4):928–934, 2006.

[WD08] J.H.B. Wijekoon and P. Dudek. Compact silicon neuron circuit with spiking and bursting behaviour. *Neural Networks*, 21:524–534, 2008.

[Wid60] B. Widrow. An adaptive "adaline" neuron using chemical "memistors". Technical Report 1553-2, Stanford University, Solid-state electronics laboratory, Stanford electronics laboratories, Stansted University, Stansted, California, October 1960.

[WMT96] S. Wolpert and E. Micheli-Tzanakou. A neuromime in VLSI. *IEEE Transactions on Neural Networks*, 7(2):300–306, 1996.

[WZPF06] D.H.B. Wicaksono, L.-J. Zhang, G. Pandraud, and P.J. French. Fly's proprioception-inspired micromachined strain-sensing structure: idea, design, modeling and simulation, and comparison with experimental results. In *J. Physics: Conference Series: International MEMS Conference 2006*, volume 34, pages 336–341, 2006.

[YMW+06] Z. Yang, A.F. Murray, F. Woergoetter, K.L. Cameron, and V. Boonsobhak. A neuromorphic depth-from-motion vision model with stdp adaptation. *IEEE Transactions on Neural Networks*, 17(2):482–495, 2006.

Part III
Neural Computation

Preface

Neural Computation (NC 2008) is the oldest running (fifth) Symposium in the BICS Series. It is a major point of contact for multi-disciplinary researchers interested in looking at the scientific and engineering challenges of understanding the brain and building truly intelligent computers. NC highlights common problems and techniques in modeling the brain, and in the design and construction of neurally-inspired information processing systems. It covers both theoretical foundations as well as the development of new models, algorithms, implementations and applications.

This Part comprises four selected chapters that represent examples of works demonstrating current progress in neural systems.

Gomes, Braga and Borges present a model of multi-level associative memories where a set of coupled generalized-brain-state-in-a-box neural networks is employed as basic building blocks. The authors report a series of experiments that show that it is possible to build a multi-level memory based on correlation and evolutionary principles. In particular, their results demonstrate the feasibility of employing genetic algorithms that allow the emergence of complex behaviours which could otherwise be potentially excluded in other learning processes.

Simões, Neto, Machado, Runstein and Gomes propose a speech compression technique based on vector quantization. A neural network with unsupervised learning is used to implement the vector quantizer. The authors introduce the idea of using a codebook to perform speech compression and the use of a 2-dimensional self-organizing Kohonen map to generate the codebook. Simulation results are used to provide some insights on the network topology, its initialization and training strategies, and codebook size. The authors also carry out a comparative performance analysis to demonstrate the superior speech quality obtained using their approach compared to another state-of-the-art compression algorithm.

Oliveira, Andreãoand Sarcinelli-Filho propose the use of a static Bayesian network as a tool to support medical decision-making in the on-line detection of Premature Ventricular Contraction (PVC) beats in electrocardiogram (ECG) records. The authors apply the Bayesian network (BN) framework to the problem of heart beat classification using two labeled databases containing representative sets of long-term ECG records. The BN is shown to produce the best results, both in terms of sensitivity and specificity, by combining information provided by two ECG channels, thus implementing a kind of data fusion. Their results confirm that the combination of different ECG channels improves the performance of the classifier, and demonstrate the viability of using BN as a tool to effectively classify this kind of a signal. In conclusion, the BN are shown to represent an efficient model for allowing the representation of both quantitative and qualitative knowledge in the same model.

Finally, Coelho and Ynoguti propose a new method for multi-class support vector machines (SVM) based on a pruning strategy. The main idea behind their method is that it seeks for the class that will receive the greatest possible number of votes, implying, when a test sample is submitted to a binary classifier, the class that doesn't receive a vote is eliminated from future comparisons. Their strategy leads to a binary search, which is known to be very fast. The authors report experimental results performed on an isolated word, speaker independent, small vocabulary speech recognition problem, which show that their proposed method exhibits similar performance to conventional SVM and "one-against-one" methods.

Amir Hussain

Genetic Algorithm Applied to Hierarchically Coupled Associative Memories

Rogério Martins Gomes, Antônio Pádua Braga, and Henrique E. Borges

Abstract Inspired by the theory of neuronal group selection (TNGS), we have carried out an analysis of the capacity of convergence of a multi-level associative memory based on coupled generalized-brain-state-in-a-box (GBSB) networks through evolutionary computation. The TNGS establishes that a memory process can be described as being organized functionally in hierarchical levels where higher levels coordinate sets of functions of lower levels. According to this theory, the most basic units in the cortical area of the brain are called neuronal groups or first-level blocks of memories and the higher-level memories are formed through selective strengthening or weakening of the synapses amongst the neuronal groups. In order to analyse this effect, we propose that the higher levels should emerge through a learning mechanism as correlations of lower level memories. According to this proposal, this paper describes a method of acquiring the inter-group synapses based on a genetic algorithm. Thus the results show that genetic algorithms are feasible as they allow the emergence of complex behaviours which could be potentially excluded in other learning process.

Keywords Associative memory · Evolutionary computation · Generalized-brain-state-in-a-box (GBSB) model · Theory of neuronal group selection (TNGS).

1 Introduction

Associative memories have been studied over the years mainly as a non-hierarchical system, however some authors have regarded them as part of hierarchical or coupled systems [19, 20, 23]. In these studies, the neocortex is considered as an associative

R.M. Gomes (✉) and H.E. Borges
CEFET-MG, Av. Amazonas 7675, Belo Horizonte, MG, CEP 30510-000, Brazil
e-mail: rogerio@lsi.cefetmg.br; henrique@lsi.cefetmg.br

A.P. Braga
UFMG, Av. Antônio Carlos 6627, Belo Horizonte, MG, CEP 31270-010, Brazil
e-mail: apbraga@cpdee.ufmg.br

A. Hussain et al. (eds.), *Brain Inspired Cognitive Systems 2008*, Advances in Experimental Medicine and Biology 657, DOI 10.1007/978-0-387-79100-5_10,

memory in which some of the long and short-range cortical connections are responsible for the storage and retrieval of global patterns. Thus, from this perspective, the cortex could be divided into various discrete modular elements with their short-range connections associated with in-module synapses and the long-range connections associated with inter-module synapses.

Based on the same principles of multi-module organization, the theory of neuronal group selection (TNGS), proposed by Edelman [5], establishes that a memory process can be described as being functionally organized in hierarchical levels in which higher levels coordinate the functionality of lower levels. According to Edelman's theory, synapses of the localized neural cortex cells generate a hierarchy of cluster units denoted as: neuronal groups (clusters of tightly coupled neural cells), local maps (reentrant clusters of coupled neuronal groups) and global maps (reentrant clusters of coupled neural maps). Edelman argues that a neuronal group is the most basic unit in the cortical area and therefore is the basic memory constructor. Still according to Edelman, these neuronal groups are a set of localized tightly coupled neurons developed in the embryo which continue their development in early childhood, i.e. they are structured during phylogeny and account for the most primitive functions in human beings.

Immediately after birth, the human brain rapidly starts creating and modifying synaptic connections between neuronal groups. According to this proposition, Edelman proposed an analogy based on Darwin's theory of natural selection and Darwinian theories of population dynamics. The term Neural Darwinism could be used to describe a physical process observed in neurodevelopment in which used synapses amongst different clusters (neuronal groups) are strengthened, while unused ones are weakened, giving rise to a second level physical structure regarded in the TNGS as a local map. Each of these arrangements of connections amongst clusters within a given local map results in a certain inter-neuronal group activity which yields a second-level memory - in other words, the second-level memory could be viewed as a correlation amongst the first-level memories. This process of coupling smaller structures through synaptic interconnections between neurons of different neuronal groups in order to generate larger ones could be repeated recursively. This process of coupling intra and inter-module neurons by strengthening or weakening the connections according to the correlation of their activities has its roots on Hebb's theory of synapses adaptation [9].

The hebbian hypothesis also paves the way for implementing this system by using correlated associative memories. Therefore, based on these principles, Gomes et al. [7, 8] presented a multi-level hierarchically coupled associative memory in which the first-level structure is built of generalized-brain-state-in-a-box (GBSB) neural networks [2]. The functionality of these recurrent neural networks, which will also be used in this paper, is tightly connected with the TNGS principles.

The GBSB model [11] can be applied in the implementation of associative memories, where each stored pattern, i.e. a memory, is an asymptotically stable equilibrium point [22]. The design of artificial neural network associative memories have been explored in the last two decades and some methods have been proposed in [10, 13, 14, 16, 17].

The algorithm used in [7] to build the first-level memories ensures that each first-level pattern is stored as an asymptotically stable equilibrium point and also assures that the network has a nonsymmetric interconnection structure.

While the first level memories are built of GBSB networks, the higher levels could emerge from a learning mechanism as correlations of lower levels memories. As a result, this paper describes a method to estimate the inter-group weights based on evolutionary computation, or more specifically, via a genetic algorithm which is also inspired by the Neural Darwinism principles.

This structure of the remainder paper is organized as follows. In Sect. 2 we present a model of hierarchically coupled GBSB neural networks. Sect. 3 illustrates the analysis carried out through a sequence of experiments showing the behaviour of the global network and its capacity of convergence to global patterns considering orthogonal and LI vectors. Finally, Sect. 4 concludes the paper.

2 Multi-level Memories

The hierarchically coupled artificial associative network was built considering symmetric connections, asynchronous updating, and local and global features emerge from Hebbian learning [19,20,23]. Notwithstanding, these synapses are expected to mimic some important characteristics inherited from biological systems [5] which had not yet been considered, such as parallelism amongst synapses in different regions of the brain, re-entrant and asymmetric connections, synchronous activation, different *bias* as well as different maximum and minimum firing rates, redundancy, non-linear dynamics and self-connection for each neuron. For this reason, based on the theory of neuronal group selection (TNGS) proposed by Edelman [4, 5] a multi-level or hierarchically coupled associative memory by means of coupled generalized-brain-state-in-a-box (GBSB) neural networks was proposed and analyzed in [7,8,21].

In this multi-level memory model, each GBSB neural network plays the part of a first-level memory. In order to build a second-level memory, a great number of GBSB networks can be coupled by means of bidirectional synapses. These new structures will then play the role of our second-level memories, the same way as the local maps of the TNGS. Hence, some global patterns could emerge as selected couplings of the first-level stored patterns.

In Fig. 1 we see an illustration of a two-level hierarchical memory devised via a coupled GBSB model, where each individual neural network A, B and C, represents a single GBSB network. In a given network, each single neuron has synaptic connections with all the others in the same network, i.e. the GBSB is a fully connected nonsymmetric neural network. Besides, some selected neurons in a given network are bidirectionally connected with a number of neurons selected from different networks [19, 23]. These inter-network connections, coined in this paper "inter-group connections", are represented by a weight inter-group matrix W_{cor}, which in turn accounts for the interconnections of the networks made through coupling.

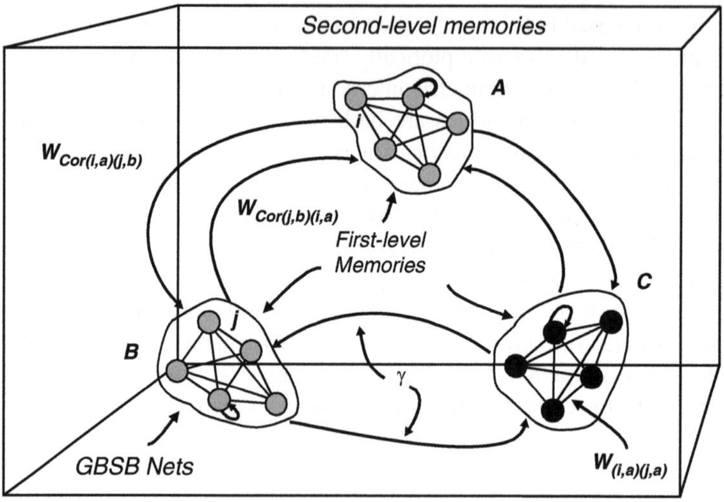

Fig. 1 Coupled neural network design

An analogous procedure could be followed in order to build higher levels in the proposed aforementioned hierarchy [1,5].

In order to observe the results of the coupling of a given GBSB network with the remaining GBSB networks, one should extend the original GBSB model [15] by adding to it a term which represents the inter-group coupling. Consequently, our general version of the multi-level associative memory model can be defined by:

$$\mathbf{x}_a^{k+1} = \varphi\left((\mathbf{I}_n + \beta\mathbf{W}_a)\,\mathbf{x}_a^k + \beta\mathbf{f}_a + \gamma \sum_{b=1,b\neq a}^{N_r} \mathbf{W}_{cor}\mathbf{x}_b^k\right) \tag{1}$$

where $\mathbf{x}_a^k$ is the vector state of the ath network at time k, $\beta > 0$ is a small and positive constant referred to as intra-group gain of the ath network, $\mathbf{f}_a$ is the bias field of the ath network, $\mathbf{W}_a$ is the synaptic weight of the ath network, N_r is the number of networks, $\mathbf{W}_{cor}$ is the inter-group weight matrix and γ is a positive constant referred to as inter-group gain between the ath and b^{th} network. To sum up, the first three terms represent N_a uncoupled GBSB networks. The fourth term of (1) represents the influence of the $(N_r - 1)$ networks on the ath network where the strength, or inter-group gain is parameterised by γ.

3 Simulation Results

Up to this point, we have presented a model of multi-level associative memories and its associated equations that allow the system to evolve dynamically towards a global pattern when one of the networks is initialized in one of the previously stored patterns as a first-level memory. In this section some simulations that show the rate of memory recovery will be presented.

Computational experiments consisting of three up to five GBSB networks connected as in Fig. 1 have been conducted and each network has been designed to present the same number of neurons and patterns stored as first-level memories. The weight matrix of an individual network was designed according to the algorithm proposed in [15]. Such algorithm ensures that the aforesaid patterns, which are in an inverse proportion to the desired ones are not automatically stored as asymptotically stable equilibrium points of the network, as a result they minimize the number of spurious states.

The second-level memories, or global emergent patterns, were built by randomly selecting a set of patterns stored as first-level memories taking into consideration linearly independent (LI) or orthogonal vectors. The convergence and capacity of the system was measured through the inter-group value (γ) and the inter-group weight matrix $\mathbf{W}_{cor(a,b)}$ calculated in accordance with a genetic algorithm strategy.

Firstly, the representation of each chosen individual was composed of real-valued variables, or genes. The aforementioned individual variables account for the γ values and the components $w_{(i,j)}$ of the inter-group weight matrix $\mathbf{W}_{cor(a,b)}$. This representation acts as a genotype (chromosome values) and is uniquely mapped onto the decision-variable (phenotypic) domain.

The next step up is to create an initial population consisting of 50 individuals, a typical value considered in the literature, and the first variable of each single one represents the γ value. The remaining variables of each individual represent every single element ($w_{(i,j)}$) of the inter-group weight matrix $\mathbf{W}_{cor(a,b)}$. γ is a randomly chosen real number uniformly distributed ranging from 1 to 2 and w_{ij} is a random real number also uniformly distributed which ranges from -0.5 to 0.5 (Fig. 2). In addition, one of the individuals of the initial population is seeded with the inter-group matrix developed in [7]. This technique aims to guarantee that the solution produced by the GA will not be less effective than the one generated by the Hebbian analysis. It is worth mentioning that the range of the γ and $\mathbf{W}_{cor(a,b)}$ values were chosen based on the values obtained in the Hebbian analysis developed in [7].

The objective function used to measure how individuals have performed a convergence to a global pattern was $\{-10, -5, -2\}$, being -10 the payoff for a complete recovery ($N_r \longrightarrow$ Number of networks), (-5) and (-2) for a partial recovery ($N_r - 1 \longrightarrow$ Number of networks minus 1 and $N_r - 2 \longrightarrow$ Number of networks

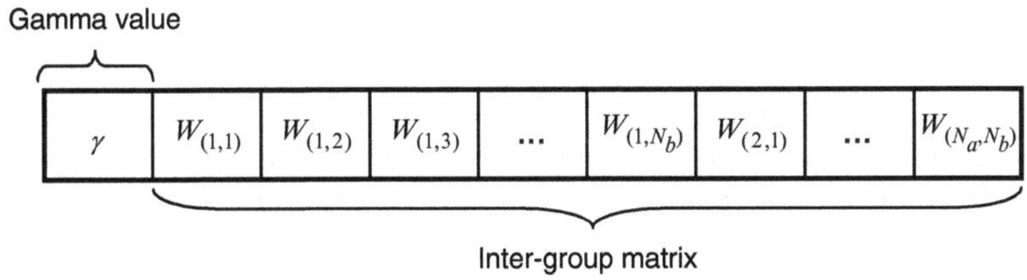

Fig. 2 Individuals - Chromosome values

minus 2, respectively) and 0 for no recovery. These values aim to enhance the chromosomes responsible for the creation of global patterns.

The fitness function used to convert the value of objective function into a measure of relative fitness was developed through a linear ranking method. The selective pressure was chosen equal 2 and individuals were assigned a fitness value according to their rank in the population rather than their raw performance. This fitness function suggests that by limiting the reproductive range, no individual generates too big an offspring, so it can prevent premature convergence from happening [3].

In the next phase, called selection, a number of individuals is chosen for reproduction, such individuals will determine the size of the offspring that a population will produce. The selection method used in this case is the stochastic universal sampling (SUS) with a generation gap of 0.7 (70%). The generation gap turned out to be suitable due to the fact that it eliminates the least fitted chromosomes.

Once the individuals to be reproduced are chosen, a recombination operation takes place. The type of crossover developed in this work is *intermediate recombination*, considering a real-valued encoding of the chromosome structure. *Intermediate recombination* is a method of producing new phenotypes around and amongst the values of the parents' phenotypes [18]. In this operation, the offspring is produced according to the equation:

$$O_1 = P_1 + \alpha(P_2 - P_1), \tag{2}$$

where α is a scaling factor uniformly chosen at random, over an interval, typically $[-0.25, 1.25]$ and P_1 and P_2 are the parents' chromosomes [18]. Each variable in the offspring is the result of the combination of the variables in the parents' genes according to the above expression added to a new α chosen for each pair of parent genes.

Now, as in natural evolution, it is necessary to establish a mutation process [6]. For real-value populations, mutation processes are achieved by either adjusting the gene value or by making a random selection of new values within the allowed range [12, 24]. In the experiment a real-value mutation is carried out at a mutation rate of $1/N_{var}$, where N_{var} is the number of variables in each single individual.

Given the fact that, by means of recombination, the new population becomes smaller than the original one by 30% resulting in a generation gap of 70%, the reinsertion of some new individuals into the old population is necessary so as to keep the size of the populations stable. Thus 90% of the new individuals are reinserted into the old population in order to replace its least fitted members.

In our simulations each network contains 12 neurons producing 4096 possible patterns from which 6 were selected to be stored as our first-level memories. This set of 6 patterns stored as first-level memories were chosen randomly considering the LI or the orthogonal vectors. In addition, in the first experiment, 3 amongst the $6^3 = 216$ possible combinations of the 3 sets of first-level memories were chosen randomly to represent second level memories.

The system was initialized randomly at time $k = 0$ in one of the networks, and in one of its first-level memories which compose a second level memory. The other networks, in their turn, were initialized in one of the 4096 possible combination of patterns, also at random. Then, we measured the number of times that a system consisting of three coupled networks converged into a configuration of triplets. The GA was tried 5 times and the algorithm was halted after 100 generations. In the end, the quality of the best members of the population were tested against the desired object.

In the first experiment a typical value of β was chosen ($\beta = 0.3$) and the number of times that a system consisting of three coupled networks converged into a configuration of triplets was measured. The rate of memory recovery in our experiments was averaged after 5 trials of 1000 iterations of the algorithm proposed in Sect. 2 for each population. The β value was chosen according to the value used in the Hebbian analysis developed in [7].

The convergence capacity of the global system can be seen in Figs. 3 and 4. They show that our model presents a mean rate of memory recovery of around 90% for LI vectors, and a rate of nearly 100% for orthogonal vectors (Table 1 - 3 coupled networks). The upper and lower limit, which represent the mean curve of the maximum and minimum convergence in all trials were close to the mean score of the system. The highest score achieved was 97.3% and 92.2%, for orthogonal and LI vectors respectively (Table 1).

In the second experiment, we analyse the capacity of convergence into global patterns in systems where three, four or five networks are coupled. Three patterns of each network (first-level memories) were chosen at random to be second-level memories.

For example, considering a system with three coupled networks as shown in Fig. 1, we assume that the stored patterns $\mathbf{P}_{(1,A)}$, $\mathbf{P}_{(4,A)}$ and $\mathbf{P}_{(6,A)}$ from network

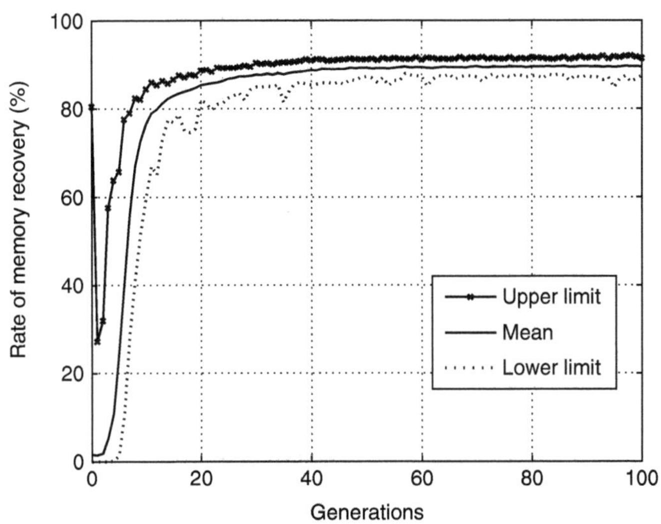

Fig. 3 Score of triplets for LI vectors

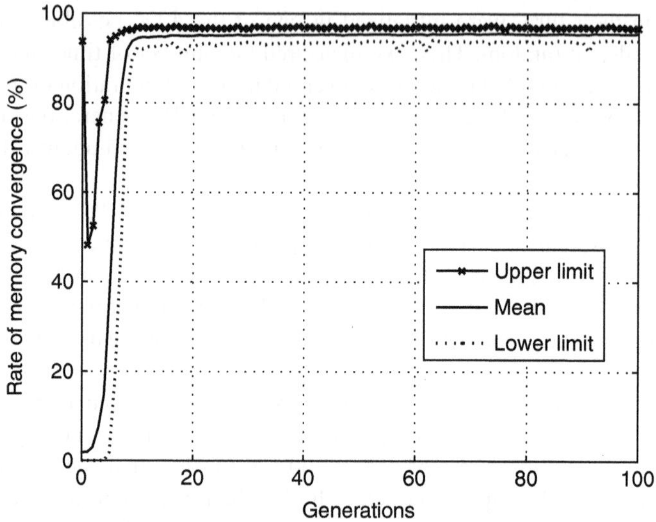

Fig. 4 Score of triplets for orthogonal vectors

Table 1 Maximum rate of memory recovery and gamma values for orthogonal and LI vectors considering 3, 4 and 5 coupled networks

	3		4		5	
	ORT	LI	ORT	LI	ORT	LI
CONV. (%)	97.3	92.2	91.4	83.9	85.18	70.9
gamma	1.42	1.55	1.53	1.55	1.64	1.55

A, $\mathbf{P}_{(2,B)}$, $\mathbf{P}_{(5,B)}$ and $\mathbf{P}_{(6,B)}$ from network B and that $\mathbf{P}_{(1,C)}$, $\mathbf{P}_{(3,C)}$ and $\mathbf{P}_{(5,C)}$ from network C were chosen as first-level memories of each network to be first and second-level memories simultaneously. Therefore, our second-level memories are a combination of these first-level memories, which are:

- second-level Memory 1: $[\mathbf{P}_{(1,A)}\ \mathbf{P}_{(2,B)}\ \mathbf{P}_{(1,C)}]$;
- second-level Memory 2: $[\mathbf{P}_{(4,A)}\ \mathbf{P}_{(5,B)}\ \mathbf{P}_{(3,C)}]$;
- second-level Memory 3: $[\mathbf{P}_{(6,A)}\ \mathbf{P}_{(6,B)}\ \mathbf{P}_{(5,C)}]$.

The procedure to apply for four, five or more coupled networks is a straightforward extension of the previous one.

A comparison of all these different couplings can be seen in Fig. 5 and 6. One can note that the rate of memory recovery which converts into global patterns decrease as more networks are coupled. Likewise, as seen in Hebbian analysis [7] the system presented a better performance regarding its capacity of memory recovery when orthogonal vectors were used.

In the experiments carried out to now, 6 first-level memory patterns were stored in each network. However, only 3 of them were chosen to compose the second-level memories. In the following experiment, considering 3 coupled networks, 1 to 6 first-level memories were chosen to compose our first and second level-memories

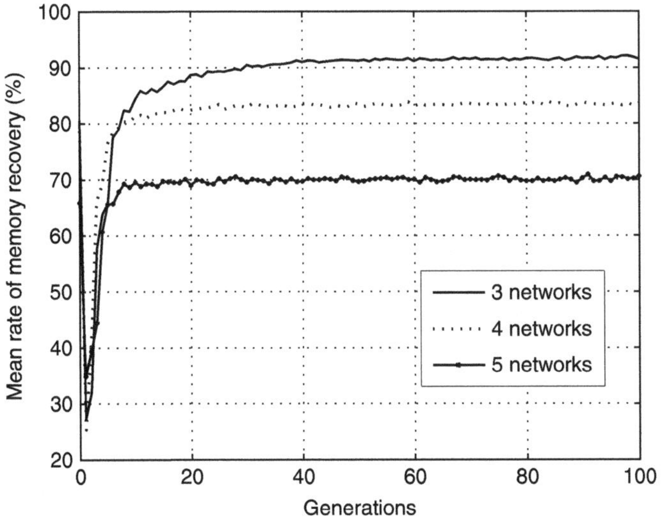

Fig. 5 Mean score of memory recovery for 3 to 5 coupled networks - LI vectors

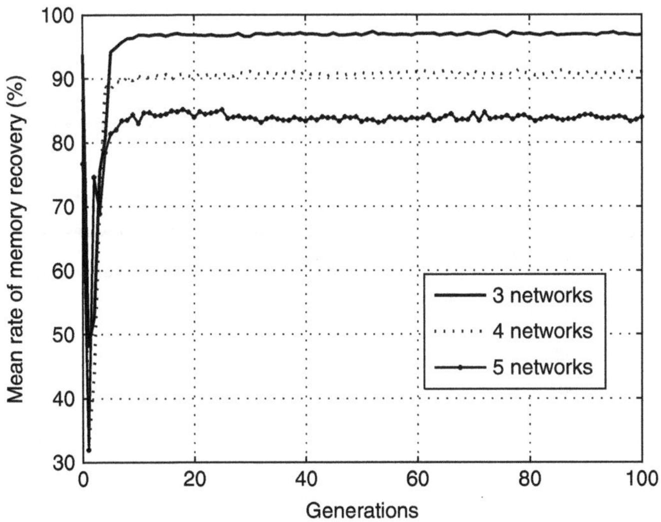

Fig. 6 Mean score of memory recovery for 3 to 5 coupled networks - Orthogonal vectors

simultaneously. Therefore, the system yielded up to 6 different sets of triplets or global memories. In Figs. 7 and 8 we plot the system's capacity to recover to the chosen global patterns (Table 2). It can be noted that the system loses its capacity of recovery when a larger number of sets of triplets is chosen to perform a second-level memory. It is also true that, despite a decrease in the recovery capacity in all

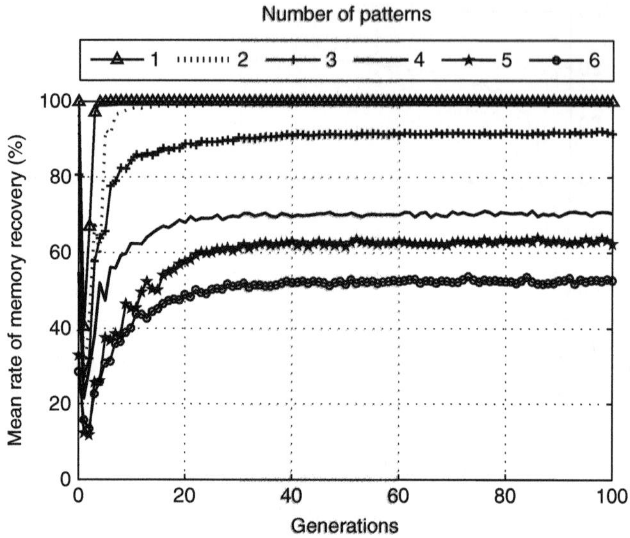

Fig. 7 Mean score of triplets for LI vectors considering 1 to 6 patterns chosen as first-level memories

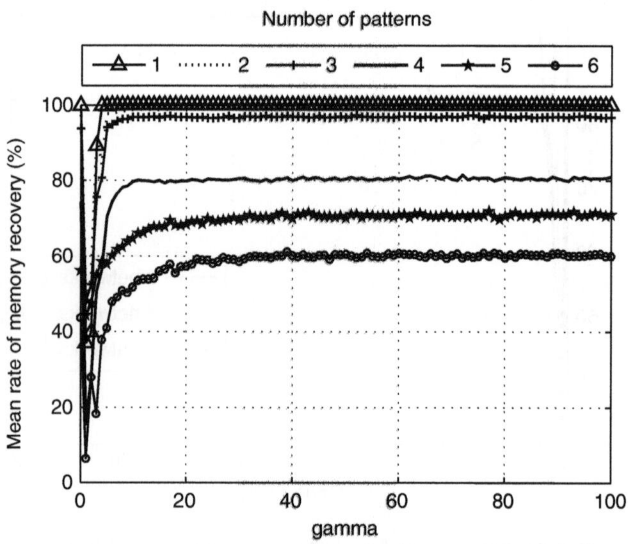

Fig. 8 Mean score of triplets orthogonal vectors considering 1 to 6 patterns chosen as first-level memories

cases, the difference between LI and the orthogonal vectors remained almost level or presented a variation of around 12% for the genetic algorithms when more triplets were selected.

Table 2 Maximum rate of memory recovery and gamma values for orthogonal and LI vectors considering from 1 to 6 patterns chosen as first-level memories

Patterns	Type	Conv. (%)	Gamma
1	ORT	100	1.49
	LI	100	1.43
2	ORT	99.4	1.44
	LI	99.3	1.49
3	ORT	97.3	1.42
	LI	92.16	1.55
4	ORT	81.6	1.49
	LI	71.2	1.42
5	ORT	72.0	1.48
	LI	64.0	1.52
6	ORT	61.2	1.63
	LI	53.7	1.39

4 Conclusions

In this paper, we have presented a model of multi-level associative memories where a set of coupled GBSB neural networks is employed as basic building blocks. Numerical computations for a two-level memory system are performed through a genetic algorithm.

It was verified that the capacity of convergence to a global pattern proved to be significant for both LI and orthogonal vectors, even though the percentage of convergence achieved for orthogonal vectors exceeded that of LI vectors, as had been expected.

The recovery of global patterns was more evident as the number of first-level memories composing the repertoire of the second-level memories increases. In fact, GA performs a compensation, reducing the effect of the *Cross Talk* or *Interference Term* as in the Hebbian analysis performed in [7], suggesting that in those cases one should have been using a genetic algorithm and orthogonal vectors.

Our experiments show that it is possible to build a multi-level memory based on correlation and evolutionary principles. Our main objective in further works will be to compare a multi-level memory model with the two-level memory model studied in this paper. We expect the simulations presented in this paper to be of use in the creation of further experiments that may lead to a better understanding of the behavior and capacity of hierarchical memory systems.

Acknowledgements The authors would like to thank Coordenação de Aperfeioamento de Pesquisa do Ensino Superior (CAPES - Brazil) and Conselho Nacional de Desenvolvimento Científico e Tecnológico (CNPQ - Brazil) for theirs financial support.

References

1. Aleksander, I.: What is thought? NATURE **429**(6993), 701–702 (2004)
2. Anderson, J.A.: An introduction to neural network. MIT Press, Cambridge, Massachusetts (1995)
3. Baker, J.E.: Adaptive selection methods for genetic algorithms. In: J.J. Grefenstette (ed.) Proceedings of the First International Conference on Genetic Algorithms and Their Applications. Lawrence Erlbaum Associates, Publishers (1985)
4. Clancey, W.J.: Situated cognition : on human knowledge and computer representations. Learning in doing. Cambridge University Press, Cambridge, U.K. (1997)
5. Edelman, G.M.: Neural darwinism: The theory of neuronal group selection. Basic Books, New York (1987)
6. Goldberg, D.E.: Genetic Algorithms in Search, Optimization and Machine Learning. Addison-Wesley Publishing Company, Reading, Massachusetts (1989)
7. Gomes, R.M., Braga, A.P., Borges, H.E.: A model for hierarchical associative memories via dynamically coupled GBSB neural networks. In: Proceeding of Internacional Conference in Artificial Neural Networks - ICANN 2005, vol. 3696, pp. 173–178. Springer-Verlag, Warsaw, Poland (2005)
8. Gomes, R.M., Braga, A.P., Borges, H.E.: Storage capacity of hierarchically coupled associative memories. In: A.M.P. Canuto, M.C.P. de Souto, A.C.R. da Silva (eds.) International Joint Conference 2006, 9th Brazilian Neural Networks Symposium, Ribeiro Preto - SP, Brazil, October 23-27, 2006, Proceedings. IEEE, Ribeiro Preto, Brazil (2006)
9. Hebb, D.O.: The Organization of Behavior. Wiley & Sons, New York (1949)
10. Hopfield, J.J.: Neurons with graded response have collective computational properties like those of two-state neurons. Proceedings of the National Academy of Science U.S.A. **81**, 3088–3092 (1984)
11. Hui, S., Zak, S.H.: Dynamical analysis of the brain-state-in-a-box (BSB) neural models. IEEE Transactions on Neural Networks **3**(5), 86–94 (1992)
12. Janikow, C.Z., Michalewicz, Z.: An experimental comparison of binary and floating point representations in genetic algorithms. In: R. Belew, L. Booker (eds.) Proceedings of the Fourth International Conference on Genetic Algorithms, pp. 31–36. Morgan Kaufman, San Mateo, CA (1991)
13. Lee, D.L., Chuang, T.: Designing asymmetric hopfield-type associative memory with higher order hamming stability. IEEE Transactions on Neural Networks **16**(6), 1464– 1476 (2005)
14. Li, J., Michel, A.N., Porod, W.: Analysis and synthesis of a class of neural networks: Variable structure systems with infinite gains. IEEE Transactions on Circuits and Systems **36**, 713–731 (1989)
15. Lillo, W.E., Miller, D.C., Hui, S., Zak, S.H.: Synthesis of brain-state-in-a-box (BSB) based associative memories. IEEE Transactions on Neural Network **5**(5), 730–737 (1994)
16. Michel, A.N., Farrell, J.A., Porod, W.: Qualitative analysis of neural networks. IEEE Transactions on Circuits and Systems **36**, 229–243 (1989)
17. Muezzinoglu, M., Guzelis, C., Zurada, J.: An energy function-based design method for discrete hopfield associative memory with attractive fixed points. IEEE Transactions on Neural Networks **16**(2), 307– 378 (2005)
18. Mühlenbein, H., Schlierkamp-Voosen, D.: Predictive models for the breeder genetic algorithm: I. continuous parameter optimization. Evolutionary Computation **1**(1), 25–49 (1993)
19. O'Kane, D., Sherrington, D.: A feature retrieving attractor neural network. J. Phys. A: Math. Gen. **26**(21), 2333–2342 (1993)
20. Pavloski, R., Karimi, M.: The self-trapping attractor neural network-part ii: properties of a sparsely connected model storing multiple memories. IEEE Transactions on Neural Networks **16**(6), 1427– 1439 (2005)
21. Reis, A.G., Acebal, J.L., Gomes, R.M., Borges, H.E.: Space-vector structure based synthesis for hierarchically coupled associative memories. In: A.M.P. Canuto, M.C.P. de Souto, A.C.R.

da Silva (eds.) International Joint Conference 2006, 9th Brazilian Neural Networks Symposium, Ribeirão Preto - SP, Brazil, October 23-27, 2006, Proceedings. IEEE, Ribeiro Preto, Brazil (2006)
22. Sussner, P., Valle, M.E.: Gray-scale morphological associative memories. IEEE Transactions on Neural Networks **17**(3), 559–570 (2006)
23. Sutton, J.P., Beis, J.S., Trainor, L.E.H.: A hierarchical model of neocortical synaptic organization. Mathl. Comput. Modeling **11**, 346–350 (1988)
24. Wright, A.H.: Genetic algorithms for real parameter qptimization. In: G.J.E. Rawlins (ed.) Proceedings of the First Workshop on Foundations of Genetic Algorithms, pp. 205–220. Morgan Kaufmann, San Mateo (1991)

Vector Quantization of Speech Frames Based on Self-Organizing Maps

Flávio Olmos Simões, Mário Uliani Neto, Jeremias Barbosa Machado, Edson José Nagle, Fernando Oscar Runstein, and Leandro de Campos Teixeira Gomes

Abstract We propose a speech compression technique based on vector quantization. A neural network with unsupervised learning is used to implement the vector quantizer. Some basic aspects related to speech signal processing are presented, as well as some general issues concerning the vector quantization problem. The idea of using a codebook to perform speech compression is introduced, and the use of a 2-dimensional self-organizing Kohonen map to generate the codebook is proposed. Simulation results are presented, giving some insights on the network topology, its initialization and training strategies, and codebook size. Finally, a comparison of speech quality obtained with our method and with a well-known compression algorithm is made.

Keywords Speech compression · Vector quantization · Self organizing maps · Clustering algorithms

1 Introduction

A database composed of fixed-dimension vectors can be stored in a compact form through the use of vector quantization. This technique replaces each sample in the original database by the best match obtained from a previously generated set called *codebook*. The degradation introduced by this form of representation is called *quantization error*. When building a codebook, one aims at minimizing the mean quantization error of the database elements.

F.O. Simões (✉), M.U. Neto, E.J. Nagle, F.O. Runstein, and L. de Campos Teixeira Gomes
Telecommunications Research Center (CPqD), Rod. Campinas–Mogi-Mirim (SP 340), km 118,5, Campinas-SP 13086-902, Brazil
e-mail: simoes@cpqd.com.br; uliani@cpqd.com.br; nagle@cpqd.com.br; runstein@cpqd.com.br; tgomes@cpqd.com.br

J.B. Machado
School of Electrical and Computer Engineering, FEEC - UNICAMP, Av. Albert Einstein - 400, Cidade Universitária Zeferino Vaz, Distrito Barão Geraldo, 13083-852 - Campinas-SP, Brazil
e-mail: jeremias@dca.fee.unicamp.br

A. Hussain et al. (eds.), *Brain Inspired Cognitive Systems 2008*, Advances in Experimental Medicine and Biology 657, DOI 10.1007/978-0-387-79100-5_11,

Speech frames in parametric form can be represented as a sequence of fixed-dimension vectors, lending themselves to compression through vector quantization.

This article proposes a speech compression strategy based on vector quantization of speech frames. The speech signal is viewed as a sequence of pitch-synchronous frames, each frame being represented as a vector of speech parameters and compressed through vector quantization.

The proposed technique uses a 2-dimensional self-organizing map [1, 2], in conjunction with the K-Means clustering algorithm, to generate cluster centroids from a speech training set. The set of centroids can be interpreted as a codebook, which is used to quantize new vectors representing the frames. Speech frames to be quantized are converted into a sequence of codebook indexes. When decoding the signal, the sequence of indexes is mapped back into a sequence of speech frames extracted from the codebook. The recovered frames are used to reconstruct the speech signal by means of an overlap-and-add operation.

2 Speech Signal Analysis

Speech is produced by the flow of air through the human vocal tract. When represented in digital form, a speech signal can be described by a sequence of samples in time.

The instantaneous characteristics of a speech signal depend on the vocal tract configuration (e.g. lips and jaws overture, tongue position and vocal chords vibration rate). During the speech production process, the speaker continuously modifies his vocal tract configuration to produce a sequence of sounds composed of basic units called *phones*. A phone can be defined as a speech segment whose acoustic characteristics match a given pattern and are different from all other patterns.

In voiced speech (Fig. 1), vocal chords vibrate almost periodically. This is the case, for example, when vowels are uttered. In this case, the speech signal is also approximately periodic. The period of the signal varies according to the vibration rate of vocal chords. Higher vibration rates correspond to shorter periods and higher voice tones, while lower vibration rates produce longer periods and lower voice tones. The frequency of the voiced speech signals is called *pitch*.

In unvoiced speech (Fig. 2), there is no vibration of the vocal chords. In this case, the speech signal has lower energy compared to voiced speech and is typically aperiodic, resembling a noise signal. This is the case for sounds such as *s*, *f* and *sh*.

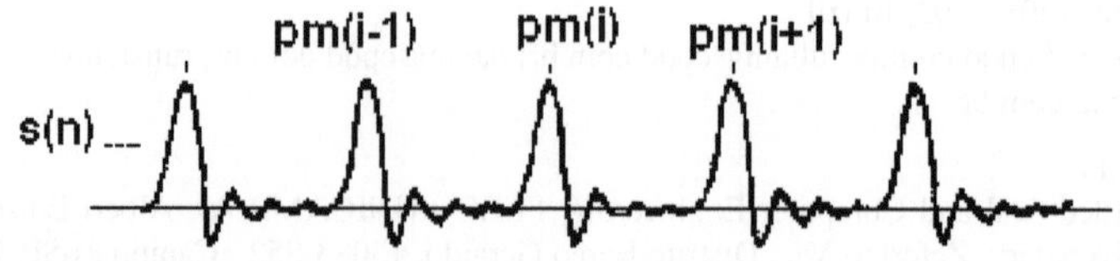

Fig. 1 Example of voiced speech signal

Fig. 2 Example of unvoiced speech signal

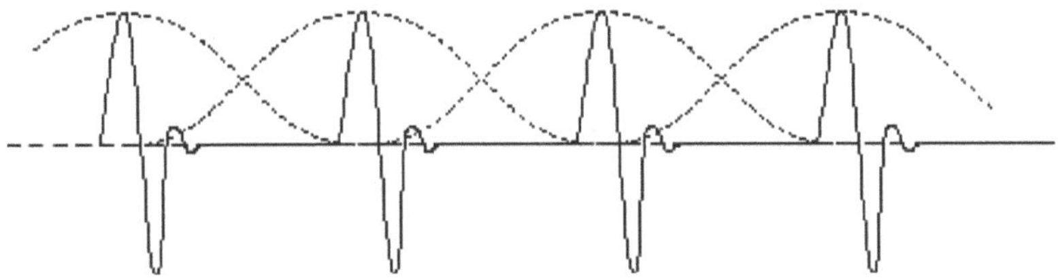

Fig. 3 Pitch synchronous speech frames

There can also be speech segments that have hybrid characteristics between voiced and unvoiced.

Even though the characteristics of speech signals vary continuously with time, it is possible to perform a discrete analysis of their acoustic features. This can be achieved by dividing the speech signal into short term segments (frames). If the frame length is short enough, we may consider that its acoustic features are approximately constant.

The analysis of speech signals may be performed for fixed-length or pitch-synchronous frames. In the first case, all frames have the same length and there is a fixed overlap between successive frames. In the second case, which is the one we adopted, pitch marks are positioned at signal peaks as show in Fig. 3. A frame is defined as a speech segment centered at a pitch mark, beginning at the preceding pitch mark and ending at the next pitch mark. Here, the samples on the right side of a frame overlap the samples on the left side of the next frame. The distance between pitch marks (referred to as *pitch period*) is related to the vocal chords vibration frequency. In unvoiced speech segments, which are not periodic, pitch marks are uniformly distributed in time (10 ms apart). In our work, the pitch marks were automatically positioned by a specific algorithm.

The analysis frame is multiplied by an asymmetric window with the same number of samples as the frame and positioned at the central pitch mark of the segment. We have used a concatenation of two half Hanning windows [3,4]: the first half corresponds to the left side of a $2N_1$ Hanning window, and the second half corresponds to the right side of a $2N_2$ Hanning window. The window samples (beginning at the origin) are given by the following expression:

$$w(n) = \begin{cases} 0,5\left[1 - \cos\left(\frac{\pi n}{N_1 - 1}\right)\right]; & 0 \leq n < N_1 \\ 0,5\left[1 - \cos\left(\frac{\pi(n+N_1-N_2+1)}{N_2}\right)\right]; & N_1 \leq n \leq N_2 \end{cases} \qquad (1)$$

A windowed frame corresponds to the samples of the original frame with an increased attenuation towards the edges.

In the process of windowing, a window is positioned in such a way that its first sample aligns with the central sample of the preceding window, and its last sample aligns with the central sample of the next window. Windows positioned in this way can be overlaped and added, allowing the reconstruction of the original signal from its original windowed frames with no distortion.

3 Vector Quantization Applied to Speech Compression

Vector quantization is a classical data compression technique used in a variety of applications, such as image compression, voice compression and speech recognition [5, 6]. It is a lossy compression technique, since it is not possible to recover all the information contained in the original signal from a limited-size codebook [7].

In vector quantization, an input data set $\{v_1, v_2, \ldots, v_N\}$ containing N k-dimensional vectors must be compressed. The task of a vector quantizer is to map this input data set into another set $\{c_1, c_2, \ldots, c_M\}$, with finite size $M < N$, also containing k-dimensional vectors. This new set is called *codebook*, and its constituent vectors are called *codevectors.*

In the coding process, each vector in the input set will be mapped into a codevector. The new vector will be a quantized (approximate) version of the original vector. We will assign a codebook index related to a windowed frame to each codevector.

Quantization introduces error in the representation of the input data. This error is referred to as *vector quantization error* [6].

If $d(.)$ is a generic measure that represents the distance between two vectors (e.g. the Euclidean distance), then the quantization error of a vector v_i is given by the distance between this vector and its quantized version $q(v_i)$:

$$Q_i = d\left(v_i, q\left(v_i\right)\right) \tag{2}$$

When building a codebook, one aims at minimizing the mean quantization error over all points of the input data set. The mean quantization error is given by:

$$Q_N = \frac{1}{N}\sum_{i=1}^{N} Q_i \tag{3}$$

In the quantization process, points that are similar to each other wil be grouped in the same cluster, and thus mapped to the same codevector. The number of output codevectors will depend on the sparseness of the input training data and also on the compression rate to be achieved. The process of codebook generation is illustrated in Fig. 4.

Once a codebook is generated, it can be used for data compression. The data to be coded will also be a set of k-dimensional vectors. In the coding phase, the codebook will be searched and, for each point in the input set, the closest codevector will be

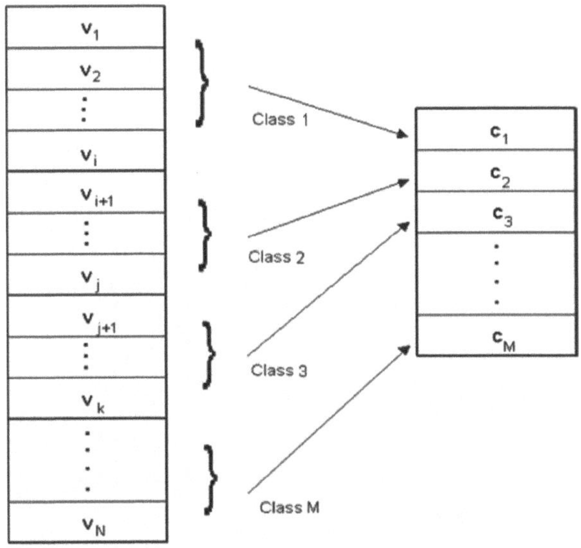

$$q(v_1) = q(v_2) = \cdots q(v_i) = c_1$$
$$q(v_i + 1) = \cdots q(v_j) = c_2$$

Fig. 4 Codebook creation: input data is grouped into class and a codevector is assined for each class

Fig. 5 Speech frames coding

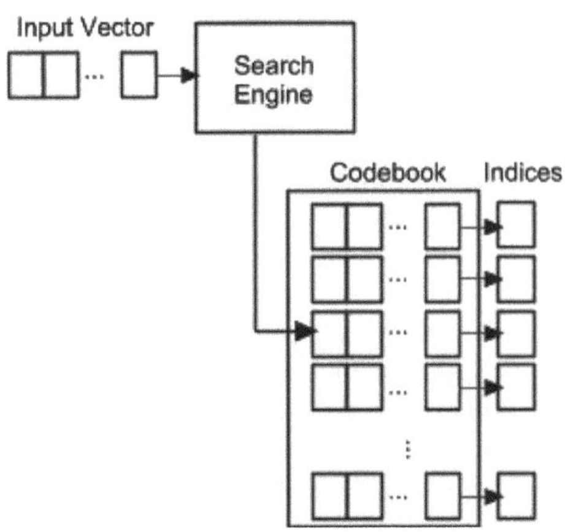

selected. So, a set of input vectors will be mapped into a set of codebook indices. In the decoding phase, the indices will be mapped back into a sequence of codevectors. The whole process is illustrated in Fig. 5 and Fig. 6.

The use of vector quantization for speech compression requires representation of speech signal as a finite sequence of fixed dimension vectors. The windowed frames of speech are variable length sequences of samples (the higher the pitch period, the

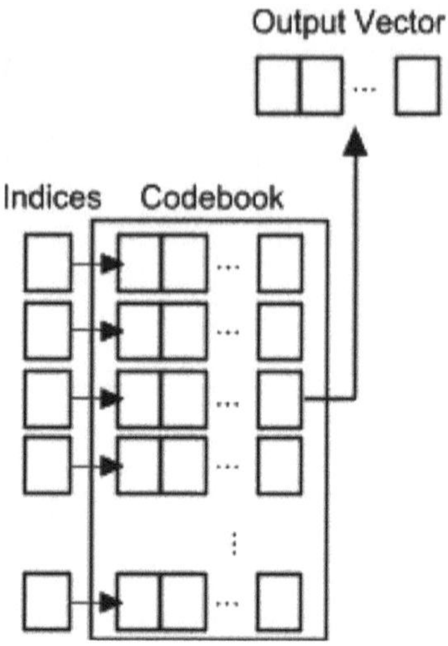

Fig. 6 Speech frames decoding

larger the number of samples). If we intend to use windowed frames as the basic units to be quantized, they must first be transformed into fixed dimension vectors. This process is called *parametrization*. Each windowed frame will be represented by a vector of parameters associated with acoustic characteristics of the speech signal. The more similar two windows are in terms of acoustic features, the closer their parameter vectors will be.

In order to implement an efficient vector quantizer, we must define a set of parameters well suited for frame discrimination. It is important that these parameters be uncorrelated, avoiding redundant information.

Once the parameter set is defined, a parameter vector is calculated for each windowed frame of the training set. These vectors are grouped into clusters, and for each cluster a centroid is calculated. The vector closest to the centroid is selected as the codevector that represents the cluster and is included in the codebook. This choice minimizes the mean quantization error of the vectors belonging to the cluster.

Different techniques can be used to generate the codebook. In the next section, we will discuss the strategy implemented in this work, based on unsupervised neural networks (2-dimensional Kohonen map).

The obtained codebook can then be used to encode speech signals. The coding process consists of the following stages:

- determination of pitch marks and analysis frames;
- determination of the energy for each frame;
- determination of the parameter vector for each frame;
- mapping of frame vectors into codebook vectors and corresponding indexes.

To recover the speech signal from the received sequence of codebook indexes, the decoder concatenates the windowed frames associated with each codevector.

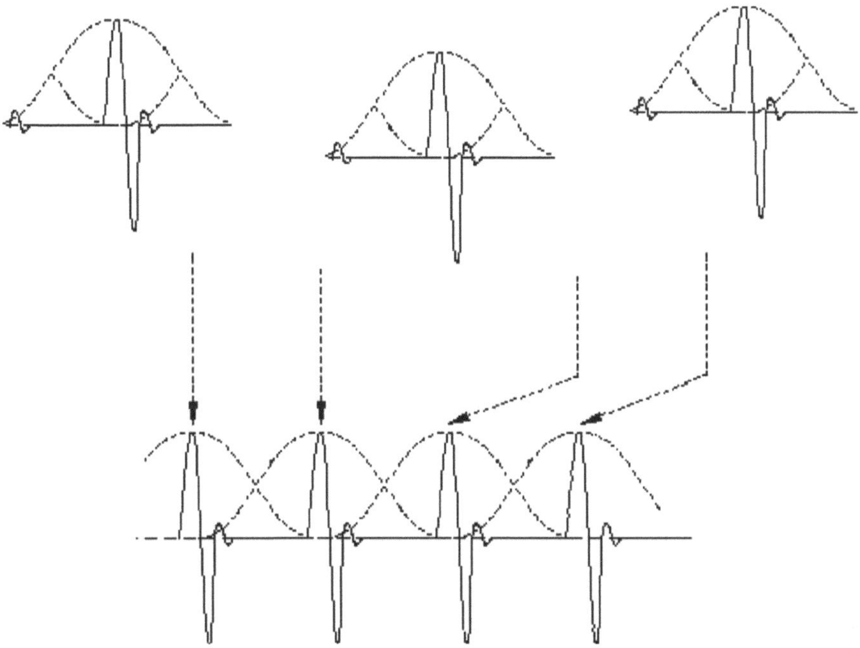

Fig. 7 Overlap and add of speech frames

A frame dictionary is thus required in the decoder, associating each codebook index with a windowed frame. This dictionary is created in the training phase.

The amplitude of each windowed frame recovered from the dictionary is adjusted so that its energy equals the energy of the original windowed frame. This prevents the occurrence of envelope discontinuities in the reconstructed signal.

The signal is reconstructed by overlapping and adding the windowed frames after gain correction. Frames are positioned in such a way that original pitch periods are preserved. The pitch and rhythm curves of the reconstructed sentence will thus match the original ones. This process is illustrated in Fig. 7.

When the windowed frames are positioned for the overlap-and-add operation, the first sample of a frame does not necessarily align with the central pitch mark of the previous frame, nor does the last sample of this frame align with the central pitch mark of the next frame. This happens because the pitch period of the frame stored in the codebook may be different of the original frame's pitch period. This difference in the superposition level of the windowed frames causes the reconstructed signal to be distorted. The smaller the mismatch between pitch periods, the smaller the distortion introduced by the overlap-and-add operation.

4 Self-Organizing Maps

Self-organizing maps (SOM) are a kind of artificial neural network initially proposed by Kohonen [1]. Its training phase consists of an unsupervised process, which comprises 4 stages [7]:

- initialization;
- competition;
- cooperation;
- adaptation.

In the initialization stage, the following initial parameters of the network are defined: dimension of the grid (1 or 2), network topology (leaf, cylinder or toroid), initial neuron weights (their values can be defined randomly or through a specific criterion).

The follow stages corresponding to the learning stage, where training data are sequentially submitted to the network. In this stage, the vector x representing one input data is presented to all neurons in the network, and the distance between this vector and each weight vector w of network neurons is calculated. The neuron whose weight vector is closest to the input is called the *best matching unit* (BMU) or *winner neuron*. This stage is called *competitive training* [8,9], and its main aim is to determine the most activated neuron for each input data. We use the Euclidean distance as a measure of similarity between x and each weight vector w.

The BMU weights are then updated in order to move the winner neuron towards the input data. The neurons in the neighborhood of the BMU are also updated with an update factor that decays according to a Gaussian distribution curve around the BMU. The activation function of the neighbourhood is given by the following equation:

$$h_{i,j}(n) = \exp\left(-\frac{d_{i,j}^2}{2\sigma^2(n)}\right), \qquad n = 0, 1, 2, \ldots \tag{4}$$

where $d_{i,j}$ is the distance between the winning neuron i and its neighbour j, and σ is the deviation adjusted at instant n. The rate at which σ is updated is given by:

$$\sigma(n) = \sigma_0 \exp\left(-\frac{n}{\tau_1}\right), \qquad n = 0, 1, 2, \ldots \tag{5}$$

where σ_0 is the initial deviation and τ_1 is a constant. So, the radius of the neighborhood decreases with time [7]. At the adaptation phase, the synaptic wheights of the BMU and its neighbours are updated towards the input data x, according to the following equation:

$$w_j(n+1) = w_j(n) + h_{i,j}(n)(x - w_j(n)) \tag{6}$$

where $w_j(n)$ is the synaptic vector of neuron j and $h_{i,j}$ is a weight that determines how closer w_j should get to the input data x.

This is known as *cooperative process*, because the winner neuron shapes the neurons in its neighborhood. Finally, in the adaptive process, the weight vectors w are updated towards each input vector x.

The input vectors are presented to the network until a stop criterion is satisfied. This criterion can be the number or iterations or epochs (an epoch implies the presentation of all the training vectors to the network), or a minimum mean quantization error between the centroids and the training vectors.

4.1 The U-Matrix

Self-organizing maps are a projection of a high dimensional data space onto a grid of neuron locations that is usually embedded in a 1 or 2-dimensional structure. The learning algorithm of a SOM is designed to preserve, in the map dimension, the neighborhood relationships that occur in the high dimensional data space.

The use of the U-Matrix [10] was proposed for the analysis of the emergent properties of the Kohonen map. The U-Matrix is a display of the distance between two neurons on grid positions of the map, giving an idea of distance relationships of the input data in the original data space.

4.2 Clustering and the K-Means Algorithm

The U-Matrix represents the distance between neuron weights, preserving the topology neighborhood of input data that generally are in a higher dimensional space. It is often difficult to determine the limits of each cluster. In several applications, the U-Matrix is sufficient to determine the clusters and their boundaries. However, in other applications, such as speech compression – addressed in this work – it is necessary to determine not only the clusters but also the neuron that best represents each cluster. The weight vector related to that neuron will be the chosen codebook vector.

To perform this task, the K-Means clustering algorithm is applied to the U-Matrix. The K-Means algorithm [11, 12] enables the creation of clusters from a set of objects. The goal is to find prototype clusters that minimize the distance between the prototype and the objects in the set.

The most popular algorithm used to deal with this problem is the iterative Lloyd algorithm [13]. It starts by partitioning the set of input samples into k initial sets either randomly or following some heuristic. Then, the mean point or centroid of each cluster is calculated, which will be the initial prototype. After that, a new partition is made by associating each point with the closest centroid. The centroids are recalculated for the new clusters, and the steps are repeated until convergence is reached.

After convergence, a set of clusters and corresponding prototypes is obtained. The prototypes are used as codevectors of the codebook.

5 Analysis Results

In this work, a Kohonen network was used together with the K-Means algorithm to perform frame grouping based on acoustic characteristics of speech frames. Different frames belonging to a speech database were presented in batch to the network. Since adjacent speech frames tend to be highly correlated, they were randomly ordered before training. The number of iterations in the batch training was high enough to ensure the convergence of the quantization errors. At the end of the training phase, each neuron in the network represented a cluster of similar speech frames.

To test the accuracy and usefulness of this method of speech quantization, the tests were divided in three phases. In the first phase, different speech parameters were tested to establish which ones are best suited for speech frame discrimination. In the second phase, an analysis of different network topologies and training strategies was made, in order to evaluate their influence in the quality of the created codebook (which is directly related to the quality of reconstructed speech signals). Finally, in the third phase a comparison was performed between results of the proposed method and those produced by a popular perceptual coding algorithm.

The same speech database was used in the training and coding/decoding phases. The speech database consisted of 100 sentences in Brazilian Portuguese uttered by a female speaker . Sentences were chosen so as to guarantee a rich phonetic content. Speech files were stored in wave format (linear PCM coding), 16 kHz sampling rate and 16 bits per sample. The complete set of sentences contained about 52,000 frames of speech.

5.1 Parameter Selection Tests

The target of this analysis was to establish which parameters are best suited for speech frame discrimination (i.e. to group frames into clusters). It is expected that better parameters lead to a better reconstructed signal and also minimize the size of the codebook for a certain quality.

Different sets of parameters (attribute vectors) were extracted from the speech frames. For each parameter set, the related vectors were grouped into clusters in order to create a speech codebook using the Kohonen network and the K-Means algorithm. The different codebooks obtained were used to reconstruct the set of 100 sentences used in the training phase. The quality of these sentences was evaluated and compared to identify the best set of parameters.

The speech parameters chosen for testing are commonly used in applications involving speech signal processing, and are widely referenced in the literature [14–17]. They are:

- Left pitch period: number of samples between the beginning of the frame and the central pitch mark.
- Right pitch period: number of samples between the central pitch mark and the end of the frame.
- Zero-crossing rate: number of times the signal passes through zero, per units of 10 ms.
- Maximum and minimum rate: number of inflections in the waveform throughout the frame, per units of 10 ms.
- Mel-cepstral coefficients [18]. This method computes the discrete cosine transform (DCT) of the modulus of the signal spectrum in decibels. The spectrum was computed using a 1024-point FFT. Before the modulus operation, the spectral coefficients were filtered by a filter bank with 24 bands distributed on the mel scale.

Different sets of mel coefficients were tested: the first six coefficients (Mceps 1-6), the first ten coefficients (Mceps 1-10) and the first twelve coefficients (Mceps 1-12).

The ITU-T standards for objective evaluation of speech quality [19, 20] were used to compare the generated speech signals. These standards make use of psychoacoustics models, which are based on characteristics of the human hearing system, and generate measurements that are analogous to those obtained by subjective evaluations [21–23]. The objective evaluation algorithm used in this work was the PESQ (Perceptual Evaluation of Speech Quality), described in the ITU-T standard P.862 [20]. The PESQMOS measurements generated by the PESQ algorithm range from 0.5 (worst case) to 4.5 (best case). The algorithm was used to evaluate reconstructed speech signals.

Table 1 presents the main results for different parameter sets. These results were obtained using the best configuration established in phase 2 of our tests (see next section). We did not normalize the mel-cepstral coefficient values since these coefficients are proportional to their variances. On the other hand, the pitch period parameters were normalized by the variance of the first mel-cepstral coefficient.

We can see that the best set is composed of 13 parameters: the first 12 mel-cepstral coefficients and the left pitch period of the frame.

5.2 *Topology and Training Strategies*

In this section, we present an analysis of the influence of network topology and training strategies on the final speech quality. A Kohonen network was employed in the simulations, using the public domain SOM (Self-Organizing Map) toolkit, available on the website of the Helsinki University of Technology [24].

Table 1 Objective measurement for different parameter sets

Mceps 1-6	Mceps 1-10	Mceps 1-12	Left Period	Right Period	Zero Cross	Max/Min	PESQMOS
		X	X				**2.358**
		X	X	X			**2.356**
		X	X		X		**2.351**
		X	X		X	X	**2.334**
		X	X			X	**2.331**
		X					**2.311**
	X						**2.306**
X							**2.165**

Table 2 Network performance as a function of the number of neurons

Number of neurons	PESQMOS
25 × 25 – 625	2.330
40 × 40 – 1,600	2.349
60 × 60 – 3,600	2.358

The values of PESQMOS for objective measurements presented here correspond to an average of 10 measurements obtained for different networks, each one trained using all 100 sentences from the database.

5.2.1 Number of Neurons

Preliminary tests using the vector quantization method showed that there is a dependence between the number of neurons in the network and the number of clusters obtained by the K-Means algorithm. It was thus necessary to evaluate the influence of the number of neurons for a fixed number of clusters. A network was trained with a single sentence containing 440 frames, and the number of output clusters generated by the K-Means algorithm was set to 440. The main objective of this analysis was to observe the network capability to generate a codebook with a different representative codevector for each input vector. In the first test, a network with 21×21 neurons was used (i.e. the number of neurons and clusters was similar); in the second test, a network with 50×50 neurons was used (i.e. the number of neurons was about six times the number of clusters). The objective measurement result was 1.816 for the first test and 3.564 for the second one. An immediate auditive inspection showed that the first speech signal was seriously degraded, while the second one presented little audible degradation.

The influence of the number of neurons when training the network with all database sentences (approximately 52,000 frames) was then analyzed. Table 2 presents objective measurements obtained for networks with different numbers of neurons. As expected, an increasing number of neurons leads to better objective measurement results.

5.2.2 Neighborhood Adjustment

The neighborhood around the winner neuron, in which synaptic adjustments are performed on training, is an important topological feature of the Kohonen map. This neighborhood affects the quality of the codebook and, consequentially, of the reconstructed speech signal. The objective of the present analysis is to evaluate the influence of the neighborhood size. Three different neighborhoods were tested: a neighborhood large enough to cover all network neurons, a neighborhood covering just the neurons around the winner neuron, and a variable neighborhood starting with full network coverage and decreasing linearly along training until only the neurons around the winner are covered. The objective measurement results are shown in Table 3. We can see that the small and the decreasing neighborhoods led to similar results, while the large neighborhood led to are covered stronger signal degradation.

5.2.3 Codebook Size

In this section, we investigate the influence of codebook size (the number of output codevectors obtained by the K-Means algorithm) in the quality of reconstructed speech signals. Theoretically, larger codebooks lead to more precise representations of the original speech frames in the database, and thus to better perceptual quality of the reconstructed signal.

Table 4 presents PESQMOS results for different codebook sizes. These results were obtained using the full database for training (about 52,000 frames), a network with 60×60 neurons and a decreasing neighborhood. As the table shows, the PESQMOS value increased with the number of codevectors up to 1,000. However, with 3,000 vectors, we observed a reduction in the PESQMOS value. This is most likely due to the fact that the number of neurons in the network is insufficient to represent 3,000 clusters. As shown earlier, there is a relationship between the number of neurons in the network and the number of clusters to be formed.

Table 3 Network performance as function of neighborhood

Neighborhood	PESQMOS
All neurons	2.241
Only neurons around the winner	2.429
Decreasing neighborhood	2.472

Table 4 Perceptual performance as a function of codebook size

Number of codevectors	PESQMOS
20	2.058
100	2.226
500	2.341
1,000	2.415
3,000	2.382

Thus, if we want to increase the number of codebook vectors in order to obtain a better reconstructed signal, we may need to increase the number of neurons as well.

The training time of the neural network grows with the number of neurons in the network, with the number of codebook vectors and with the number of speech frames used to generate the codebook. For each of the examples above, the training time needed was lower than 30 minutes using a Pentium IV microcomputer with a clock of 3 GHz and 1 GB of RAM. The training time grows approximately linearly with the size of the codebook and with the number of neurons.

5.3 Comparison with MPEG1-Layer 3

In order to evaluate the potential of the proposed technique for the compression of speech signals, we conducted informal tests to compare our method with the MPEG1-Layer 3 coder (MP3). MP3 is a compression technique vastly used in portable players and the Internet. The distortion level of an MP3-encoded signal is a function of the bit rate of the compressed bitstream. Higher bit rates (lower compression) produces higher quality signals, while lower bit rates (increasing compression) produce lower quality signals.

To compare our method with MP3, a codebook with 500 codevectors was used. The network was trained with a 100-sentence database and was generated by a Kohonen network with 60×60 neurons and a decreasing neighborhood. The use of this codebook resulted in a compression rate of 100x.

The MP3 codec was configured to operate with a constant bit rate of 8 kbps, resulting in a compression rate of 30x. Compression rates of 100x render the speech unintelligible. The employed coder was supplied with CoolEdit Pro 2.0 (Syntrillium Software Corporation).

Table 5 presents the objective measurement results.

The results show that the proposed method has a compression rate about 3 times higher than MP3's, while providing lower levels of signal degradation. It is important to remark that the two methods produced different kinds of degradation. While the vector quantization technique introduced signal discontinuity effects, the MP3 technique at this bit rate resulted in a stifled signal, indicating the attenuation of high frequency components.

Table 5 Comparison between the proposed vector quantization technique and an MPEG1-Layer 3 codec

Compression type	PESQMOS
Without compression	4.500
Vector quantization (100x compression rate)	2.343
MPEG1 - Layer 3 (30x compression rate)	1.904

6 Conclusion

In this work, we present a method for speech compression based on vector quantization. The method uses pitch-synchronous frames as basic speech units. The frames are represented in parametric form and grouped in clusters, according to their acoustic characteristics.

A Kohonen network in conjunction with the K-Means algorithm was used to perform frame clustering. K-Means was used because better results were obtained when the number of neurons in the network was larger than the number of clusters, allowing more than one neuron to represent the same cluster.

Experimental results showed that the proposed method was successful in grouping speech frames coherently. Subjective and objective tests were performed to evaluate the quality of the reconstructed signals. Objective results obtained with the PESQ algorithm were presented.

Frame parametrization is an important step in frame grouping. Among the analyzed coefficients, the mel-cepstral ones seemed to preserve relevant acoustic characteristics of the frames. Better results were obtained using a higher number of mel coefficients (12). The quality of reconstructed speech signals increased with the size of the codebook and the number of frames used to train the neural network (speech database).

The vector quantization method proposed here showed promising results for applications involving speech compression. Speech sentences coded with the proposed method and with an MPEG-1 Layer 3 codec were compared using an objective speech quality algorithm (PESQ). The vector quantization method here led to better results.

In the next steps of this work, other speech parameters will be analyzed, such as linear prediction coefficients (LPC), linear prediction cepstral coefficients, spectral coefficients and mel/bark spectral coefficients. Growing self-organizing maps will also be studied as an alternative to improve neuron grouping and avoid the use of K-Means algorithm. Finally, better strategies of initializing neuron synapses, using real frame samples from the database, will also be studied.

References

1. KOHONEN, T.: Self-organized formation of topologically correct feature maps. In: Biological Cybernetics, (43):59–69 (1982).
2. RUNSTEIN, F. O.: Sistema de reconhecimento de fala baseado em redes neurais artificiais. PhD thesis, FEEC/Unicamp (1998).
3. MAKHOUL, J.; WOLF, J.: Linear prediction and the spectral analysis of speech. In: Bolt, Beranek, and Newman Inc., pages 172–185 (1972).
4. BOLL, S.; WOLF, J.: Suppression of acoustic noise in speech using spectral subtraction. In: IEEE Trans. on Acoustics, Speech, and Signal Processing, 27(2):113–120 (1979).
5. GERSHO, A.; GRAY, R. M.: Vector quantization and signal compression. Kluwer Academic Publishers, 2nd edition (1992).
6. GRAY, R. M.; NEUHOFF, D.: Quantization. In: IEEE Trans. on Inf. Theory, 44(6) (1998).

7. HAYKIN, S.: Neural networks, a comprehensive foundation. Prentice Hall, 2nd edition (1999).
8. KOHONEN, T.: Improved versions of learning vector quantization. IJCNN International Joint Conference on Neural Networks, (1):545–550 (1990).
9. KOHONEN, T.: The self-organizing map. Proceedings of the IEEE, (78):1464–1480 (1990).
10. ULTSCH, A.; SIEMON, H. P.: Kohonen's self-organizing feature maps for exploratory data analysis. In: Proc. INNC'90, Int. Neural Network Conf., 305–308 edition (1990).
11. FORGEY, E.: Cluster analysis of multivariate data: efficiency vs. interpretability of classification. Biometrics, (21):768 (1965).
12. MACQUEEN, J.: Some methods for classification and analysis of multivariate observations. In: Proc. Fifth Berkeley Symp. Math. Statistics and Probability, (1):281–296 (1967).
13. LLOYD, S. P.: Least squares quantization in PCM. In: IEEE Trans. Information Theory, (28):129–137 (1982).
14. ANDERSON, T. R.: Phoneme recognition using an auditory model and a recurrent self-organizing neural network. In: ICASSP'92: IEEE International Conference on Acoustics, Speech and Signal Processing, (2):337–40 (1992).
15. CAWLEY, G. C.; NOAKES, P. D.: The use of vector quantization in neural speech synthesis. In: IEEE Service Center, IJCNN'93, International Joint Conference on Neural Networks, volume III, Piscataway, NJ, USA (1993).
16. HERNANDEZ-GOMEZ, L. A.; LOPEZ-GONZALO, E.: Phonetically-driven CELP coding using self-organizing maps. In: ICASSP'93, International Conference on Acoustics, Speech Propagation and Signal Processing, volume II, Piscataway, NJ (1993).
17. KITAMURA, T.; TAKEI, S.: Speaker recognition model using two-dimensional mel-cepstrum and predictive neural network. In: Proceedings ICSLP'96. Fourth International Conference on Spoken Language Processing, volume III, New York, USA (1996).
18. DAVIS, S.; MERMELSTEIN, P.: Comparison of parametric representations for monosyllabic word recognition in continuously spoken sentences. In: IEEE Trans. on ASSP, 28(4):357–366 (1980).
19. ITU-T. P.861: Objective quality measurement of telephone-band (300-3400 Hz) speech codecs, February (1998).
20. ITU-T. P.862: Perceptual evaluation of speech quality (PESQ), an objective method for end-to-end speech quality assessment of narrow band telephone networks and speech codecs, February (2001).
21. ITU-T. 85: A method for subjective performance assessment of the quality of speech voice output devices, June (1994).
22. ITU-T. P.830: Subjective performance assessment of telephone-band and wideband digital codecs, February (1996a).
23. ITU-T. P.830: Subjective performance assessment of telephone-band and wideband digital codecs, February (1996b).
24. SOM TOOLBOX. Available from: <http://www.cis.hut.fi/projects/somtoolbox/>.

The Use of Bayesian Networks for Heart Beat Classification

Lorena Sophia Campos de Oliveira, Rodrigo Varejão Andreão, and Mário Sarcinelli-Filho

Abstract This work proposes to use a static Bayesian network as a tool to support medical decision in the on-line detection of Premature Ventricular Contraction beats (PVC) in electrocardiogram (ECG) records, which is a well known cardiac arrhythmia available for study in standard ECG databases. The main motivation to use Bayesian networks is their capability to deal with the uncertainty embedded in the problem (the medical reasoning itself frequently embeds some uncertainty). Indeed, the probabilistic inference is quite suitable to model this kind of problem, for considering its random character; as a consequence, random variables are used to propagate the uncertainty embedded in the problem. Some topologies of static Bayesian networks are implemented and tested in this work, in order to find out the one more suitable to the problem addressed. The results of such tests are discussed in details along the text, and the conclusions are highlighted.

Keywords Bayesian Networks · Decision-Support Systems · PVC Detection · Uncertainty · Medical Informatics

1 Introduction

Since the early nineties Bayesian networks (BN) have been explored in the development of computer systems for medical applications. In fact, Bayesian network is a promising tool for acquiring expert knowledge, since it handles the uncertainty in a natural way and deals with missing data through inference methods [11].

Most medical applications adopting BN are related to diagnosis, treatment selection, planning, and prognosis prediction [11], issues frequently related to a high level decision making. A remarkable example, in such a context, is the growing

L.S.C. de Oliveira, and M. Sarcinelli-Filho (✉)
Graduate Program on Electrical Engineering, Federal University of Espírito Santo, Av. Fernando Ferrari, 514, 29075-910, Vitória, ES, BRAZIL
e-mail: mario.sarcinelli@ele.ufes.br

R.V. Andreão
Instituto Federal do Espírito Santo, Av. Vitória, 1729, 29040-780, Vitória, ES, BRAZIL

A. Hussain et al. (eds.), *Brain Inspired Cognitive Systems 2008*, Advances in Experimental Medicine and Biology 657, DOI 10.1007/978-0-387-79100-5_12,

usage of artificial intelligence tools in e-health applications [3,4]. In such a context, a very promising research field is the remote monitoring in cardiology. The vital signals of a monitored patient, specially the electrocardiogram (ECG), are recorded 24 h a day by a health center and his/her condition is followed up. Thus, automatic ECG analysis can be carried out to identify risky events and generate proper alarms. The particular interest for ECG is due to its efficiency in the diagnosis of cardiac arrhythmias and the great incidence of cardiac diseases in industrialized countries [8].

Most systems developed for ECG analysis use heuristic rules, fuzzy logic, neural networks or statistical approaches as decision tool [1, 2]. An example of a heuristic approach for PVC classification is presented in [1], where the limitation of the heuristic rule is overcome by using regions of certainty related to the possible values of a given variable. On the other hand, statistical approaches are built after a learning phase based on a set of selected examples. Thus, the classification capability of this group of approaches is highly dependent on the information previously learnt. However, they embed a certain potential of evolution (through the use of a new set of examples in the learning phase), making them very attractive.

Since the last decade, neural networks are in evidence to cope with the problem of arrhythmia classification [6, 9]. However, they are considered as black boxes, because the medical expert can only have access to their inputs and outputs, and the network itself can not be easily interpreted and configured by him. Another characteristic associated to the use of neural networks is their lack of flexibility to adapt themselves to new examples, since the learning procedure demands a large set of examples.

In this context, this work evaluates the performance of the Bayesian network framework as a tool to assist the cardiologist in the diagnosis of premature ventricular beats. To the extent of the authors' knowledge, such framework has not been used yet for the particular problem of beat-classification, which gives more importance to the work. Different network configurations are tested, including one for which information coming from more than one ECG lead is fused, aiming at exploring the complementarities among ECG leads, as the medical expert frequently do.

2 Methodology

People often need to deal with uncertainty in order to make a decision, which is particularly important in the medical domain. The Bayesian network is a formalism that allows representing and reasoning over problems involving uncertainty [11]. The Bayes' theorem plays an important role in such a framework, being used to compute $P(A|B)$, the conditional probability of the occurrence of an event A given the evidence B, considering that the event B has been observed.

Such theorem states that

$$P(A|B) = \frac{P(B|A)P(A)}{P(B)},$$

where $P(A|B)$ is the *a posteriori* probability of A given B, $P(B|A)$ is the *a posteriori* probability of B given A, $P(A)$ is the *a priori* probability of A, and $P(B)$ is the *a priori* probability of B. Based on this theorem, the BN works on the causal relations between random variables. It is a graphical representation in which nodes represent random variables with some uncertainty associated to them, and arcs indicate the direct causal relations between the connected nodes [12]. Thus, the BN can model the knowledge of a specialist in an intuitive way.

Therefore, the BN basically consists of a set of variables and a set of arcs linking them, forming a direct acyclic graph. For the discrete case, each variable has a limited group of mutually exclusive states, and the causal relation between a discrete random variable A and its parents $b_1, \ldots, b_n$ is given by a table of conditional probabilities $P(A|b_1, \ldots, b_n)$. For a continuous random variable A, by its turn, the conditional probability is given by a probability density function (pdf).

Another characteristic of a Bayesian Network is that even if two nodes are not directly connected, there is dependence among them, because whenever new evidence is included in the network the probabilities are redistributed to all nodes.

This approach has been employed in intelligent systems designed to generate possible solutions for several types of problems, like medical diagnosis.

2.1 ECG and PVC

Electrocardiography is the most commonly used exam in cardiology, standing out for being fast, cheap and non invasive [7] (compared to other exams available in cardiology). It is simply a record of the heart electrical activity gotten from different leads (the electrocardiogram - ECG), and the characterization of a cardiopathy corresponds to specific modifications in the shape of the waveform associated to each beat. An illustration of part of an ECG record is shown in Figure 1, where the elementary waveforms and time intervals are identified.

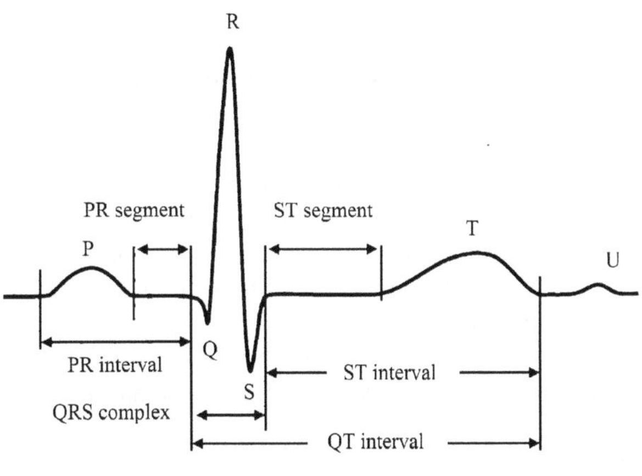

Fig. 1 Heart beat observed in an ECG record, with elementary waveforms and time intervals identified [1]

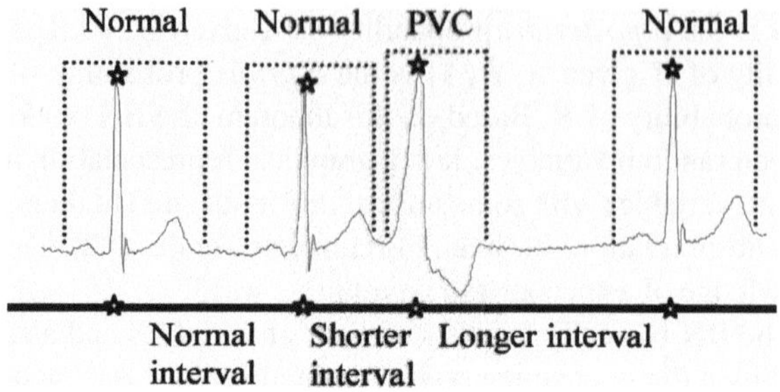

Fig. 2 An ECG segment showing a PVC beat (the third one)

This work deals with a particular type of arrhythmia perceivable in the ECG signal, the premature beats (PVC), which are caused by the advanced electric activation of the ventricles, compared to the normal activation coming from the sinus node (heart's natural pacemaker). In the ECG signal, this arrhythmia is characterized by a premature QRS complex followed by a compensatory pause. In addition, the P wave is not observed before the premature QRS complex, whose wave is wider than in the normal case (see Figure 2).

The automatic classification of PVC, using the theory of Bayesian networks, is the objective of this work. Several BN structures have been implemented and tested, aiming at identifying the network configuration with better performance, as it is discussed hereinafter in the chapter.

2.2 *Designing a Bayesian Network for PVC Classification*

There are at least four steps involved in the design of a Bayesian Network, namely 'Identify Random Variables and Evidences', 'Identify Rules and Causal Relations', 'Process the Signal and Estimate the Parameters of the Bayesian Network' and 'Validate the Results' (see Figure 3).

Identify Random Variables and Evidences: To perform the appropriate diagnosis of any disease, the first step it is to identify all the random variables and the evidences or observations necessary to deliver the most likely cause of that set of evidences (this is what the medical expert does).

In the case of PVC beat classification, the medical expert, in general, classifies each beat looking for the sinus rhythm and the shapes of the beat waveform in the ECG. The analysis is locally performed, since the main interest is to distinguish normal and abnormal beats. Moreover, most times the medical expert has several leads (or channels[1]) at his hands.

However, the evidences are usually not obtained directly from the ECG signal. Actually, it is generally necessary to process the signal in order to extract the

[1] The word channel will be used hereinafter, instead of the word lead.

Fig. 3 Steps involved in the design of a Bayesian Network

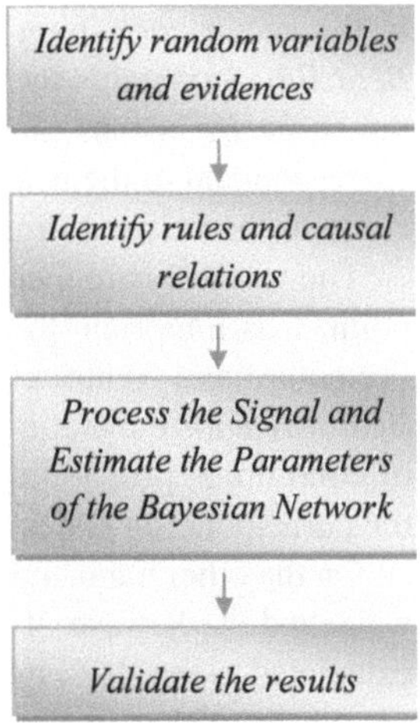

evidences. In this work, the evidences corresponding to the distance between the peaks of two consecutive QRS-complexes and the QRS-complex shape are provided by a framework based on the hidden Markov model, developed by Andreão, Dorizzi and Boudy [1]. Therefore, such framework corresponds to the layer number zero of the system here proposed to identify the occurrence of a PVC beat.

Identify Rules and Causal Relations: After defining the random variables (the nodes of the Bayesian Network), the rules and causal relations are established through the arcs connecting pairs of nodes.

The second step is then to use the BN to combine the evidences through a structure composed of some types of random variables (hidden versus observable, discrete versus continuous), modeled by probability functions [5, 10, 13]. Such probability functions are obtained after a training phase, where the expert's knowledge is learnt (e.g. through using labeled databases).

In this work, only observable nodes are used, i.e., during network training the evidences of all nodes are available through the labels of each heart beat. Moreover, the database is composed of ECG signals from two different channels, where each channel is assigned to an ECG lead. Consequently, the evidence of each channel can be combined in the network, similarly to what is done by the medical expert before classifying each beat.

The network is formed by discrete and continuous variables. The discrete variables are the nodes corresponding to the *PVC Beat*, the *Premature Beat* and the *Ventricular Beat*. On the other hand, the continuous variables are *RRC1* and *RRC2*, which correspond to the R-R interval (time interval) between the two last beats detected in channels 1 and 2, respectively, and *LLC1* and *LLC2*, which represent the likelihood that the QRS complex belongs to a normal beat, obtained from the last detected beat in channels 1 and 2, respectively.

In some of the implementations tested here it is adopted a suffix *BA* or *BP* for the nodes *RR* and *LL*, which mean, respectively, previous beat and following beat. The suffix *BA* means that the corresponding vector is a one-beat delayed version of the vector containing the beat under analysis. This way, the vector '*Premature Beat_BA(i)*' contains the value of the variable '*Premature Beat(i-1)*'. By its turn, the suffix *BP* means that the vector to which it is associated contains the values correspondent to the beat following the current one. Thus, the vector '*LLC1_BP(i)*' contains the values correspondent to the vector '$LLC1(i+1)$'.

The variables are then connected according to their causal dependences. At this point, it is important to stress that a PVC beat is detected every time the beat is Premature and Ventricular. Therefore, the dependencies between the nodes *PVC Beat*, *Premature Beat* and *Ventricular Beat* are quantitatively represented by a table of conditional probabilities with binary alternatives *V* (True) and *F* (False), since the corresponding nodes are discrete.

On the other hand, a beat is said to be Premature when the interval between its R wave and the R wave of the previous beat is shorter than the normal one, which is calculated as the R-R interval. Furthermore, a beat is Ventricular when the morphology of the QRS-complex is larger than the normal one, what is computed from the likelihood of the QRS-complex associated to a normal beat. These four continuous random variables, *LLC1*, *LLC2*, *RRC1* and *RRC2*, are qualitatively represented by Gaussian probability density functions (pdf).

The causal relations quantified through the values of the conditional probabilities are illustrated in the directed graph of Figure 4, which corresponds to the Bayesian network implemented here. Notice, however, that such network is just a general model, which was something changed for each case tested here, as detailed in the next section.

Process the Signal and Estimate the Parameters of the Bayesian Network: The statistical behavior of the continuous variables, represented by the R-R interval and the QRS likelihood, was studied using a training set of examples in order to estimate their probability distributions accordingly. Thus, the histogram of the QRS-complex likelihood and the R-R interval, for the Normal and Ventricular beat types, were built, and four Gaussian functions were adopted to approximate the histograms.

The values of the R-R interval and the QRS-complex likelihood have been normalized, for a better discrimination between the normal, the premature and the ventricular beats. Actually, there is a great variability in the beat morphology among individuals, and the effect of this variability can be reduced through normalization. For doing that, we have taken the R-R interval between the last two detected normal beats to normalize the current R-R interval, and the same has been done for the QRS-complex likelihood, considering the last normal QRS-complex.

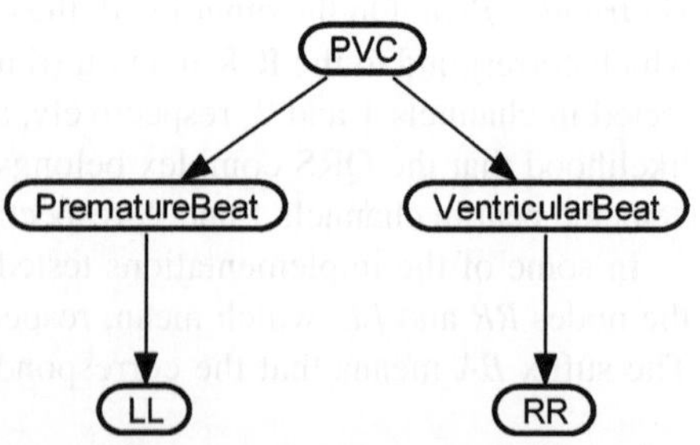

Fig. 4 General kernel of the Bayesian networks adopted in this chapter

Table 1 Representation of the Confusion Matrix for the classification of Normal and PVC beats

	Classification of the Bayesian network		
Database Labels	Normal	PVC	Total
Normal	a true positives	b false negatives	$a + b$ total amount of labeled Normal beats
PVC	c false positives	d true negatives	$c + d$ total amount of labeled PVC beats
Total	$a + c$ total amount of beats classified as Normal beats	$b + d$ total amount of beats classified as PVC beats	$a + b + c + d$

The estimation of the BN parameters is accomplished through a learning phase using a set of examples labeled by a physician. Since there is only observable nodes, the learning algorithm based on the classical junction-tree method [12] has been adopted, which maximizes the probability of the observations given the model.

Validate the Results: The validation of the use of the BN in the problem of beat classification is performed through the use of ECG databases composed of long-term ECG records containing a representative set of labeled beats. The results obtained by the BN are compared to the labels assigned by a cardiologist, in order to get some performance indexes for the proposed classifier.

The network successes and mistakes are used to build a Confusion Matrix (see the structure of such matrix in Table 1), which allows visualizing the errors and checking the quality of the classification obtained (false negatives and false positives). From such matrix, four performance indexes, sensitivity, specificity, positive predictivity and negative predictivity, are calculated, whose characterization is as follows:

- the sensitivity is defined as $a/(a + b)$, and represents the capability of the network of recognizing true positives. In this case, this index gives the probability that the classification is 'Normal' when the label of the physician is also 'Normal';
- the specificity is defined as $d/(c+d)$, and represents the probability that the system classifies a beat that is not 'Normal', thus expressing the capability of the system to identify true negatives. In the present case, such index represents the probability of a classification as 'PVC' when the label of the physician is also 'PVC';
- the positive predictivity is defined as $a/(a + c)$, and, in the present case, indicates the degree of certainty of the hypothesis 'Normal Beat' when the diagnosis offered by the system is positive (in this case, 'Normal');
- the negative predictivity is defined as $d/(b + d)$, and, in the present case, indicates the degree of certainty of the hypothesis 'PVC Beat' when the diagnosis offered by the system is negative (in this case, 'PVC Beat').

3 Experiments

Several topologies of static Bayesian networks were implemented, tested and validated in this work, using the ECG records available in the MIT-BIH (1997) database. The main kernel from which the tested Bayesian networks were generated is the one presented in Figure 4. From such basic structure some other structures were derived and tested, aiming at comparing results and identifying the more suitable structure for the problem being addressed.

3.1 Results for Different Topologies of Static Bayesian Networks Using the MIT-BIH Database

The ECG records available in the database were split in two sets, one for training the network considered and the other for testing it. The first phase, the training one, corresponds to the adjustment of the quantitative part of the network, i.e., the adjustment of the probability tables. The second phase, the test one, by its turn, is adopted for checking the reliability of the result of the classification process. Finally, it is worthy mentioning that in all cases the network implementation was performed using the BN Toolbox for MATLAB® (2002).

3.1.1 Bayesian Network with Empirical Estimation of the Probability Tables

As a first step, the probability tables associated to all nodes of Figure 4 were empirically estimated. The estimation associated to the discrete nodes was performed through a try-and-error procedure, resulting in the values presented in Table 2. Such values were estimated using the same set of beats reserved for training the Bayesian networks. The results of the test phase for the network thus implemented are shown in Table 3. As it can be observed from the table, the negative predictivity value is not good, meaning that the structure thus implemented is a bad classifier for PVC beats.

Then, a Bayesian network using the Bayesian inference for both training and testing (thus being able to adapt its parameters) will be adopted hereinafter.

Table 2 The probabilities empirically estimated for the discrete nodes

PVC	Probability
V	40%
F	60%

PVC	Premature Beat	%	PVC	Ventricular Beat	%
F	F	95%	F	F	95%
V	F	5%	V	F	5%
F	V	5%	F	V	5%
V	V	95%	V	V	95%

3.1.2 Bayesian Network Having only Observable Nodes

Considering the same structure of Figure 4, this new approach has the objective that the network itself learns and changes its parameters, in the training phase. This means to implement the Bayesian learning paradigm, with the goal of improving the network performance, for what the algorithm '*learn_params_em*' [12] was adopted. The learning procedure considers just observable nodes, so that to each node in the network it was assigned an exact value. For the testing phase, all the nodes in the network were considered as observable nodes, except for the node '*PVC*', since this is the hypothesis being classified.

Then, the parameters of the network were adjusted, based on the same training dataset, and the results obtained, using the same test dataset, are presented in Table 4. From Tables 3 and 4, one can notice that the negative predictivity value obtained in Table 4 is much better, making clear that the system performance improves meaningfully when using the Bayesian learning, in the training phase, to estimate the value of each variable.

Aiming at improving this result, the next step is to implement a Bayesian network having observable and non-observable variables.

3.1.3 Bayesian Network with Observable and Non-observable Nodes

In this new simulation, it was considered the same Bayesian network of Figure 4, whose training was also the same presented in the previous experiment, what means that the learning procedure considered only observable nodes. However, there is a difference in the testing phase: only the nodes '*RR*' and '*LL*' are observable.

The classification results for this approach are presented in Table 5. Comparing the results presented in the last two tables, one can notice that the Bayesian network with observable and non-observable nodes leads to better classification. The good results obtained in Table 5 also confirm that the methodology based on Bayesian networks is a powerful tool for learning and classification. Therefore, from the results so far obtained we can state that we have obtained a classifier that exhibits good performance, from the comparison of the results obtained with the Bayesian network with the labels defined by the medical specialists.

Table 3 Results obtained with the Bayesian network with probabilities empirically estimated

Index	Value
Sensitivity	89,78%
Specificity	82,16%
Positive predictivity	98,83%
Negative predictivity	32,34%

Table 4 Results obtained for the Bayesian network with Bayesian learning and considering all nodes as observable nodes

Index	Value
Sensitivity	99,35%
Specificity	77,82%
Positive predictivity	98,69%
Negative predictivity	87,75%

Table 5 Results considering Bayesian learning and observable and non-observable nodes in the testing phase

Index	Value
Sensitivity	99,69%
Specificity	79,53%
Positive predictivity	98,79%
Negative predictivity	93,92%

3.1.4 Some Remarks About Using Bayesian Networks

The use of Bayesian networks represents an approach able to deal with the uncertainty through its probabilistic representation, working with incomplete and random data in its inference engine.

So far, the learning and classification capabilities of such network have been verified, the best results being obtained when using observable and non-observable nodes (Table 5). The training method adopted was the *'learn_params_em'* [12], while the testing algorithm implemented was the *junction-tree* [12].

The next step of this work is to use a strategy similar to a fusion of different ECG channels. To do that, two ECG channels are used, as well as information about the previous beat (the last classified beat), aiming at improving the performance of the classifier, mainly in terms of the specificity. In order to accomplish such task it was necessary to use another ECG database, the QT database, because the records from the MIT-BIH database had problems during the processing of the channel two in the layer zero (beat segmentation) of our system. The idea is just to study how the strategy of channel fusion affects the system performance.

3.2 Results for Different Topologies of Static Bayesian Networks Using the QT Database

In the experiments reported in the sequel 82,509 labeled heart beats obtained from the QT database were used, 75% of them selected as the training set (58,744 beats) and 25% (23,765 beats) employed as the test set. It is worth noting that only 5.5% of the total amount of beats corresponds to premature ventricular contraction (PVC) beats.

3.2.1 Bayesian Network Using Channel Fusion and Previous Beat Information

Different Bayesian networks topologies were implemented and evaluated employing two ECG channels and information about the beat previous to the one currently being analyzed. The best results (presented in Table 6) were obtained with the Bayesian network topology of Figure 5.

Table 6 Classification results for the Bayesian Network of Figure 5

Index	Value
Sensitivity	99,98%
Specificity	78,15%
Positive predictivity	99,72%
Negative predictivity	98,33%

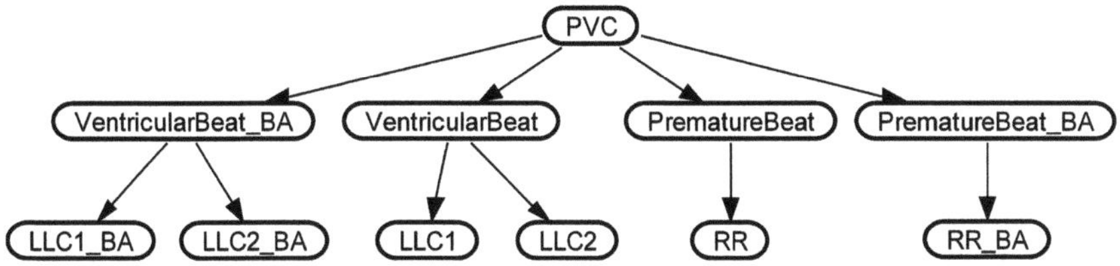

Fig. 5 Bayesian Network using channel fusion, i.e., employing information from channels 1 and 2 (C1 and C2), plus the information about the last beat (BA nodes)

Despite the fact that it was used two different ECG databases, the results of Tables 5 and 6 are very close. Specifically speaking, the negative predictivity value is quite similar to the positive predictivity value. Nevertheless, this comparison is not conclusive since the databases are different.

The next step is to build another network, now using information about the next beat, in order to verify which of the two networks, the one using information about the last beat or the one using information about the next beat, effectively improves system performance.

3.2.2 Bayesian Network Using Channel Fusion and Next Beat Information

In this experiment, a Bayesian Network was built using information coming from the two ECG channels once more, but now including information about the next beat, which means that the classification of a heart beat depends on the information about the next beat (BP), thus delaying the classification result by one beat.

Different topologies have been evaluated and the best results were achieved considering the one presented in Figure 6. The corresponding results for beat classification are shown in Table 7.

Comparing the Bayesian Networks using information about the previous beat and about the next beat (Figures 5 and 6), it is observed that the performances are quite similar. However, considering the fact that the goal is to improve specificity (since it is important for PVC classification and it is the only value below 90 %), the best topology is the one of Figure 6. On the other hand, if the goal is to improve negative predictivity, the best topology is the one of Figure 5.

The next step is to assess the impact of the channel fusion in the system performance by excluding the nodes related either to the previous or the next beat.

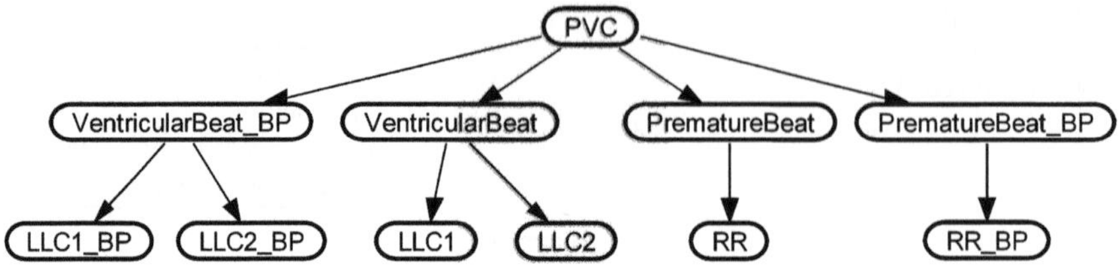

Fig. 6 Bayesian Network using channel fusion, i.e., employing information from channels 1 and 2 (C1 and C2), plus information about the next beat (BP nodes)

Table 7 Classification results for the Bayesian Network of Figure 6

Index	Value
Sensitivity	99,95%
Specificity	78,48%
Positive predictivity	99,72%
Negative predictivity	95,56%

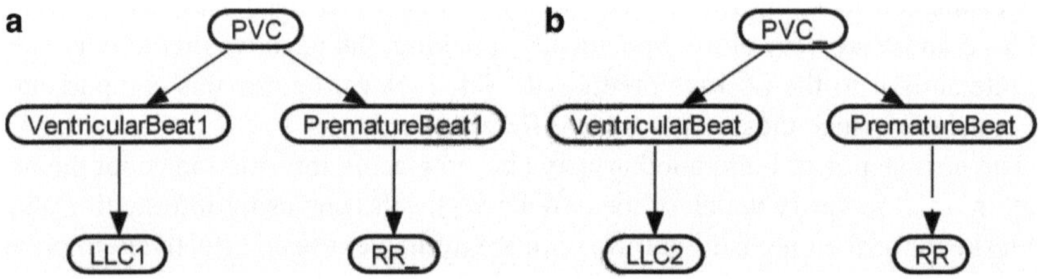

Fig. 7 Bayesian Network using only one channel: (**a**) channel 1 (C1) (**b**) channel 2 (C2)

3.2.3 Bayesian Network Using a Single ECG Channel

Two Bayesian Network topologies were implemented and tested, one using only information provided by channel 1 (Figure 7(a)) and the other using only information provided by channel 2 (Figure 7(b)). The classification results for both of them are presented in Table 8.

At a first glance, it is clear that the network using information from channel 1 performs better than the other one. The main reason for that is the signal-noise ratio, which is worse for channel 2. Moreover, the specialists suggest that channel 1 should be normally used for diagnosis while channel 2 provides redundancy to confirm the diagnosis, whenever possible. Nevertheless, the results using only one channel are worse than those using channel fusion, what confirms that channel fusion is the best solution.

In order to estimate the contribution of the channel fusion to the system performance, another Bayesian Network was built, based on the topologies of Figures 6 and 7, now discarding the information about the next or the previous beat, as discussed in the sequence.

Table 8 Results for the Bayesian Networks of Figures 7 (**a**) and (**b**), respectively

Index	Value	Value
Sensitivity	99,85%	99,98%
Specificity	75,50%	53,31%
Positive predictivity	99,69%	99,40%
Negative predictivity	86,69%	96,99%

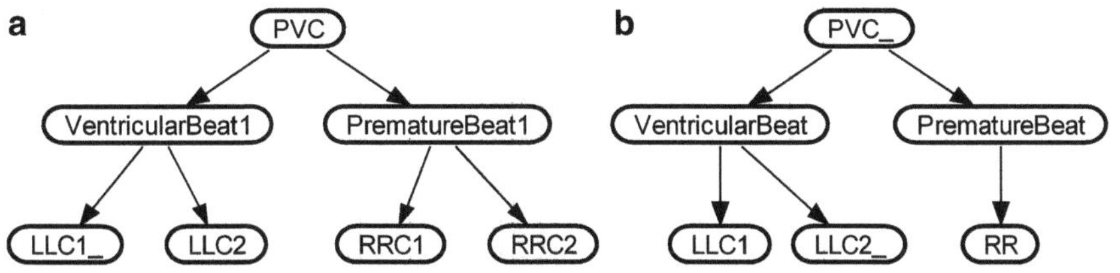

Fig. 8 Bayesian Network using channel fusion. (**a**) RR interval from both channels ($RRC1$ and $RRC2$). (**b**) RR interval just from channel 1 (RR)

Table 9 Classification results for the Bayesian Network of Figure 8 (a) and (b), respectively

Index	Value	Value
Sensitivity	99,99%	99,96%
Specificity	72,19%	78,15%
Positive predictivity	99,64%	99,72%
Negative predictivity	98,64%	96,33%

3.2.4 Bayesian Network Using only Channel Fusion

Two Bayesian Network topologies using channel fusion were implemented and tested (see Figures 8(a) and 8(b)). One topology uses two nodes RR and the other one just one node RR. The difference is because the RR interval is obtained in the segmentation phase, which may be influenced by the bad quality of the signal in channel 2, thus generating many segmentation errors for such channel. Then, we have decided to compare the network using the RR interval information from channel 1 with the network combining the RR interval from both channels, thus generating the topologies of Figure 8. The classification results for such topologies are presented in Table 9.

Comparing the performances of the topologies of Figure 8, it is observed that for negative predictivity the topology using *RR* intervals from both channels (second column of Table 9) is better than the one using only the *RR* interval from channel 1 (third column of the same table). On the other hand, considering only the specificity, the ability of detecting PVC beats, the topology of Figure 8 (b) performs better (third column of Table 9).

It is worth noting that Bayesian Network using channel fusion and next beat information (Figure 6) and the Bayesian Network employing only channel fusion (Figure 8(b)) present specificity values very close (the difference is of 0,33%). From that, it can be concluded that implementing the Bayesian Network of Figure 8(b) with less nodes in the particular problem of PVC detection, one can have a simpler topology, less computational effort and classification performances very similar to more complex topologies (loosing just 0,33% in terms of specificity).

Finally, it is important to stress that the best topologies are those which take into account the information provided by both channels, a result that is fully compatible with the results associated to information fusion.

3.2.5 Discussing the Results Obtained with the QT Database

The best Bayesian Network topology, in terms of sensitivity and specificity, is the one which combines information coming from the two ECG channels.

Although it has presented a small number of false positives (normal beats detected as PVC beats), the specificity (number of true PVC detected) has not improved enough to be as good as the state of the art for that database [1,3]. The main reason for that is the small number of PVC examples available in the database used (approximately 5.5%), compared to the normal ones, what makes the network training for this class of beats a difficult task. However, the performance results obtained so far validate the use of Bayesian Network as a tool for heart beat classification, which is the main contribution of this work.

4 Conclusion

In this work the Bayesian network framework (BN) has been applied to the particular problem of heart beat classification. According to the physician approach when classifying premature ventricular beats (PVC), different BN structures were implemented and tested using the MIT-BIH and QT databases, two labeled databases containing representative sets of long-term ECG records. The BN which presented the best results, in terms of sensitivity and specificity, was the one that combined information provided by two ECG channels (thus implementing a kind of data fusion). However, it was evaluated using only the data available in the QT database.

The results obtained confirm that the combination of different ECG channels improves the performance of the classifier, and demonstrate the viability of using Bayesian networks as a tool to classify this kind of signal as well. Indeed, the Bayesian networks represent an efficient model, for allowing the representation, in the same model, of quantitative and qualitative knowledge.

Another statement based on the results here presented is that the training of the Bayesian networks implemented was restricted because of the low number of PVC beats available in the QT database. This statement is clearer when the better result obtained is compared to the equivalent result obtained using the MIT-BIH database, in which a greater number of PVC beats is available. Thus, in order to improve the performance of the network it is necessary to change the database adopted, to get more examples of PVC beats for training the classifier and thus getting better results in the classification step. However, the available databases do not have more labeled PVC beats than the one used here. As an example, we can mention the AHA database (2008 edition), which contains just 4,600 beats labeled as PVC beats, although containing 200,000 beats labeled as normal beats. Thus, the sequence of this work will be developed using the MIT-BIH database, which contains more PVC beats than the QT database, in absolute numbers, although corresponding to a lower percentage of the total amount of labeled beats.

References

[1] Andreão R V, Dorizzi B, Boudy J (2006) ECG signal analysis through hidden markov models, IEEE Trans Biomed Eng 53: 1541–1549

[2] Andreão R V (2004) ECG beat segmentation through a markovian approach: application to the detection of ischemic episodes (Ph. D. Thesis). National Telecommunications Institute, Evry (written in French)

[3] Andreão R V, Pereira Filho J G, Calvi C Z (2006) TeleCardio – telecardiology to serve the domiciliary patient. In: Proceedings of the X Brazilian conference in health informatics. Florianópolis, Brazil (written in Portuguese)

[4] Boudy J, Delavault F, Muller M et al (2006) Telemedicine for elderly patient at home: the TelePat project. In: Proceedings of the international conference on smart homes and health telematics. Belfast

[5] Buntine W L (1991) Theory refinement on bayesian networks. In: Proceedings of the 7th conference on uncertainty in artificial intelligence. San Francisco

[6] Farrugia S, Yee H, Nickolls P (1991) Neural network classification of intracardiac ecgs. In: Proceedings of the IEEE and INNS international joint conference on neural networks. Singapoore

[7] Gawande A (2002) Complications: a surgeon's notes on an imperfect science. Henry Holt and Company, New York

[8] Kadish A, Buxton A E, Kennedy H L et al (2001) ACC/AHA clinical competence statement on electrocardiography and ambulatory electrocardiography, J Am Coll Cardiol 38:2091–2100

[9] Kuppuraj R N (1993) A neural network system to classify simulated ecg rhythms. In: Proceedings of the IEEE biomedical engineering conference, New Orleans

[10] Lauritzen S L, Spiegelhalter D J (1988) Local computations with probabilities on graphical structures and their application to expert systems (with discussion). J R Stat Soc Ser B 50:425–448

[11] Lucas P (2001) Bayesian networks in medicine: a model-based approach to medical decision making. In: Proceedings of the EUNITE workshop on intelligent systems in patient care. Vienna

[12] Pearl J (1988) Probabilistic reasoning in intelligent systems: networks of plausible inference. Morgan & Kaufman Publishers Inc., San Mateo

[13] Shachter R D (1986) Evaluating influence diagrams. Oper Res 34:871–882

A Histogram Based Method for Multiclass Classification Using SVMs

Sandro Tomassoni Coelho and Carlos Alberto Ynoguti

Abstract SVMs were primarily proposed to deal with binary classification. In this work an alternative $O(\log_2(n))$ method for multiple classes classification using SVMs is proposed. Experimental results showed that it can be 23 times faster than the one vs one method, and 1.3 times faster than the one vs all classic methods, with the same error rate. Tests were performed on a speaker independent, isolated word speech recognition scenario.

1 Introduction

The technique of Support Vector Machines, first proposed in the late seventies [1], has now receiving increasing attention.

The main idea behind this method is to use a hyperplane as the decision surface in such a way that the separation margin between positive and negative examples is maximized. Traditional techniques such as multilayer perceptron neural networks try to minimize the empirical risk, (frequency of errors made by the learning machine on the training samples set). On the other side, the SVM technique searches for structural risk minimization, that implies the realization of the best generalization performance by matching the machine capacity to the available training data for the problem at hand. Therefore, the goal of this technique is to find, among the systems with the minimum training error, the simpler one (the one with the least complexity).

SVMs are characterized by the use of kernels, lack of local minima, and sparse solution. The use of kernels make it possible to map the input data into a high dimensional space, easing the classification task.

Originally, the SVM technique was proposed to perform binary classification, so extension methods are necessary to make it possible to deal with multiple classes.

S.T. Coelho and C.A. Ynoguti
Instituto Nacional de Telecomunicações, Av. João de Camargo, 510,
Santa Rita do Sapucaí -MG, Brazil
e-mail: sandrot@inatel.br; ynoguti@inatel.br

A. Hussain et al. (eds.), *Brain Inspired Cognitive Systems 2008*, Advances in Experimental Medicine and Biology 657, DOI 10.1007/978-0-387-79100-5_13,

There are two ways to achieve this goal: the first is by combining several binary classifiers ("one against one", "one against all", DAG, ECOC, MOC, among others); the second form considers all the classes in the optimization problem formulation. In general, the first form is preferred, because it's simpler to solve several binary classification problems than a single problem with several classes. In this work, only the former approach is considered.

This work describes a new such technique, that was proven to be faster than the existing ones, with the same performance in terms of error rate. In the next section, the main techniques are presented in order to provide a comparison background for the proposed one.

2 SVMs for Multiple Classes

2.1 *One vs One*

This method was introduced by Knerr [8], and later, the *Max Wins* strategy was proposed by Friedman [10].

If n is the number of classes, one binary classifier is trained for each of the possible two classes combinations. This procedure will generate $n(n-1)/2$ binary classifiers.

The most popular method to test the system after all $n(n-1)/2$ binary classifiers are constructed, is to submit the unknown sample x to all classifiers. Then x is predicted in the class with the largest number of votes (*max wins strategy*).

The advantages of this method are it's easy understanding and implementation, besides a good performance. The disadvantage is the huge number of binary classifiers ($n(n-1)/2$), which means a great memory and computational load ($O(n^2)$).

2.2 *One Against All*

This was probably the earliest implementation for SVM multi-class classification [12]. If n is the number of classes, this method constructs n SVM models in the following way: the ith SVM is trained with all examples in the ith class with positive labels, and all other examples with negative labels.

Again, in the testing step, the unknown sample x is submitted to all of the n classifiers. In general, only one of the classifiers will give a positive value for the separate class, and this is the classification criterion. In some cases, it is possible that more than one classifier give a positive output for the separated class; in this case, the one with the highest output is selected.

The advantage of this method is the small number of classifiers, leadig to memory savings and faster classification. However, the training stage is very time consuming, because each SVM must be trained with all training samples.

2.3 DAGSVM

Platt e co-authors [14] suggested a tree based algorithm named DAG (*Directed Acyclic Graph Support Vector Machine*). In this method, the training step is the same as the one against one method by solving $n(n-1)/2$ binary classifiers. However, in the testing step, this method uses a directed acyclic graph with $n(n-1)/2$ nodes and n leaves. Each node is a binary SVM of ith and jth classes. Given a test sample x, starting at the root node, the binary decision function is evaluated. Then it moves either left or right depending on the output value. The process is repeated for each level, until a leaf is reached, which indicates the predicted class.

To illustrate this method, let us consider a simple example with 4 classes, shown in Figure 1: in the initial node, all the four classes are active (the active classes are indicated by the numbers in italics at each node side). The test sample x is initially submitted to the SVM corresponding to the classes 1 and 4. If this SVM decides that the test sample is not from class 1 then, for the next level, the SVM corresponding the classes 2 and 4 will be selected; otherwise, the SVM corresponding to the classes 1 and 3 will be selected. This process goes until one leaf is reached.

An advantage of this method is the low computational cost in the test step, when compared with the previous methods. In fact, a benchmark conducted by Hsu and Lin [11] indicated that the "one against one" and the DAGSVM methods are more suitable for practical use than other methods.

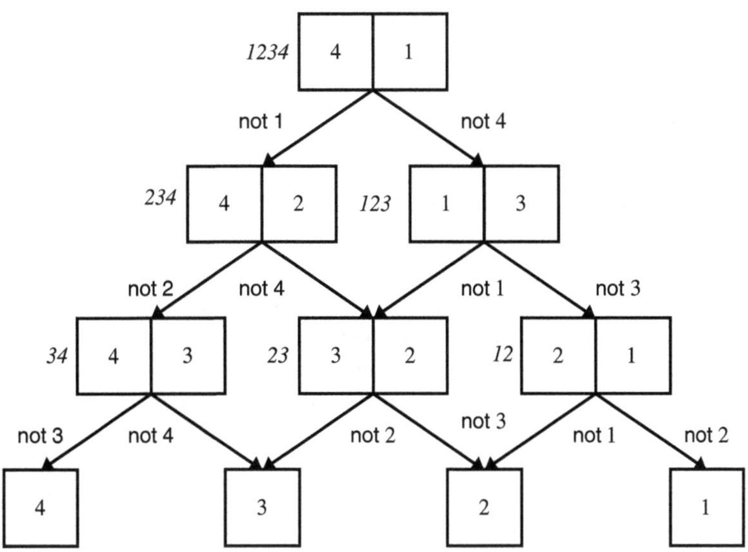

Fig. 1 Example of the DAGSVM method. The numbers in italics indicate the classes that are still active

Another benchark test, conducted by Platt [14] suggests that the performance of the DAGSVM is the same or a little better than the "one against one" and "one against all" methods.

3 The Histogram Pruning Method

This is the contribution of this work, and is based in the "one against one" method: the training step is the same, and modifications were made in the test step in order to lower down it's computational cost.

The idea of this technique arose by observing the behavior of the "one against one" method: if n is the number of classes, then the winner class can receive no more than $n-1$ votes. Furthermore, in our tests, with $n = 50$ classes, it was verified that the winner class received all the $n-1 = 49$ possible votes in the great majority of the cases, as shown in Table 1.

From the Table 1, it can be seen that in 97.66% of times, the winner class received 49 out of 49 possible votes and in 2.16% of times, the winner class received 48 out of 49 possible votes. The occurence of less than 48 ($n-2$) votes for the winner class was very low.

With these results in mind, the following strategy was outlined: if a test sample x is submitted to the classifier for the classes say i and j, then if this classifier decides that x is more likely to be from class say i than class j, then class j will not receive the maximum allowed number of votes ($n - 1$). In this case, class j will not be considered anymore. In other words, this strategy seeks for the class that will receive the maximum allowed number of votes ($n - 1$); if a given class fail in some of the contests, it will be automatically considered out of fight.

An example will help make things clear. Let's suppose that a problem involves the classification of a test sample x into one of 6 possible classes. In the first stage, x is submitted to the classifiers say 1 and 2, 3 and 4, 5 and 5. Let's suppose that the winner classes in this stage were 1, 3 and 5.

For the second stage, only classes 1, 3 and 5 survive. Since each SVM is a binary classifier, one of these classes must compete twice. Say that in this case, class 3 was chosen to play this role. Now suppose that in this stage, winner classes were 1 and 5.

In the third and final stage, the two winner classes of the previous stage (1 and 5) compete to each other, and the winner one (say class 1) is the predicted class.

To improve the performance in terms of error rate, instead of considering only the classes that would receive the maximum allowed number of votes ($n - 1$), classes that would receive $n - 2$ votes are also considered. To do this, no elimination is performed until stage 2. After this stage, classes with no votes are eliminated. The same example given above will be solved in order to provide a comparative view.

Table 1 Number of votes received by the winner class

Votes	49	48	47	46
% results	97.66	2.16	0.22	0.04

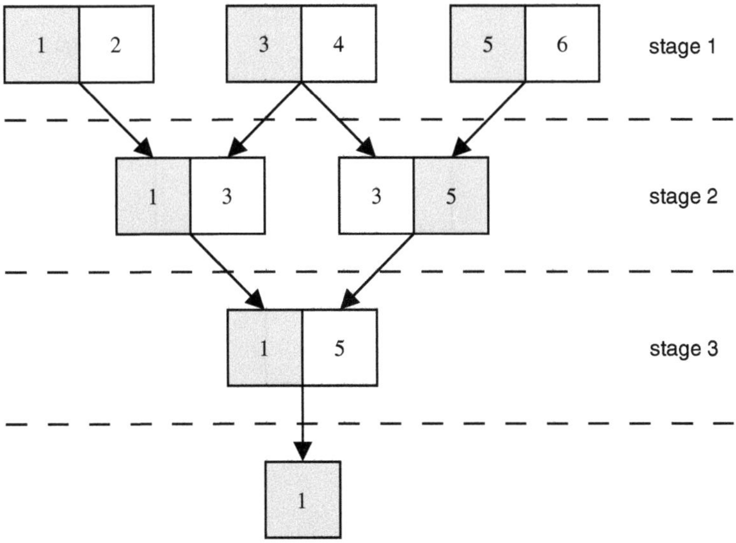

Fig. 2 The histogram pruning method. Winner classes are shaded

Table 2 Situation after stage 1

Class	Votes	Active
1	1	yes
2	0	yes
3	1	yes
4	1	yes
5	0	yes
6	0	yes

Table 3 Situation after stage 2

Class	Votes	Active
1	2	yes
2	0	no
3	2	yes
4	1	yes
5	1	yes
6	0	no

In the first stage, all the classes are active, and the test sample x is submitted say to the classifiers 1 and 6,2 and 4, and 3 and 5. Let's suppose that in this stage, classes 1, 3 and 4 receives a vote. Then, the situation after stage 1 is shown in Table 2.

In the second stage, all the classes are still active, and the test sample x is submitted say to the classifiers 1 and 2,3 and 4, and 5 and 6. Let's suppose that in this stage, classes 1, 3 and 5 receives a vote. Then, the situation after stage 2 is shown in Table 3.

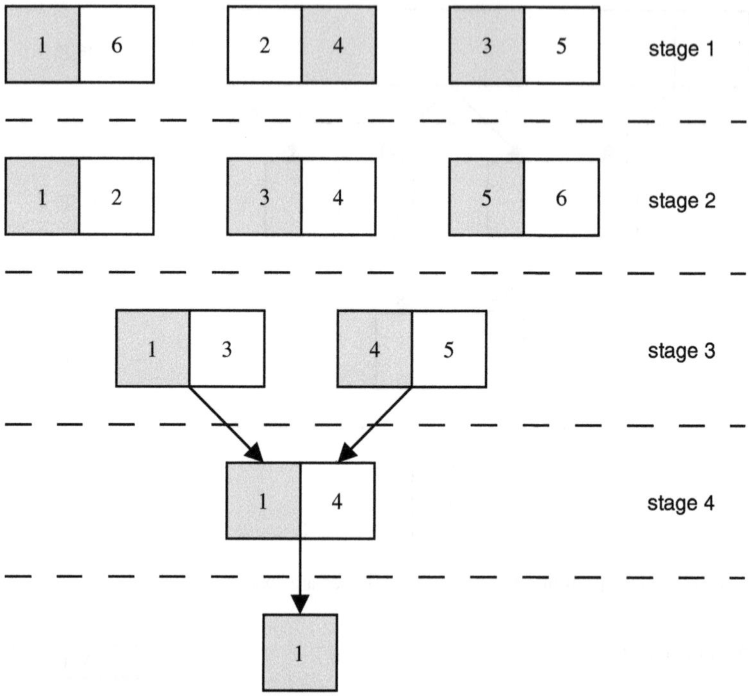

Fig. 3 The modified histogram pruning method. Winner classes are shaded

For the stage 3, classes 2 and 6 are discarded, because they have received no vote. The remaining of the algorithm is the same as the previous case. In Figure 3 this process is shown in graphical mode.

For this method, at each stage, the number of classes is half the number of classes of the previous stage and therefore, it's computational cost is the same as the binary search ($O(log_2(n))$). In our tests, when the pruning process of the looser classes starts at stage 1, the execution time of the algorithm was half the time of the DAGSVM method; when the pruning process starts at stage 2, the execution time was similar to the DAGSVM method, but the recognition rate was better.

Of course, a classification error in one of the classifiers can lead to an error in the final classification. However there are two things to be considered: a) the DAGSVM suffers the same problem (and it's performance is very good), experimental results, that will be presented in the sequel, showed that the performance of this algorithm is the same as the one against one method for this problem. In fact, this is true because, in the great majority of the cases, the winner class received all $(n-1)$ of possible votes.

Table 4 Number of SVMs for each method, in the training and testing steps. n is the number of classes

Method	training	testing
"one against one"	$\frac{n(n-1)}{2}$	$\frac{n(n-1)}{2}$
DAGSVM	$\frac{n(n-1)}{2}$	n
Proposed	$\frac{n(n-1)}{2}$	n
Proposed (modified)	$\frac{n(n-1)}{2}$	$n+\frac{n}{2}$
"one against all"	n	n

Table 5 Complexity for each method

Method	training	testing
"one against one"	$O(n^2)$	$O(n^2)$
DAGSVM	$O(n^2)$	$O(\log_2(n))$
Proposed	$O(n^2)$	$O(\log_2(n))$
Proposed (modified)	$O(n^2)$	$O(log_2(n))$
"one against all"	$O(n)$	$O(n)$

4 Memory and Computational Requirements

In this section, the memory and processing requirements of the proposed method will be compared to some of the existing ones, considering the task of classification among n classes. The required number of SVMs is shown in Table 4, and in the Table 5, the computational requirements are presented.

From the analysis of the Tables above, it's clear that the proposed method have memory requirements and computational load similar to the DAGSVM method.

5 Experimental Results

In this section, experimental results of several tests comparing the performance of the proposed method with some of the traditional ones are shown. Before presenting the results, in the next section, the classification problem will be established.

5.1 The Classification Problem

The tests were performed for the task of isolated word, speaker independent, small vocabulary, speech recognition.

The database consists of 50 different words, pronounced by 69 adult speakers, 33 male and 26 female (a total of 3450 utterances). Recordings were made in an office environment, with small background noise, with 8kHz sampling frequency and 16 bits resolution.

Table 6 Training, validation and testing subsets

subset	male	female	total	%
training	22	13	35	50.72
validation	03	01	04	5.80
testing	18	12	30	43.48

Table 7 Tests with different number of voting sessions

Number of sessions	1	2	3
Recognition rate (%)	90,53	90,69	90,61
Training time	2 hs	2 hs	2 hs
Total testing time	5,2 hs	9,9 hs	14,2 hs

Speakers were divided into training, validation and testing subsets according to Table 6.

All utterances were parameterized using 12 mel-cepstral coefficients [16] and, because of the nature of the classifiers, all utterances were parameterized using the same number of frames (49 frames). With these settings, each utterance is then parameterized with a feature vector of dimension $d = 12 \times 49 = 588$.

Summarizing the information above, the problem consists in the classification of feature vectors with 588 dimensions in one of the 50 possible classes.

For the next sessions, training and testing times were obtained with simulations under the Matlab platform, running on a Pentum 4 2GHz, 512 MB RAM machine.

5.2 *Initial Tests with the Proposed Method*

In the initial tests, the goal is to determine the optimum number of full voting sessions before the pruning step. Intuitively, the greater is the number of full voting sessions, the better is the performance. The results of these tests, presented in Table 7, show a slightly different behavior.

Two facts can be observed by looking at Table 7: the testing time is proportional to the number of full voting sessions and, the performance does not necessarily increase with the increase of the number of voting sessions. It seems that there is no gain using more than two full voting sessions before starting the pruning process, because of the results from Table 1: for the most of times (97.66%), the predicted class won with all $n-1$ possible votes, and in some cases (2.16%), with $n-2$ votes. Also, the gain for two full voting sessions over one full voting session is quite small.

Based on these results, the strategy considering only one full voting session was chosen. In the next section, the results of this method are all related to this number of full voting sessions.

5.3 *Comparison with Other Methods*

In this section, some tests results are shown in order to compare the proposed method with "one against one", DAGSVM and "one against all" methods, in terms

Table 8 Comparison of the proposed method (D) with the "one against one" (A), DAGSVM (B), "one against all" (C) methods. Testing time is referred to the mean time for one utterance classification

	1×1	DAG	$1 \times A$	Hist
recognition rate (%)	90.53	90.29	92.18	90,53
training time (h)	2	2	24	2
testing time (s)	84.7	7	4.8	3.67
number of SVMs	1225	50	50	50

of computational load (for training and testing), memory requirements (number of SVMs) and classification peformance (in terms of recognition rate).

For all tests gaussian RBF kernels were used for all SVMs, with the c and σ parameters being set for the best performance. The results are presented in Table 8:

From the results of Table 8, it can observed that the the performance of the proposed method is similar to "one against one" and DAGSVM methods; the "one against all" method achieve a slightly better recognition rate.

On the other hand, the proposed method achieved the best performance in terms of recognition time, being the fastest of them all.

6 Conclusions

In this paper, a new method for multi-class support vector machines, based on a pruning strategy, is proposed. It can be viewed as a variation of the "one against one" method.

The main idea behind this method is that it seeks for the class that will receive the greatest possible number of votes. So, when a test sample is submited to a binary classifier, the class that doesn't receive a vote is eliminated from future comparisons. This strategy leads to a binary search, which is known to be very fast.

Experimental results, performed on a isolated word, speaker independent, small vocabulary speech recognition problem, showed that the proposed method has performance similar to "one against one" and DAGSVM methods.

In terms of recognition rate, and can be twice as faster than the DAGSVM method and have a performance that is the same to the "one aginst one" method.

References

[1] Vapnic, Vladimir, *Estimation of dependencies based on empirical data [in Russian]*, Nauka, Moscow, Russia, 1979.

[2] Vladimir Vapnik, *Nature of Stastistical learning Theory*. Springer, New York, 1995.

[3] N. Cristiannini and J. Shawe-Taylor, *An Introduction to Support Vector Machines*. Cambridge University Press, Cambridge, U.K., 2000.

[4] V. Vapnik , *Stastistical learning theory*. John Wiley, New York, USA, 1998.
[5] Sandro T. Coelho, *Reconhecimento de fala usando Máquinas de Vetor de Suporte*. Master's thesis, Inatel, 2005.
[6] Christopher J. C. Burges, "A Tutorial on Support Vector Machines for Pattern Recognition," *Data Mining and Knowledge Discovery*, v. 2, no. 2, p. 121–167, 1998.
[7] Simon Haykin, *Redes neurais: princípios e prática*. Bookman, Porto Alegre, 1986.
[8] Knerr and L. Personnaz and G. Dreyfus , "Single-layer learning revisited: A stepwise procedure for building and training a neural network," *In F. Fogelman-Souli´e and J. H´erault, editors, Neurocomputing: Algorithms, Architectures and Applications*, p. 41–50, Springer-Verlag, 1990.
[9] U. Krebel, "Pairwise classification and Support Vector Machines", *In B. Sch¨olkopf, C. J. C. Burges, and A. J. Smola, editors, Advances in Kernel Methods Support Vector Learning*, p. 256–268, MIT Press, Cambridge, 1999.
[10] J.H.Friedman, "Another approach to polychotomous classification", *Technical report, Standford University, Dept.of Statistics*, 1996.
[11] C.-W. Hsu, C.-J.Lin , "A comparison of methods for multi-class support vector machines," *IEEE Transactions on Neural Networks*, v. 13, p. 415–425, 2002.
[12] C. Hsu and C. Lin, "A comparison of methods for multi-class support vector machines", *Technical report, Department of Computer Science and Information Engineering*, National Taiwan University, Taipei, Taiwan, 2001.
[13] Tom G. Dietterich and Ghulum Bakiri, "Error-correcting output codes: A general method for improving multiclass inductive learning programs," *In Proceedings of the Ninth National Conference on Artificial Intelligence*, p. 572–577, Anaheim, CA, 1991. AAAI Press.
[14] J. Platt and N. Cristianini and J. Shawe-Taylor, "Large margin DAGs for multiclass classification," *In S. A. Solla, T. K. Leen, and K.-R. Muller, editors, Advances in Neural Information Processing Systems.*, MIT Press, Cambridge, 2000.
[15] José Antônio Martins, *Avaliação de Diferentes Técnicas para o Reconhecimento de Fala*. PHD's thesis, UNICAMP, 1997.
[16] Davis, S. & Melmertstein, P. Comparison of parametric representations for monossylabic word recognition in continously spoken sentences. *IEEE Transactions on Acoustics, Speech and Signal Processing*, ASP-28(4):357–366. August, 1980.

Part IV
Models of Consciousness

Models of Consciousness
Ron Chrisley and Rob Clowes

Until recently, many believed the subjective nature of consciousness rendered it an unsuitable topic for scientific investigation at all, let alone the object of detailed computational modelling. But more and more, neuroscientists, psychologists, philosophers and artificial intelligence researchers have been devising methods for investigating the processes that are responsible for consciousness – the "what it's like" to be something. The six papers in this section vary in the way they approach this investigation, but all focus on modelling aspects of cognitive systems that they take to be essential to the presence and character of conscious, phenomenal states.

Much work in modelling consciousness has to do with modelling emotion and affect. Ricardo Sanz, Carlos Hernández, Jaime Gómez and Adolfo Hernando, in their paper entitled "A functional approach to emotion in autonomous systems", take an engineering approach to this task, asking what features of naturally conscious systems might be useful in the design of robust autonomous systems, particularly in the context of ASys, a framework based on an integrated control architecture. They propose that "emotional mechanisms may play a critical role in sustainment of function in changing environments." For them, the key point is that emotions provide a control broadcast infrastructure that is fast, global and externalisable. Going beyond superficial models of emotion, that see emotions as mechanisms for mere input handling, social affective coordination, or action selection, Sanz and his colleagues propose instead that emotions perform more sophisticated control-theoretic/computational functions, such as state-space reduction for better meta-control, and modulation of control structures at the component level, providing value-centric functional reorganisation.

Chella's contribution, "A robot architecture based on higher order perception loop", focuses on consciousness in the sense of *self*-consciousness. While basic perceptual experience involves a first-order loop from actual sensor values to anticipated sensor values to action to actual sensor values again, Chella argues that self-consciousness involves second-order perceptual loops from meta-perceptive sensor readings of first-order loops, to anticipations of the state of the first-order loop, to action, to meta-perceptive sensor values again. Like first-order loops, second-order loops may be either synchronic (allowing higher-order inferences about the current scene) or diachronic (anticipating changes in the first-order relation over time in response to changing scene). An example of a synchronic second-order loop is when a subject sees a stick placed in water; the first-order loop results in an experience of the stick as bent; the second order loop allows the subject to make sense of the difference between how the stick appears, and how it actually is. This reflective capacity differs from move conventional, symbolic architectures for reflection in that it is not strictly hierarchical: the second order loop typically alters the parameters of the first order loop. Chella gives a brief description of the implementation of his architecture on Panormo, a robot that guides visitors to the Botanical Gardens at the University of Palermo. The self-order loops allow Panormo to notice when its reliance on laser-based navigation is untrustworthy, allowing it to switch to visual-based navigation, and back.

Sulamita Frohlich and Carlos A. Franco's paper, "The consciousness circuit – An approach to the hard problem", concerns a hypothesis about how consciousness might be reduced to a functional property of the brain. Their paper considers a means of representing consciousness based on their Consciousness Orthogonal Graphic Model. Their approach develops out of a critical consideration of Lamme's (2004) model that proposed a two factor decomposition, where consciousness is considered a function composed of an attended/non-attended term and an unconscious/conscious term. While being sympathetic to the general approach, Frohlich and Franco point out that this unfortunately defines consciousness in two terms, one of which is itself consciousness. To improve on this situation they propose their own two term orthogonal scale in which the dimensions are now on the one hand, following Lamme, level of attention; and on the other, following Woolf & Hameroff (2001), a term which the authors call the quantum. This second term is developed with respect to the hypothesis that the level of consciousness depends on quantum activity in the microtubules, such that more activity signals greater awareness. Another way to think about this model is that the two dimensions are level of attention and something which can loosely be thought of as level of awareness (but defined in terms of quantum effects). This allows for the possibility that a subject could be highly aware of, but not attending to, a given mental content or visa versa. The authors develop an analysis of certain mental states based upon their graphic model and argue that this approach might lead in the future to new ways of measuring consciousness.

In his article, "Computational consciousness: building a self-preserving organism", Allan Barros asserts that the word 'consciousness' currently has no definition and therefore proposes an operational concept he calls 'computational consciousness'. Barros' idea is based upon Damasio's thought that consciousness is related to "the perception that an organism has of itself" (Damasio, 2000). Unlike Damasio the possible 'organisms' that Barros allows can regarded as computationally conscious can be as simple as a single simulated neuron with a number of connections to some internal function and some external sensors. He proposes: "consciousness arises when new information, which [is] important for the organism to self-preserve itself, [modifies] the current state of the whole or of part of the organism." Although this concept of self-preservation is not fully clarified, a computationally conscious organism is implicitly defined as one capable of learning given some internal state.

The contention is that "more complex organisms can be thought of using this model" and some attempt is made to relate the model to phenomena such as perceptual masking, attention and emotions.

L. Andrew Coward's contribution, "The hippocampal system as the cortical resource manager: A model connecting psychology, anatomy and physiology", is an extensive, in-depth model of how the hippocampus may be performing the function of regulating learning, or "information recording", within the cortex. Locating the model within a useful picture of the architecture of the brain as a whole, and relying on a wide range of physiological, anatomical and psychological data, Coward persuasively makes the case for a model of hippocampal function that parsimoniously does justice to these data. Although his initial focus is on the maintenance of cortical receptive fields, he shows how an account of such can be extended to phenomena more obviously related to higher conscious states, such as episodic and semantic memory, navigation, and the cognitive functions of sleep and dreaming.

The contribution by Marilda Spindola, Giovani Carra, Alexandre Balbinot and Milton A Zaro, "Cognitive measure on different profiles", aims at developing methodologies for using brain recording techniques to aid educationalists. With reference to a version of faculty psychology recently popularized by Pinker (1997; 2002) and especially Gardner's (1993) theory of multiple intelligences, they propose that future educative work might be enhanced by relating the various postulated specialized intelligence systems to an ERP based measure. The study examines two groups of participants (three subjects each); those who work in scientific-technological areas and those who work in social-humanist areas. For these groups different intelligence profiles are assumed. From this the authors find that the ERP readings for participants shown 3D and 2D graphic representations evince substantially different readings between the groups and thus that ERP might be used in helping further examination and / or registration of these differences. The authors hope this work might point the way toward new methodologies for examining learning and new instruments for studying learning types.

References

Damasio, A. R. (2000). *The Feeling of What Happens: body, emotion and the making of consciousness*: Vintage.

Gardner, H. (1993). Frames of Mind: The Theory of Multiple Intelligences. In (Second ed.).

Lamme, V. (2004). Separate neural definitions of visual consciousness and visual attention; a case for phenomenal awareness. Neural Networks, 17(5–6), 861–872.

Pinker, S. (1997). *How the mind works*: The Penguin Press, Penguin Book Ltd.

Pinker, S. (2002). *The Blank Slate*.

Woolf, N., & Hameroff, S. (2001). A quantum approach to visual consciousness. Trends in Cognitive Sciences, 5(11), 472–478.

A Functional Approach to Emotion in Autonomous Systems

Ricardo Sanz, Carlos Hernández, Jaime Gómez, and Adolfo Hernando

Abstract The construction of fully effective systems seems to pass through the proper exploitation of goal-centric self-evaluative capabilities that let the system teleologically self-manage. Emotions seem to provide this kind of functionality to biological systems and hence the interest in emotion for function sustainment in artificial systems performing in changing and uncertain environments; far beyond the media hullabaloo of displaying human-like emotion-laden faces in robots. This chapter provides a brief analysis of the scientific theories of emotion and presents an engineering approach for developing technology for robust autonomy by implementing functionality inspired in that of biological emotions.

1 Introduction

The central tenet of engineering research is the development of technology for achieving some desired level of performance in artificial systems: 220 volt in a wall socket, 240 k/m in a car, 80 Hz beat in a pacemaker, etc.

Once the development of the base technology—electrical engineering, mechanical engineering, embedded electronics—lets reach this performance level, a second aspect gains in importance: maintaining this performance. The maintenance of a certain level of performance is obvious in the pacemaker or wall socket, where pumping rate and voltage must be maintained at certain values; it may be less obvious in the car with its continuously varying road conditions, but is clear for the speed which in modern vehicles is to be maintained by the cruise control system. The need for performance keeping may be not so easy to pinpoint; e.g. for the brakes, where it is not so clear what is the variable to be kept within a certain range, although, if carefully analysed, you may find one: the braking power.

R. Sanz, C. Hernández, J. Gómez, and A. Hernando
Autonomous Systems Laboratory (ASLab-UPM), Universidad Politécnica de Madrid, José Gutiérrez Abascal 2 28045 Madrid
e-mail: ricardo.sanz@upm.es; carlos.hernandez@upm.es; jd.gomez@upm.es; adolfo.hernando@upm.es

A. Hussain et al. (eds.), *Brain Inspired Cognitive Systems 2008*, Advances in Experimental Medicine and Biology 657, DOI 10.1007/978-0-387-79100-5_14,

Fig. 1 The IST ICEA project is focused in the extraction of integration patterns among the cognitive, emotional and autonomic systems of the rat. These are evaluated on technical systems including physical and simulated rats

1.1 Sustainable Performance

The preservation of performance levels shall be understood in relation with the concepts of *resilience* and ultimately *robust autonomy*, which relates to the capability of systems to keep their proper functioning despite the uncertain and sometimes hampering and even hazardous dynamics of the environment where they perform. This idea of sustaining performance obviously maps in different ways to different kinds of applications that have more or less resilient structures in changing environments.

In our research of robust autonomy, we focus on two domains of artificial systems: large-scale industrial plants and mobile robotics. While in the case of industrial plants sustaining performance maps crystal clear to maintaining the production rate and keep it within a quality range, this mapping in a mobile robot scenario is far from evident.

In order to achieve *robust autonomy*, one approach could be to build systems physically robust, so as to minimise the effect of external disturbances from the environment. However this approach is always cost prohibitive—i.e. let us think of building a bulldozer-like field robot so that it does not need to avoid obstacles—not to say usually unrealisable—consider a thermodynamically isolated kiln with no need for a temperature control. Discarded this somewhat outdated "brute force" stratègy, engineers now must look for a more "intelligent" approach, where such a term does not only refers to smartness by their part, but also to the capability of the built systems to take advantage of information (Sanz et al. 2000).

To effectively and efficiently address the problem of robust autonomy, the material and energy flows from/to the environment ought to be accompanied by the corresponding flows of information that enable the system to manage the uncertainty in the operational conditions. Therefore artificial systems must build informational structures which implement relevant information about the environment, themselves and the interaction between both. Consequently, the systems are operationalised;

they can cope successfully with the inherently dynamic environment, maximising their effectivity when they pursue the performance goal they were designed for. The construction of maximally effective systems seems therefore to pass through the proper exploitation of goal–centric self-evaluative capabilities that let the systems teleologically self-manage.

In the context of the EU-funded project ICEA[1], this search for resilience is focused on the way that cognitive, emotional and autonomic subsystems are integrated into a single control architecture: that of the mammal brain. The special focus on emotions is due to the fact that emotions, conceived here as internal states and processes, seem to play this kind of role of valence–centric modification in biological systems. Several brain subsystems are being investigated in this direction, especially basal ganglia, amygdala and hippocampus, because of their involvement in basic emotional and cognitive processes: the hippocampus plays a key role in spatial cognition and memory, the amygdala is a main centre for emotion and basal ganglia are involved in valenced decision making.

1.2 Emotion for Engineering Sustainable Performance

Most research in applying emotion to technical systems has focused on the involvement of the display of emotions (Darwin, 1872) in social interaction and communication, and its application for the improvement of human-artificial systems interfaces (see Figure 2).

Fig. 2 iCat is the Philips research platform for studying human-robot interaction topics, intended to stimulate research in this area by building a research community through supporting a common hardware and software platform (from philips.com)

---

[1] www.iceaproject.eu

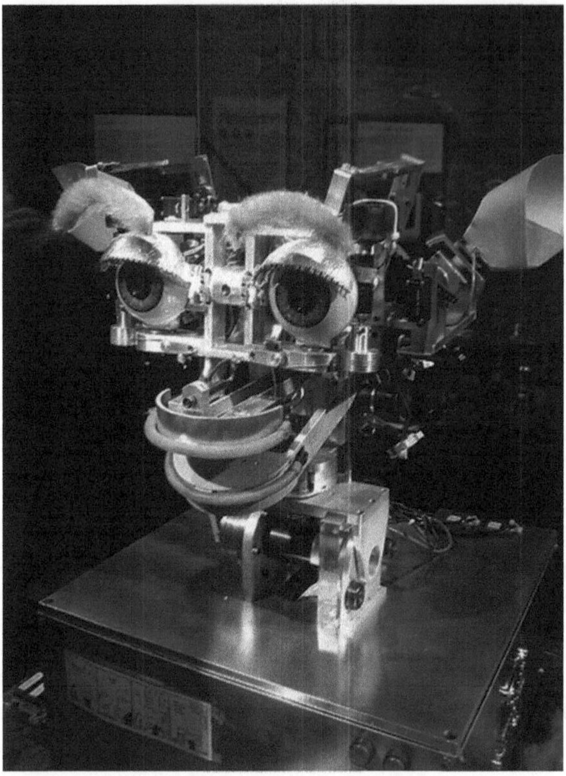

Fig. 3 The MIT robot Kismet is a paradigmatic example of research in the display of emotions by robots (Breazeal, 2000)

However, beyond the obvious capability for displaying human-like, emotion-laden faces in robots (see Figure 3), emotional mechanisms may play a *critical role in the sustainment of function in changing environments*. From a layman's understanding of this hypothesis, we could take the example of the fear one would experience upon discovering an intruder in one's home. That strong emotion elicits a different state in humans: we enhance our attention to sensory systems (i.e. hearing), the discharge of adrenaline prepares our motor system for faster responses, increases our heart rate, etc. Everything is aimed towards the maintenance of a high-level function which we could label as "survival". From a more formal, scientific perspective, it is commonly agreed that biological emotions, as considered in the previous example, are an evolved mechanism for adaptation related to a certain appraisal of internal and external events (Botelho, 2001). This appraisal seems to assign a certain value to stimuli, external or internal, related to the goal tree of the system (i.e. survival, reproduce, search food, avoid predators) and helps reconfigure the system accordingly. Let us take a National Geographic example. When in the mating epoch, a gazelle may feel hungry. That feeling (a bodily emotion) helps reconfiguring its behaviour toward a goal that has gain priority. Going further with this example, suppose that, when searching for food, the gazelle hears the possible presence of a lion or other predator: now the emotion of fear enhances her sensory system, to confirm the presence of the predator and locate it, and shifts her brain into survival mode after the decision-reconfiguration typically called *fight or flight*.

This description of core functionality of emotions in biological systems parallels the functional needs in artificial systems—presented in the previous section—in the pursue of robustness and sustainable performance. In this chapter we will analyse some of the most relevant theories on emotions and coalesce them into a theoretical framework—the *ASys Framework*—for addressing the technical issues of building similar mechanisms into artificial systems to achieve greater levels of resilience and performance.

2 A Review of Emotion

A common, somewhat folk, theory of human emotions say that they are, to a large extent, subjective and non deterministic. This accounts for the obvious fact that apparently identical stimuli may raise different emotions in different humans, and the same individual may experiment different emotions in response to similar stimuli. Obviously, from a purely systemic perspective, there must be some disparity between systems if the behaviour differs, but the ascription of the variety in emotional response by dilettantes have gone from subtle details in otherwise apparently identical stimulus to the very possibility of human freedom and non-determinism.

Starting at the last decades of 19th century, with the work of William James (James, 1884), and through the 20th century, the scientific community has managed to turn the ancestral and pre-scientific ideas about emotions into theory-laden models, were the sustaining theories formal or not. Nowadays emotions are studied just as another natural phenomenon of living systems, such as digestion or homeostasis, and, despite the diversity of approaches and theories, there is a consensus that emotions are an evolved adaptivity mechanism related to situation assessment and decision making (Panksepp, 1998).

Some researchers (Ventura and Pinto-Ferreira, 1999) have also pointed out a division regarding the scientific analysis of emotion which comprises an external, social perspective of emotion as it is involved in communication between individuals through the display by them of emotional states and attitudes, and on the other hand a internal point of view considering how emotion is involved in decision making processes.

We can summarise the usual understanding of emotions into several related aspects (Bermejo Alonso, 2006):

A1 – how emotional behaviour is *triggered* by event surrounding the agent;
A2 – how emotion is *manifested* (displayed) by/within the agent; and
A3 – how emotion is *felt* by the agent.

This heterogeneity notwithstanding, it is necessary to think about basic physiological principles going down to neural-hormonal mechanisms that make a particular event "emotional". Through this section we will summarily analyse the scientific principles and theories about emotions abstracted so far.

2.1 *Classical Models*

The classical theories of James-Lange or Cannon-Bard address the causality of the relation between A1-A2-A3. The model of James-Lange states that in animals the triggering (A1) are experiences in the world, the autonomic nervous system then creates physiological events such as increased heart rate or muscular tension (A2), and then emotions come as conscious feelings (A3) which come about as a result of these physiological changes (rather than being their cause as the Cannon-Bard model postulates).

Plenty of models can be found in the literature among which we would like to distinguish—for their closeness to the ASys emotion model—the model of Arnold (1960), due to its relation with the shaping of action tendencies; the model of Frijda (1987), due to its perspective of emotions constituting forms of action readiness; and the models of Plutchik and Kellerman (1980) and James (1884) in relation with the bodily basis of adaptive mechanisms.

2.2 *Damasio's Model*

When relating emotion to the rest of the evolved adaptive mechanisms in humans, from the hormonal system to the more cognitive ones such as consciousness, one of the most complete models is that of Damasio. The somatic marker hypothesis (Damasio, 1999) and related machinery can be used to provide a deeper understanding of emotional system organisation (see Figure 4), and accounts for all A1-A2-A3, with special relevance for the last two aspects. This structure is among the architectures explored in the beforehand mentioned ICEA project in order to provide a coherent picture of the integration of cognitive, emotional and autonomic aspects in mammals.

Damasio's is a concrete proposal in line with what is needed from a scientific and technical approach to emotion and cognition (Ortony et al. 1988). Architectural approaches go beyond the three behavioural and phenomenal aspects mentioned before (triggering, display and feeling) addressing what are the physiological mechanics for all this functioning.

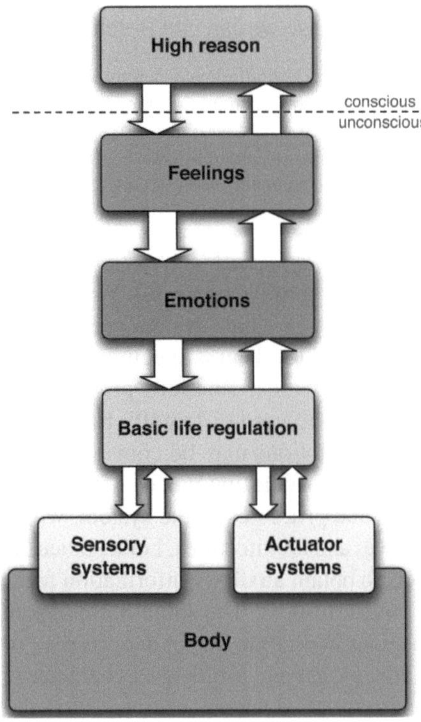

Fig. 4 Damasio proposes a hierarchical process for emotion–raising distinguishing between basic autonomic mechanisms, emotion, feeling and conscious awareness of emotion

3 Assessing Models

Most data concerning emotion comes from experimentation on animals and humans. Data coming from these sources are widely heterogeneous; from single neuron firing patterns, columnar behaviours or activation levels of a whole brain area to hormone concentrations or verbally reported psychological data. This heterogeneity in the level of resolution and abstraction of the data is surely a factor for the difficulty of building up a unified theory of emotion which could address such a broad variety of data. For example, the neurophysiological and hormonal response to fear in mammals is quite well known (Fendt and Fanselow, 1999) together with the behavioural response of rats in experiments of fear conditioning (Hatfield et al. 1996). However, there still remains missing a unified theory covering the gap between the low level, populated with neurons and hormonal mechanisms, and the higher levels of elicited behavioral responses.

Some theoretical models of emotion and associated computational implementations are being explored as a promising tool for integrative understanding of this emotive-cognitive mechanisms. The main value this approach offers is the possibility of having a precise, more rigourous methodology to grasp the core concepts and architecture.

In this sense we can cite Botelho (2001):

> We present a preliminary definition and theory of artificial emotion viewed as a sequential process comprising the appraisal of the agent global state, the generation of an emotion-signal, and an emotion-response. This theory distinguishes cognitive from affective appraisal on an architecture-grounded basis. Affective appraisal is performed by the affective component of the architecture; cognitive appraisal is performed by its cognitive component.

Emotions *affect all levels of operation in a system*, from basic life regulation to conscious, cognitive processes. We use the term *transversal* to indicate this fact. The concrete way in which system operation is affected is specific to each level. Within each of these levels, it is specific to each organ, component and process.

In other words, emotions provide a common *control broadcast* infrastructure which may be used differently by each of the processes in the system. In natural systems, emotions may be conveyed by neural firings and hormones (i.e. the bodily signal broadcasting mechanisms). These mechanisms must be shared by many organs and processes in the system, which will interpret signals according to their purposes and architectures. For instance, a cognitive process may interpret hormonal levels to obtain auxiliary information for making a decision regarding what the system must do next. The same hormones may be interpreted concurrently by other processes in order to detect danger, risk, or a need to obtain food, for example.

This global and multi-level character of emotions explains some distinctions of emotion-relative phenomena present in the literature, such as Damasio's (Damasio, 1999, 2004):

- state of emotion,
- state of feeling an emotion,
- state of a feeling of an emotion made conscious.

One particular way in which emotions are transversal is by broadcasting a summarised picture of the system state to many of its components and processes. This means not only a summary of how its components find themselves, but also a certain sense of affordance of the current scenario relative to the current system situation, processes and objectives.

This is useful to the system in order to adapt to its scenario of operation, mainly for three reasons:

- Emotions are fast, and are available before other more cognitive information.
- Emotions, being to some extent global, contribute to co-ordination and focus of large quantities of system processes and components, which is a factor for preserving system cohesion (López, 2007).
- Emotions can be externalised and hence used for behavioural organisation (co-operation and competition are examples) in multi agent environments (societal behaviour being the clearest example).

3.1 *The 'Emotional' Facial Mimicking*

This last aspect, that of externalisation of emotional states, has rendered emotional expression one of the main topics of emotion research (Figure 5) (Darwin, 1872; Ekman, 1982).

There have been plenty of efforts made towards the implementation of emotion in machines as inspired by biosystems (Trapl et al. 2002) but in most of the cases they have neglected addressing the core issues and have instead just focused on mimicking shallow, observable manifestations of emotion (e.g. making robot faces *a la* Ekman). But this work, outside the psychological arena, is irrelevant in theoretical and operational terms. All that is expected is some improvement in social capability by facial displaying for human emotions. This is hopeless because the functional value of the display of an emotional state in a social interaction is based on the

Fig. 5 A big amount of research on emotion has been focused on the expression of facial emotion neglecting the inner functional aspects of it

activation in the receptor of the display of a behavioural model of the displayer so as to maximise effectiveness of interaction—it is more a question of better exploiting the human capabilities than of concrete robot competencies.

Mimicking faces is thus useless unless the operational state of the displayer is what is captured in the model going to be activated in the receptor. Clearly this is not the case of human vs. robot architectures (i.e. the mental model of the receptor will be a model of a human whereas the robot is not a human at all from an architectural nor functional point of view). This issue has been widely addressed in the field of human-computer interaction and the mental models community (Gentner and Stevens, 1983). What is important then is the raising of mental states in the receptor (i.e. the activation of mental models) that are relevant for the interaction. This can only be done if emotion is tightly tied to the inner operational mechanisms of the agent displaying the emotion (Conde, 2005).

To conclude with this subject, we shall summarise that facial expression of emotion is an externalisation of an emotional state to help reconfigure a multi-agent organisation taking into account individual agent operational states.

3.2 *What Emotions Are and Are Not: Ideas to Solve the Puzzle*

After this succinct analysis, the puzzle of emotion, from an engineering point of view, can be reduced to three core aspects:

Three questions:

- **What is the function that emotional mechanisms do play?**
- **What is the general form of an emotional mechanism?**
- **What is the best strategy for emotion implementation?**

The analysis of the several models of emotion produce some conclusions regarding *what emotions are not*:

- Emotions are not just sophisticated input handling, i.e. *not just reacting to bears.*
- Emotions are not just sophisticated action generation for social affective behaviour, i.e. *not just showing embarrassment.*
- Emotions are not just mechanisms for re-goaling, i.e. *not just deciding to change from eating to doing sex.*

A deeper analysis abstracting from the biological mechanics into the functional structure renders some conclusions about *how emotions work*:

- Emotions do generate *synthetic compact states* (performing state space reduction) for the effective tuning and use of evolutionary meta-controllers.

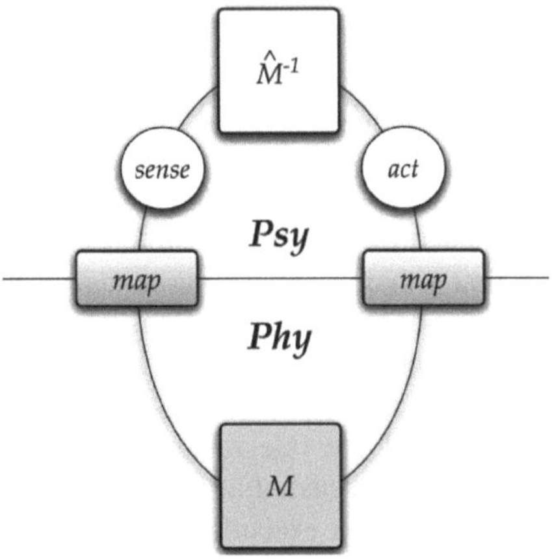

Fig. 6 The mind as model-based controller vision is a central concept in the ASys Framework. It helps bind the dynamics of the mental (Psy) to the dynamics of the physical (Phy). This should not be understood as a dualistic position, it is not. The mind itself is part of the phenomena of the physical (Landauer, 1992) and we just focus on two aspects of this physicality, stressing the informational nature of minds

- Emotions do change the *control structures* at the component functional level (patterns and roles) of subsystems.
- Emotions operate in a global controller configuration approach rendering a *transversal structural feedback* architecture.

In the following section we will develop these three core ideas about emotions in the context of a technical framework intended for the engineering of maximally autonomous systems by applying bioinspired functional concepts.

4 The ASys Framework

The ASLab ASys Project is a long-term research project focused in the development of technology for the construction of autonomous systems. What makes ASys different from other projects in this field is the extremely ambitious objective of addressing *all the domain of autonomy*. We capture this purpose in the motto "engineering any-x autonomous systems". The *ASys Framework* is both a theoretical framework for understanding all the relevant issues and a software-intensive technological framework that enables the technically sound creation of autonomous systems, where autonomy is understood in its broadest sense and not in the severely restricted sense of the term *autonomous intelligent systems* that is usually equated to *mobile reactive robots*.

One of the central topics in the ASys Framework is the pervasive model-based approach. A truly autonomous system will be continuously using models to perform its activity. An ASys system will be built using models of it. An ASys can exploit its own very models for driving its behaviour. Model-based engineering and model-based behaviour then merge into a single phenomenon: *model-based autonomy*. We equate this conceptualisation with *cognition*.

The ASys Framework hence establishes that a system is said to be cognitive if it exploits models of other systems in its interaction with them. Models and knowledge are then equated and the ASys Framework provides a link between the ontological and epistemological aspects of mind.

On the technical side, the ASys Framework follows a principled approach to autonomous system mind construction, the *cognition as model-based behaviour* being the first principle, so as to ground a systematic engineering approach that shall end in rendering machine consciousness (Sanz et al. 2007).

These principles establish guidelines for the systematic, formally grounded development of a real-time control framework based on the control and software principles of the Integrated Control Architecture (Sanz et al. 1999), which will be furtherly discussed at the end of this section. This will render a methodology, a toolset and an execution framework for the engineering of robust autonomous systems based on the implementation of cognitive mechanisms up to the level of consciousness (Sanz et al. 2005).

4.1 Emotion, Consciousness and Control in ASys

The mechanisms of emotion impinge on the behavioural capability of the agent so as to prepare it for future action. This makes emotion a core capability for sophisticated self-management control architecture where outer control loops (emotion) determine the functioning of inner control loops (homeostasis) so as to maximise survivability. Damasio's model on consciousness lays out another control loop atop of these two (see Figure 4) rendering a high-level reasoning capability. Emotions realise meta-controllers.

As was the case with emotions, there are plenty of models of consciousness that try to address the relation of physiology and the three core aspects of consciousness: world-awareness, self-awareness and qualia. We can distinguish as maximally relevant for our work, due to their abstract, general nature, the Global Workspace model of Baars (1997) and the information integration model of Tononi (2004).

The integrated control model of consciousness (Sanz et al. 2007), also part of the ASys Framework, is based on the provision of self-awareness by means of model-based perceptual mechanics.

The ASys perspective on cognition/emotion goes beyond Damasio's approach of putting emotions/feelings as additional layers in hierarchical controllers. Emotion is no longer another layer in the architecture but a transversal mechanism that crosses across all layers. This is indeed a well known fact in the studies of emotion. Emotions do appear from the subconscious plane to the conscious surface, affecting all levels in the cognitive structure, from the physiological up to the cognitive, social, self-conscious level.

This implies (see Figure 7) that emotional mechanics are part of each level of the control hierarchy. The level of focus of the analysis is what determines the

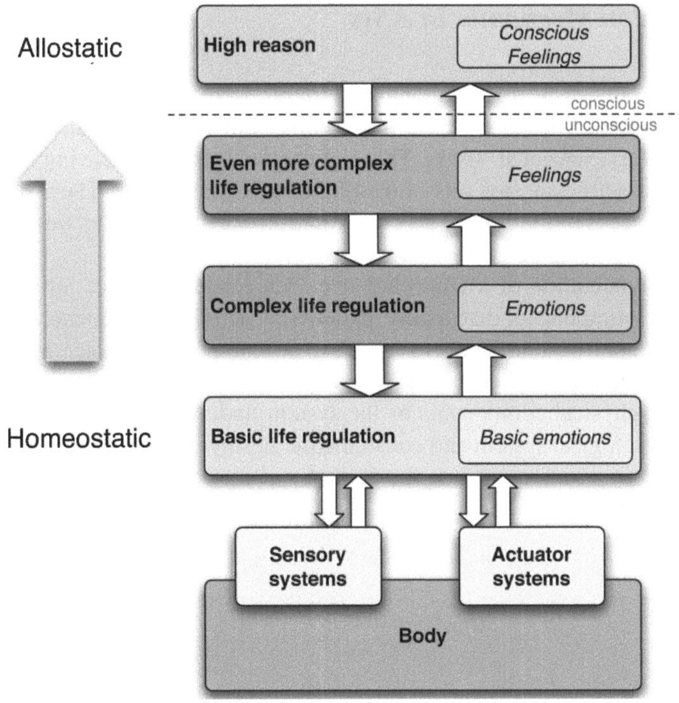

Fig. 7 Damasio's layering of emotion appears as labelling of the transversal emotional mechanisms across a layered architecture for control

labelling used for this mechanism: basic emotion, emotion, feeling, conscious feeling, etc. In the language of information technology we would say that emotion is an *aspect*—in the computer science interpretation (Filman et al. 2004)—of the different systems that constitute the body and mind of an autonomous agent. From a functional perspective we can also observe that the goals pursued by such control structures go from the purely homeostatic mechanisms for life survival to the higher-level, socially-originated allostatic mechanisms for social behaviour. From hunger to embarrassment, emotions do share the meta-control capabilities over basic behavioural structures.

Artificial implementations of emotions are not developed yet to the same degree as the natural. However, large, distributed, fault-tolerant systems include mechanisms which already play a similar role (Aström et al. 2001). Fault detection, damage confinement, error recovery and fault treatment are based on broadcast messages and other mechanisms shared and used by system components in analogous ways to the natural counterparts.

4.2 Emotion Mechanics in ASys

The ASys Framework for autonomous systems is based on an architecture for software-intensive, distributed, real-time control called the Integrated Control Architecture (ICa) (Sanz et al. 1999). This architecture is based on the implementation of patterns of activity across sets of distributed real-time agents. These patterns respond to the needs of the control task that can follow a multilayered, multi-objective control strategy (Alarcon et al. 1994).

The implementation of a controller over ICa renders a collection of interacting software components that realise patterns of activity as sequences of service requests. The software component model is of extreme importance in the implementation of such controllers because it provides a common modelling framework for both the physical components of the system under control, which are but the organs in a biological system and constitute the fleshly infrastructure, and the mental components, which constitute the control superstructure. Figure 8 shows three such components in a simple, layered control structure: an organ, a controller and a meta-controller.

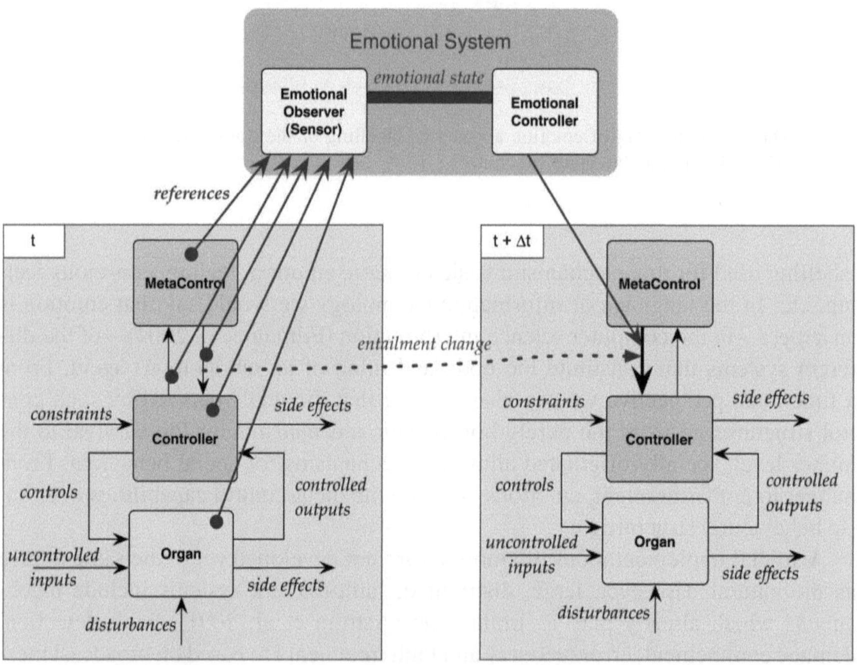

Fig. 8 The figure depicts how an emotional system following these ideas would perform reconfiguration in an modularised control architecture. The emotional system changes the functional organisation to adapt it to new operating conditions

An emotional system incorporated to ICa, according to the previous conceptualisation of emotion in the ASys Framework, would provide the structural mechanisms for control pattern adaptation to the current state of affairs. Examples of this kind of architecture are already available in the realm of control systems, for example the previously mentioned fault-tolerant controllers and sliding mode controllers.

The primary effect of the emotional system is the change in the functional organisation of the control system of the body. In Figure 8 the emotional system changes the functional organisation from time t to $t + \Delta t$, concretely in this example the output of the meta-controller to the controller—e.g. the goal or reference, in control jargon—. Both the emotional observation and control are done in terms of a value system for the agent. This happens in a multi-scale, multilayer organisation that constitutes the integrated global controller of the agent.

Now we can provide the ASys Framework answers to the three core aspects of emotion mentioned before:

Three answers:

- *What is the general form of an emotional mechanism?*
  A self-reorganising meta-controller.
- *What is the function that emotional mechanisms do play?*
  Provide value-centric functional reorganisation.
- *What is the best strategy for emotion implementation?*
  Functional modularisation of control functions and integration over a common infrastructure.

5 Summary

The chapter has reviewed some of the common approaches to emotion understanding, with an special emphasis on Damasio's model of emotion and feeling.

It has also been analysed the extended functional role that emotions can play in complex adaptive controllers and how the different aspects of emotion—triggering, emotional states, bodily effect—are addressed from this perspective.

This understanding has been put in the context of the ASys Framework, a theoretical and technical framework for the implementation of autonomous systems. This framework is based on the construction of modular, component–based control systems following the architectural guidelines of the Integrated Control Architecture (ICa)—a software architecture based on distributed real-time objects.

This framework is being applied to the modelling and understanding of autonomic-emotional-cognitive integration aspects in the rat brain and the implementation of embedded controllers in the context of the IST ICEA project.

Acknowledgements Authors would like to acknowledge the support coming from the European Community's *Seventh Framework Programme FP6/2004-2007* under grant agreement IST 027819 ICEA—*Integrating Cognition, Emotion and Autonomy*—and the Spanish *Plan Nacional de I+D* under grant agreement DPI-2006-11798 C3—*Control Cognitivo Consciente*.

References

Alarcon I, Rodriguez-Marin P, Almeida L, Sanz R, Fontaine L, Gomez P, Alaman X, Nordin P, Bejder H, de Pablo E (1994) Heterogeneous integration architecture for intelligent control systems. Intelligent Systems Engineering 3(3):138–152

Arnold M (1960) Emotions and Personality. Cambridge University Press

Aström K, Albertos P, Blanke M, Isidori A, Schaufelberger W, Sanz R (eds) (2001) Control of Complex Systems. Springer

Baars BJ (1997) In the theatre of consciousness. global workspace theory, a rigorous scientific theory of consciousness. Journal of Consciousness Studies 4:292–309

Bermejo Alonso J (2006) A state of the art on emotion. Tech. Rep. R-2006-002, UPM Autonomous Systems Laboratory, Universidad Politécnica de Madrid

Botelho L (2001) Machinery for artificial emotions. Cybernetics and Systems: An International Journal 32:465–506

Breazeal C (2000) Sociable machines: Expressive social exchange between humans and robots. PhD thesis, Department of Electrical Engineering and Computer Science, MIT

Conde RP (2005) Mapeo facial de emociones sintéticas. Master's thesis, Universidad Politécnica de Madrid

Damasio A (1999) The feeling of what happens: body and emotion in the making of consciousness. Harcourt Press, New York

Damasio A (2004) Emotions and feelings: a neurological perspective. In: Manstead A, Fridja N, Fischer A (eds) Feelings and emotions, Cambridge University Press, chap 4, pp 49–57

Darwin C (ed) (1872) The expression of the emotions in man and animals. D. Appleton and Company

Ekman P (ed) (1982) Emotion in the human face. Cambridge University Press

Fendt M, Fanselow MS (1999) The neuroanatomical and neurochemical basis of conditioned fear. Neuroscience & Biobehavioral Reviews 23(5):743–760

Filman RE, Elrad T, Clarke S, Aksit M (eds) (2004) Aspect Oriented Software Development. Addison Wesley

Frijda N (1987) The emotions. Cambridge University Press

Gentner D, Stevens AL (eds) (1983) Mental models. Lawrence Erlbaum Associates, Hillsdale, NJ

Hatfield T, Han JS, Conley M, Gallagher M, Holland P (1996) Neurotoxic lesions of basolateral, but not central, amygdala interfere with pavlovian second-order conditioning and reinforcer devaluation effects. J Neurosci 16(16):5256–5265

James W (1884) What is an emotion'. Mind 9:188–205

Landauer R (1992) Information is physical. In: Workshop on Physics and Computation, PhysComp '92., pp 1–4

López I (2007) A framework for perception in autonomous systems. PhD thesis, Departamento de Automática, Universidad Politécnica de Madrid

Ortony A, Clore GL, Collins A (1988) The Cognitive Structure of Emotions. Cambridge University Press

Panksepp J (1998) Affective neuroscience: the foundations of human and animal emotions. Oxford University Press, New York, URL http://www.loc.gov/catdir/enhancements/fy0635/98015955-d.html

Plutchik R, Kellerman H (eds) (1980) Emotion: Theory, research and experience. Volume I: Theories of emotion. Academic Press, New York

Sanz R, Matía F, Puente EA (1999) The ICa approach to intelligent autonomous systems. In: Tzafestas S (ed) Advances in Autonomous Intelligent Systems, Microprocessor-Based and Intelligent Systems Engineering, Kluwer Academic Publishers, Dordretch, NL, chap 4, pp 71–92

Sanz R, Matía F, Galán S (2000) Fridges, elephants and the meaning of autonomy and intelligence. In: IEEE International Symposium on Intelligent Control, ISIC'2000, Patras, Greece

Sanz R, López I, Bermejo-Alonso J, Chinchilla R, Conde R (2005) Self-X: The control within. In: Proceedings of IFAC World Congress 2005

Sanz R, López I, Rodríguez M, Hernández C (2007) Principles for consciousness in integrated cognitive control. Neural Networks 20(9):938–946, DOI http://dx.doi.org/10.1016/j.neunet.2007.09.012

Tononi G (2004) An information integration theory of consciousness. BMC Neuroscience 5:42

Trapl R, Petta P, Payr S (eds) (2002) Emotion in Humans and Artifacts. MIT Press

Ventura R, Pinto-Ferreira C (1999) Emotion–based agents: three approaches to implmentation. In: Velasquez JD (ed) Workshop on Emotion–based agent architectures, Seattle, U.S.A.

A Robot Architecture Based on Higher Order Perception Loop

Antonio Chella

Abstract The paper discusses the self-consciousness of a robot as based on higher order perceptions of the robot itself. In this sense, the first order perceptions of the robot are the immediate perceptions of the outer world of the robot, while higher order perceptions are the robot perceptions of its own inner world. The resulting architecture based on higher order perceptions has been implemented and tested in a project regarding a robotic touristic guide acting in the Botanical Garden of the University of Palermo.

Keywords Machine consciousness · Robotics · Perceptions

1 Introduction

One of the major topics towards robot consciousness is to give a robot the capabilities of *self-consciousness*, i.e., to reflect about itself, its own perceptions and actions during its operating life. The robot *self* grows up from the content of the agent perceptions, recalls, actions, reflections and so on in a coherent life long *narrative*.

A robot system, able to build an internal model of the environment and to generate suitable predictions, has been proposed by Holland and Goodman [15]. The system is based on a neural network that controls a Khepera minirobot and it is able to build a model of environment and to simulate perceptual activities in a simple environment. Following the same principles, Holland et al. [16] presented the robot *CRONOS*, a very complex *anthropomimetic* robot whose operations are controlled by the program *SIMNOS*, a 3D simulator of the robot itself and its environment.

Haikonen [11, 12] proposed a feedback loop in which the model of the environment is implicitly learned in terms of weights of an associative neural network. The loop is in turn the basic block of the *Haikonen cognitive architecture* for robot brains.

A. Chella (✉)
Dipartimento di Ingegneria Informatica, Università di Palermo, Viale delle Scienze, I-90128 Palermo, Italy
e-mail: chella@unipa.it

A. Hussain et al. (eds.), *Brain Inspired Cognitive Systems 2008*, Advances in Experimental Medicine and Biology 657, DOI 10.1007/978-0-387-79100-5_15,

Hesslow [13], from the standpoint of neuroscience, discussed in details the role of inner simulations in relations with sensorimotor and cognitive functions. A similar approach has been proposed by Grush [10] and inspired to the Kalman filter as a model for perception process.

A seminal theoretical founded attempt to give self reflection capabilities to an artificial reasoning system is described by Weyhrauch [30]. Weyhrauch proposed the *First Order Logic (FOL)* system, able to perform logic inferences and to reflect about its own inferences.

McCarthy [22] stressed the fact that a robot needs the ability to observe its own mental states. He proposed the *mental* situation calculus as a formalism that extended the situation calculus in order to represent *mental* situations and actions. Minsky [24] described a system based on several interacting agents at different levels, in which the tasks of higher levels agents is the *self-reflective* processing. Sloman and Chrisley [29] followed a similar approach in the design of the *Cognition and Affect (CogAff)* architecture.

McDermott [23] made a distinction between *normal access* to the output of a computational module and and *introspective access* to the same module. The first one is related with the output of the processing algorithms of the module, while the second one is related with the higher-order access inside of the processing module according to self model. According to McDermott, this second modality is at the basis of phenomenal consciousness, i.e., consciousness arises through the use of a self model.

The relationships between higher-order access and phenomenology has been longly debated in the literature about higher-order theories of consciousness, see e.g. [5, 27]. Aleksander and Morton [3] discussed the relationships between their kernel architecture based on the five axioms of consciousness, and the higher-order theories of consciousness.

From the neuroscience point of view, Rao and Ballard [25, 26] proposed a neural model of the visual cortex based on a hierarchy of neural networks. Their model combines bottom-up signals coming from input with top-down expectations coming from the higher levels.

From a control theory point of view, Sanz and colleagues [28] proposed functional principles for the design of suitable hierarchical control systems towards an effective and functional conscious machine.

In this paper, a model of robot self-consciousness is proposed and based on higher order perceptions of the robot during time, in the sense that first order robot perceptions are the immediate perceptions of the outer world of a self reflective agent, while higher order perceptions are the perceptions during time of the inner world of the agent.

The first order robot perception loop, at the basis of the proposed robot architecture, is based on tight interactions between the robot brain, body and environment. To model higher order perceptions in self reflective agents, we introduce the notion of *second-order* perception loop. According to the proposed model, a second order perception loop describes the perception of the agent at a previous time of the robot. In this sense, while the object of perceptions for the first order perception loop is the external world out there, the object of perception for the second-order loop is the first order perception loop itself.

The architecture has been tested on an effective robot architecture implemented on *Robotanic* [7], an operating outdoor autonomous robot *MobileRobots* Pioneer 3-AT using differential drive and equipped with a laser scan range finder, a sonar array, a stereo camera and a Global Positioning System.

2 First Order Perception Loop

The first order perception loop model is described in Fig. 1; see Chella [6] for a detailed description. The loop takes into account the schema proposed by Hesslow and by Grush.

The robot vision system receives in input the robot position, speed and so on from the proprioceptive sensors and it generates the scene anticipations, i.e., the expectations about the perceived scene. The perception loop is then closed by the perceptive sensors that acquire the effective scene by means of the video camera.

Macaluso et al. [9,21] describe a robot based on the perception loop in which the process of scene anticipations is performed by a 3D computer graphics simulator. The simulator generates the expected 2D image scene on the basis of the robot movements.

In the proposed model, the mapping between the anticipated and the perceived scene is achieved through a *focus of attention* mechanism implemented by means of *attractor neural networks* with delayed connections. A sequential attentive mechanism is hypothesized that suitably scans the perceived scene and, according to the hypotheses generated on the basis of the anticipation mechanism, it predicts and detects the interesting events occurring in the scene [8]. Hence, starting from the incoming information, such a mechanism generates expectations and it makes contexts in which hypotheses may be verified and, if necessary, adjusted.

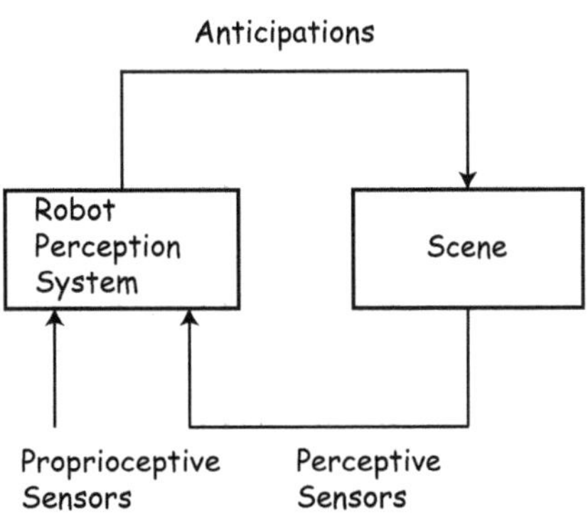

Fig. 1 The basic perception loop

The focus of attention mechanism selects the relevant aspects of the acquired scene by sequentially scanning the image from the perceptive sensors and by comparing them in the generated anticipated scene. The attention mechanism is crucial in determining which portions of the acquired scene match with the generated anticipation scene: not all true (and possibly useless) matches are considered, but only those that are judged to be relevant on the basis of the attentive process.

The match of a certain part of the acquired scene with the anticipated one in a certain situation will elicit the anticipation of other parts of the same scene in the current situation. In this case, the mechanism seeks for the corresponding scene parts in the current anticipated scene. We call this type of anticipation *synchronic* because it refers to the same situation scene.

The recognition of certain scene parts could also elicit the anticipation of evolutions of the arrangements of parts in the scene; i.e., the mechanism generates the expectations for other scene parts in subsequent anticipated situation scenes. We call this anticipation *diachronic*, in the sense that it involves subsequent configurations of image scenes. It should be noted that diachronic anticipations can be related with a situation perceived as the precondition of an action, and the corresponding situation expected as the effect of the action itself. In this way diachronic anticipations can prefigure the situation resulting as the outcome of a robot action.

Two main sources of anticipation are taken into account. On the one side, anticipations are generated on the basis of the structural information stored in the robot by design. We call *phylogenetic* these kind of anticipations. On the other side, anticipations could also be generated by a purely Hebbian association between situations learned during the robot operations. We call *ontogenetic* this kind of anticipations. Both modalities contribute to the robot conscious perception process.

Ontogenetic anticipations are acquired by *online* learning and *offline* learning. During the normal robot operations, when something unexpected happens, i.e., when the generated anticipation image scene does not match the scene acquired by the perceptive sensors, the robot vision system learns to associate, by an Hebbian mechanism, the current image scene with the new anticipation image through the previously described attentional mechanism (Fig. 2).

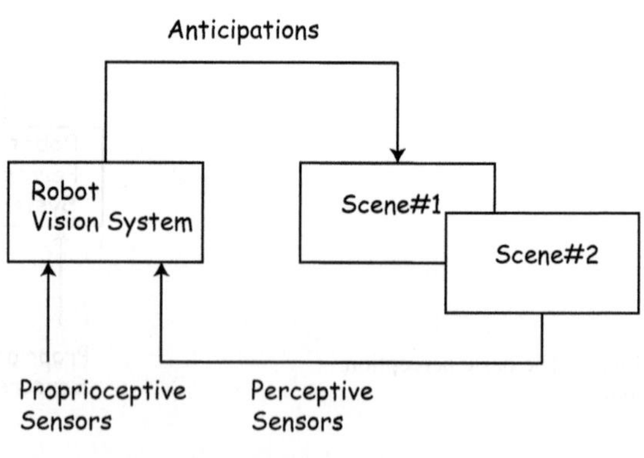

Fig. 2 Online learning of the perception loop

Fig. 3 Offline learning of the perception loop

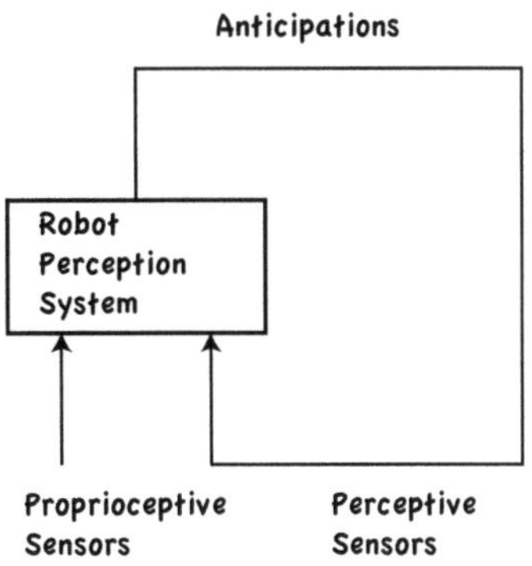

In the offline learning, the perception loop is employed to allow the robot to imagine future sequences of actions to generate and learn novel anticipations. The signal from perceptive sensors is related to the perception of a situation of the world out there. In this mode, the robot vision system freely generates anticipations of the perceptive sensors, i.e., it freely imagines possible evolutions of scenes and therefore possible interactions of the robot with the external world, without referring to a current external scene. In this way, new anticipations or new combinations of anticipations may be found and learned offline by the robot itself through the synchronic and diachronic attentional mechanisms (Fig. 3).

3 Neural Networks Implementation of the Perception Loop

The perception loop generates the anticipations of the scene by means of the focus of attention mechanism that produces a sequence p of the parts of the expected scene:

$$p = \{k_1, k_2, \ldots, k_l\} \tag{1}$$

where each k_i is a part of the expected scene described by means of suitable graphics primitives. For example, Chella et al. [8] adopted *superquadrics* as 3D primitives describing the parts of the objects present in the scene. The part of the scene k_i may be viewed as a point attractor of an energy function. In this way, a sequence of fixed point attractors models the scene: starting from an initial state representing a part of the scene imposed, for instance, from the external input, the system state trajectory is attracted in turn to the nearest stored parts of the scene.

The implementation of the focus of attention mechanism by means of an *attractor neural network* [4, 17] characterized by the corresponding energy function, appears to be a natural choice: each part of a scene is an activation pattern learned by the network. The implementation of the sequence of scanned parts of the scene is built by means of time delayed connections that learn the corresponding temporal sequences of parts [18, 19]. This modification allows the attractor neural network both to recognize and to generate all the anticipated parts of the scene.

The choice of time-delay attractor neural networks offers several advantages. It is based on the well-studied energetic approach; the learning phase is fast, since it is performed at "one shot". Furthermore, it allows for a uniform treatment of both the recognition and the generation of the parts of the scene.

For the sake of simplicity, we adopted the *binary unit* version of the attractor neural network; the coding of the parts of the scene in terms of the binary activation pattern of the network has been computed by a *coarse coding* algorithm [14].

The general expression of the energy function of an attractor neural network is:

$$E_1(t) = -\sum_{i=1}^{m}\sum_{j=1}^{m} T_{ij}k_i(t)k_j(t) \text{ with } j \neq i \tag{2}$$

where m is the number of binary units of the network, $\mathbf{T}$ is the connection matrix storing the attractors representing the anticipated parts of the scene, and $k(t)$ is the part representing the current activation pattern of the network. The number m of units depends on the number l of parts of the scene according to the *low memory load* condition [4]:

$$l < \alpha_c m \tag{3}$$

where $\alpha_c \simeq 0.3$. The connection matrix $\mathbf{T}$ is:

$$T_{ij} = \frac{1}{m}\sum_{\nu=1}^{l} k_{\nu_i}k_{\nu_j} \text{ with } j \neq i \tag{4}$$

where k_ν is the ν-th object part.

As previously stated, in order to generate anticipations and recognize a scene made up by several objects also made up by parts by means of the focus of attention, a sequential operation in the corresponding attractor neural network is implemented by introducing time-delayed connections among units. These connections store the time sequence of parts in the scene; the resulting energy term is:

$$E_2(t) = -\sum_{d=1}^{s}\sum_{i=1}^{m}\sum_{j=1}^{m} D_{ij}^{d}k_i(t)k_j(t-d\tau) \text{ with } j \neq i \tag{5}$$

where τ is the time delay among two subsequent parts, s is the amplitude of the time window of interest, $\mathbf{D}^d$ is the delayed synapses connection matrix related to the time delay $d\tau$, $k(t)$ and $k(t-d\tau)$ are respectively the current and the past $d\tau$-th parts in the focus of attention scan.

The connection matrix $\mathbf{D}^d$ is given by:

$$D_{ij}^{d} = \frac{1}{m}\sum_{\xi=1}^{h} k_{(\xi+d)_i}k_{\xi_j} \text{ with } j \neq i \tag{6}$$

where k_ξ and $k_{(\xi+d)}$ are respectively the ξ-th and the $(\xi+d)$-th part of the current scan of the focus of attention; h is the length of the considered scan.

The global external input to the network is modeled by the energy term:

$$E_3(t) = -\sum_{i=1}^{m}\sum_{j=1}^{m} F_{ij} k_i(t) I_j(t) \text{ with } j \neq i \tag{7}$$

where $\mathbf{F}$ is the external input connection matrix, $\mathbf{I}(t)$ is the actual activation pattern input of the network coming from the perceptive sensors of the scene.

The connection matrix $\mathbf{F}$ is given by:

$$F_{ij} = \frac{1}{m}\sum_{\nu=1}^{h} k_{\nu_i} L_{\nu_j} \text{ with } j \neq i \tag{8}$$

where L_ν is the input corresponding to the part k_ν.

The global energy function is the sum of (2), (5), (7):

$$E(t) = E_1(t) + \lambda E_2(t) + \varepsilon E_3(t) \tag{9}$$

where λ and ε are the weighting parameters of the time delayed synapses and the external input synapses, respectively.

The normal operation of the perception loop is implemented by setting the parameters of the energy function $E(t)$ to $\lambda < 1$ and $\varepsilon > 0$. In facts, the task of the perception loop is the generation of anticipations and the recognition of sequences of the parts representing the input scene perceptions. To accomplish this task it is necessary to consider the input term $E_3(t)$ in order to make the transitions among parts happen, as driven from the external input. When $\lambda < 1$, the term $\lambda E_2(t)$ is not able itself to drive the activation pattern transition among the parts corresponding to the items of the focus of attention scan, but when the term $\varepsilon E_3(t)$ is added, the contribution of both terms will make the transition happen. The neural network therefore recognizes the scene as the corresponding focus of attention scan "resonates" with one of the scans previously stored for the corresponding scene.

When there is a "mismatch" between the anticipated and the perceived parts in the focus of attention scan, as depicted in Fig. 2, the global energy function $E(t)$ reaches low values. In this case the connection matrices $\mathbf{T}$, $\mathbf{D}^d$ and $\mathbf{F}$ are updated according to standard Hebbian learning. This is the case of previously described *online* learning.

In the *offline* learning, as depicted in Fig. 3, the generation of anticipations is implemented by setting the parameters of the energy function $E(t)$ to $\lambda > 1$ and $\varepsilon = 0$. In this learning mode, only the the connection matrices $\mathbf{T}$ and $\mathbf{D}^d$ are updated according to standard Hebbian learning. In fact, the task of the perception loop is to generate suitable parts in the scene representing the scan of the focus of attention. This choice of parameters allows the transitions among parts in the scene to occur "spontaneously" with no external input. Referring to Eq. (9), it can be shown that an attractor is stable for a significant long time period due to the $E_1(t)$ term, so that the output knoxel is easily observed. As $\lambda > 1$, the term $\lambda E_2(t)$ after some $d\tau$ is able to destabilize the attractor and to carry the activation pattern of the network toward the following attractor of the sequence representing the next part of the stored focus of attention scan of the scene The neural network therefore visits in sequence all the parts of the stored parts in anticipated scene.

4 Many Perception Loops

In a real operating robot, we may have many perception loops in action (Fig. 4). They may be related with different sensor modalities, e.g., laser, video camera, sonar, and so on.

Moreover, perception loops related with the same sensor modality may consider different aspects and operations, e.g., a vision based perception loop may consider some kind of objects while another vision based modality perception loop may consider free space. From this point of view, the perception loops play the role of *trackers* in the sense introduced by Kuipers [20] as the basic block of conscious perception.

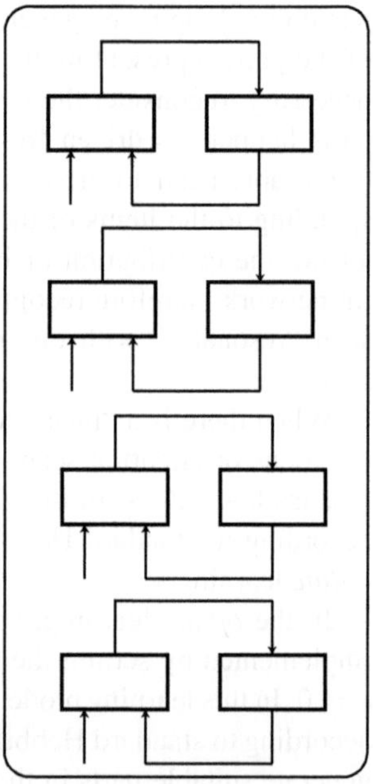

Fig. 4 Many perception loops

5 Higher Order Perception Loop

As stated in the Introduction, we hypothesize that sources of self-consciousness are higher order perceptions of a self-reflective agent. More in details, the *second-order* perception loop at time $t + \delta$ describes the perception of the agent at time t, i.e., the perception loop at a previous δ time of the robot.

The proposed second order perception loop (Fig. 5) works as follows: the robot vision system receives in input the robot position, speed and so on from the proprioceptive sensors as in the first order loop, and it generates anticipations about the operations of the first order perception loop, i.e., the expectations about the inner parameters of the loop itself. The second order loop is then closed by higher order *metaperceptive* sensors that acquire the effective inner loop parameters.

Consider the bent stick example proposed by McDermott [23]: a straight stick partially put into water is perceived as it would be bent because of the water, but the stick is still straight. It appears to be bent only because of the physical properties of the water. According to McDermott, the normal access to the perception module perceives the bent stick, while an introspective access make it appears to be straight.

In the proposed system, the stick partially immersed into water would be perceived as a bent stick from the first order perception loop, while the second order loop has access to the parameters of the first order loop and it could generate the expectations that the bent is not really bent; it only appears in this way because of water.

The second order perception modality considered so far concerns the fact that while the first order perception loop perceives an external scene, the second order

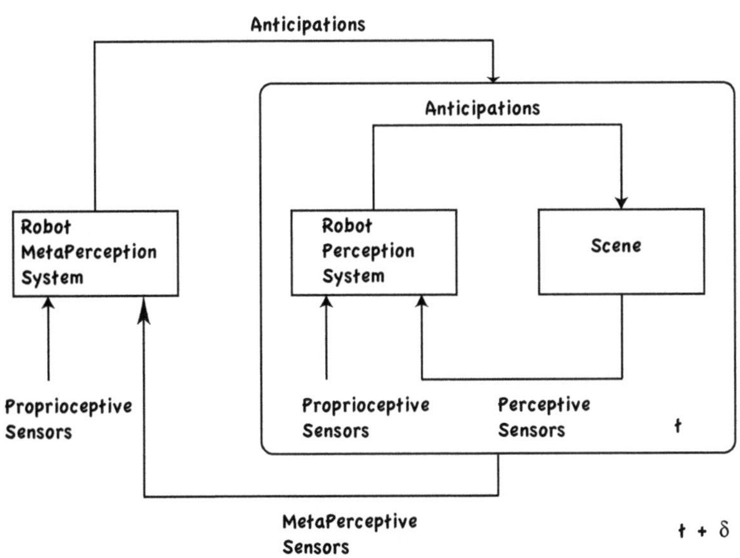

Fig. 5 The robot synchronic second order perception loop

loop, having access to the internal parameters of the perception loop, it is able to make higher order inferences about the scene, as in the bent stick case.

This type of second order anticipations is *synchronic* because it refers to the same situation scene.

The second order modality could also elicit the anticipation of evolutions of the parameters of first order loop in time; i.e., the mechanism generates the expectations for sets of parameters in subsequent operations of first order loop when perceiving dynamic scenes (Fig. 6).

We call this kind of anticipations *diachronic*, in the sense that it involves subsequent configurations of parameters of the first order loop. In particular, diachronic anticipations can prefigure the perception loop as resulting as the outcome of a robot action.

The *attractor neural networks* implementing the second order loop are similar to the ones previously described in Sect. 3, where the generic k_i weights of the first order loop are input patterns of the second order loop.

The outlined procedure may be generalized to consider higher order perception loops: they correspond to the robot's higher order perceptions loops of the of lower order at previous δ times (Fig. 7). We propose that the union of first order, second order and higher order perception loops is at the basis of the robot self-consciousness. The robot recursively embeds higher order models of its own perception loops during its operating life.

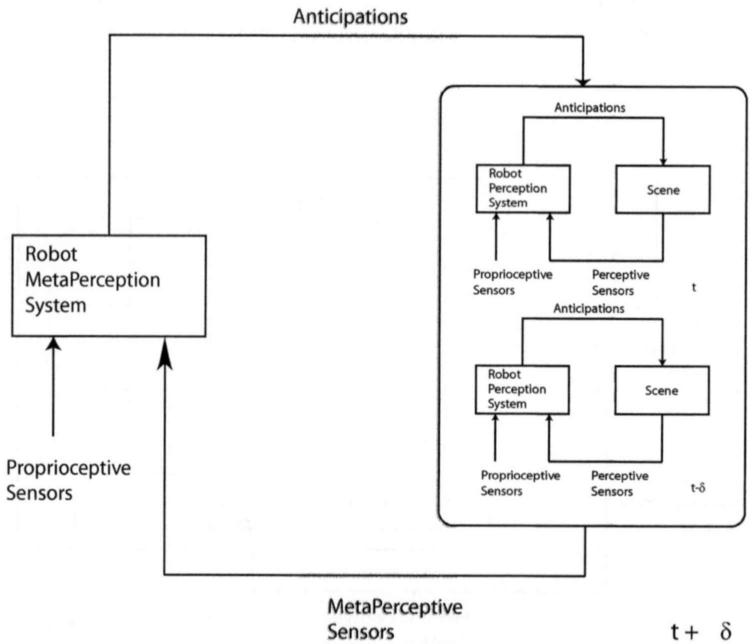

Fig. 6 Diachronic second order perception loop

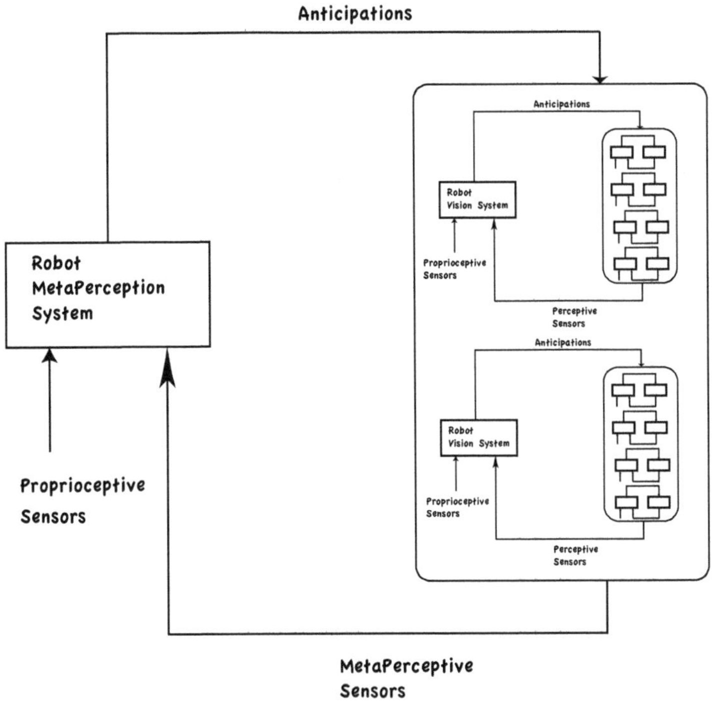

Fig. 7 The robot "self"

The robot *self* is therefore generated and supported by the dynamics of the perception loops, in the sense that the system generates dynamically first order, second order and higher order perception loops during its operations, and this mechanism of generation of higher order loops is responsible for the robot self-consciousness.

It should be remarked that higher and higher order perception loop may embed the whole past history of the robot itself. In this way the robot *identity* is related with the robot capabilities to remember and reason about its own past perceptions.

6 The Architecture at Work

The described architecture has been tested in *Robotanic*: an outdoor robotic tour guide acting in the Botanical Garden of the University of Palermo. The robotic platform is a *MobileRobots* Pioneer 3-AT using differential drive and equipped with a laser scan range finder, a sonar array, a stereo camera and a Global Positioning System (GPS). The Botanical Garden of the University of Palermo is a public outdoor environment, nearly 20000 sq.m. wide. It is characterized by several pathways sometimes bounded by short walls or plants. Furthermore, the paths are often covered by foliage or mud. A typical route takes 30 min to complete it (including speaking).

Fig. 8 Snapshots of tours in the botanical garden

The botanical garden environment is highly dynamic as visitors are all day walking inside the botanical garden, gardeners work alongside the plants and even workers trucks happen to pass by.

The robot has two main goals, namely: i) to guide tourists along the garden and ii) to ensure the safety of people, environment and of itself. During a route, the robot stops near exhibits of interest and it presents a description of them.

Live demos for the proposed system were performed in the Botanical Garden since September 2007. The robot performed several tours, covering more than 3 km in a day. Fig. 8 shows some snapshots taken during one of the tours.

Several perception loops are in action, related with different sensor modalities, as the vision based perception loop, the laser rangefinder perception loop, the odometer and the GPS loop.

Odometric errors and sensor uncertainty underpinned any effort to build a map of the whole environment, leading us to choose a mapless navigation approach. A tourist map (Fig. 9) is employed to figure out where exhibit nodes and navigation aid nodes are placed. The latter were introduced to avoid the robot being trapped in local minima. Each node location was computed by averaging a set of GPS data taken during one whole day, to take into account satellite variations.

In the current implementation the robot has no exhibit recognition perception loops, so the robot stops at fixed positions in order to illustrate the exhibits. This approach then relies on the pose localizer perception loop.

The perception loop based on laser modality is the main source of sensor data: it is used for more than 80% of the time, while the perception loop based on video camera becomes dominant only near the garden basin, as explained below.

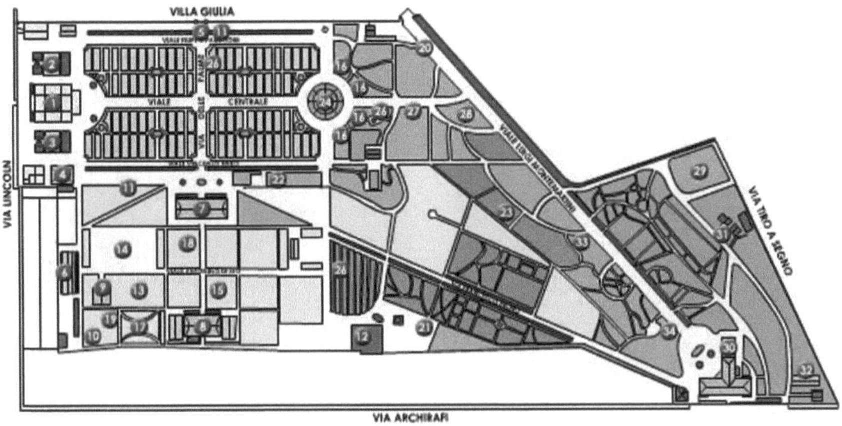

Fig. 9 Botanical garden map

Fig. 10 Basin place (*left*) and its anticipation by the vision based perception loop (*right*)

The robot reports problems when trying to get very near an exhibit, with an average pose error of about 3 ms an and average angular error of about 13 degs respect to the desired goal.

The perception loop related with GPS tracks 5 satellites on average, dropping down to 3 when under heavy foliage or rising to 9 when in open space. We have no ground truth to estimate the pose error, but experiments showed that the error is not higher than 5 ms. This is an acceptable error, as the robot is able to correctly move towards each goal.

One of the most critical sections of the botanical garden is near a basin placed in the middle of a large place and surrounded by short plants and flowers (Fig. 10). Here, the higher order perception loop covers a main role.

In facts, the laser based perception loop is unable to detect the plants as obstacles, leading to an erroneous free space in front of the robot. This causes an increasing discrepancy between the anticipations of the laser based perception loop and the corresponding anticipations of the vision based perception loop (Fig. 11).

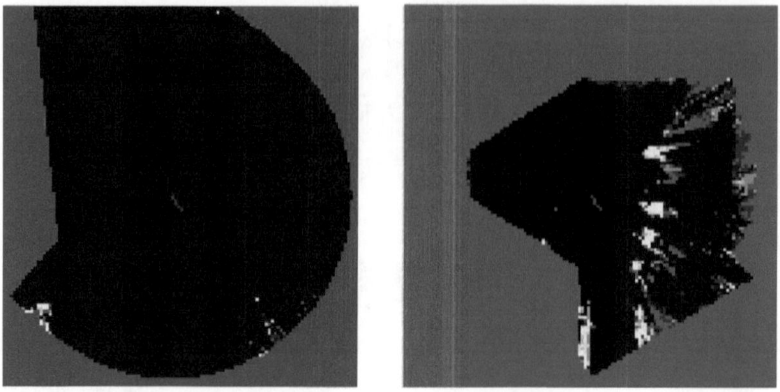

Fig. 11 Laser-based (*left*) and vision-based (*right*) map near the basin

This discrepancy allows the higher order perception loop to operate in order to solve the discrepancy by suitably changing parameters of the laser and the vision based perception loop. In this way, similarly to the example of the correction of the perception of the bent stick, the laser perception is corrected by the vision perception. As a consequence, obstacles near the basin are now allocated and the contour of the basin is clearly individuated (Fig. 11 right). As soon as the robot managed to get around the basin, the laser based perception loop resumed consistency and it is again used as the main navigation sensor.

7 Discussion

The robot operations generally resulted in "a fun experience" according to tourists attending the tours in the botanical garden, especially children.

The perception loop alone is currently not capable of dealing with entities not previously stored in the attractor neural networks: e.g., a moving person in front of the robot. To deal with this problem, *Robotanic* integrates the described perception loop with a standard set of reactive behaviours based on laser scans that take control of the robot in order to let the system to reactively cope with unexpected situations.

We have observed that the proper operating situation for *Robotanic* is when the number of persons attending the tour is up to 6-8 and the garden paths near the robot are not crowded, which is a typical situation in the botanical garden. In this case, all the people attending the tour stay behind the robot during its tour and the rest of the path is nearly clear. The robot performs its own tour without problems, also dealing with persons in front of it or with persons partially occluding plants.

When the number of persons attending the tour is higher, it may happens that tourists displace themselves all around the robot and also in front of it. In this case, the camera occlusions and the repeated activations of the obstacle avoidance behaviour may allow the robot to enter in a deadlock.

When the garden path is crowded, typically when some school-class visit the botanical garden, it may happens that the focus of attention scan is ineffective because perception loop is unable to find in some cases a satisfactory match between the anticipations and the perceived parts of the scene.

In both cases, *Robotanic* stops its tour waiting to recover from the fault after some time.

8 Conclusions

We claim that self-consciousness is based on perception of the inner world. In this sense, to model self consciousness, we adopt the same perception loop adopted for the perception of the external world, but now the higher order perception loop is "oriented" towards the inner world of the robot.

Considering the agent reasoning system, higher order perception loop may correspond to symbolic meta-predicates, i.e., symbolic predicates describing the robot perceiving its own situations and actions. These meta-predicates form the basis of the introspective reasoning of the robot, in the sense that the robot may reason about its own actions in order to generate evaluations about its own performances. Moreover, the robot equipped with the representation of self may generate more complex plans, in the sense that the robot motivations, i.e., its long term goals, may now include also the higher order perception loops.

There are analogies between the system described here and that proposed by Weyhrauch. Namely, in both cases there is the possibility of exploiting metalevel representations, and both systems associate some form of analogue representation to the symbolic formalism (the simulation structure in Weyhrauch's system), the perception loops in the proposed system.

The described architecture has points in common with the *Real-time Control System (RCS)* architecture proposed during the years by Albus and collaborators [1, 2]. In facts, the described perception loop structure has some similarity with the RCS node and also the reported architecture have similarities with the hierarchy typical of the RCS implementations.

However, the main difference between the two architectures is that RCS is a strictly hierarchical architecture while our perception loops have no a priori defined activation or computational resources. The activations of RCS nodes are defined by the architecture designer according to a fixed hierarchy of levels. Instead, in the proposed architecture, the higher order perception loop dynamically allocates computational resources, i.e., the task activation priorities and the allocated memory, according the ongoing of the robot mission. According to the received feedback from the *metaperceptive* sensors, the higher order perception loop may change the perception loops parameters and, consequently, their computational resources and their activation priorities.

In this way, when the robot operations are no more satisfactory, e.g., in the case of the robot operations near the garden basin, the feedback allows the higher order loop to change the perception loop parameters in order to affect their activation priority, memory allocated, and other computational parameters. In this way, our architecture is able to dynamically reconfigure itself in runtime.

A limitation of the proposed architecture is that the described continuous generation of perception loops at different orders poses problems from the computational point of view, in the sense that the physical memory of the robot may be easily filled up with data structures describing the parameters of perception loop. Some mechanism that lets the robot to summarize its own past experiences will be investigated. One possibility is to make the representations much more *blurred* as the levels grow: higher order perception loops could be less detailed than lower level ones.

Acknowledgements Author would like to thank Irene Macaluso, Lorenzo Riano and Rosamaria Barone for the discussions about the topic of the paper and for the implementation work of the described architecture.

References

[1] Albus, J., Barbera, A.: RCS: A cognitive architecture for intelligent multi-agent systems. In: Proc. 5th IFAC/EURON Symposium on Intelligent Autonomous Veichles IAV 2004. Lisbona, Portugal (2004)
[2] Albus, J., Meystel, A.: Engineering of Mind. Wiley-Interscience (2001)
[3] Aleksander, I., Morton, H.: Computational studies of consciousness. In: R. Banerjee, B. Chakrabarti (eds.) Progress in Brain Research, vol. 168, pp. 77–93. Elsevier Science, Amsterdam, The Netherlands (2008)
[4] Amit, D.: Modeling Brain Function. The World of Attractor Neural Networks. Cambridge University Press (1988)
[5] Carruthers, P.: Language, Thought and Consciousness: an essay in philosophical psychology. Cambridge University Press, Cambridge, UK (1996)
[6] Chella, A.: Towards robot conscious perception. In: A. Chella, R. Manzotti (eds.) Artificial Consciousness, pp. 124–140. Imprint Academic, Exeter, UK (2007)
[7] Chella, A., Barone, R.: Panormo: An Emo-Dramatic Tour Guide. AAAI Spring Symposium on Emotion, Personality and Social Behaviour pp. 10–16 (2008)
[8] Chella, A., Frixione, M., Gaglio, S.: A cognitive architecture for artificial vision. Artificial Intelligence **89**, 73–111 (1997)
[9] Chella, A., Macaluso, I.: The perception loop in *CiceRobot*, a museum guide robot. Neurocomputing **72**, 760–766 (2009)
[10] Grush, R.: The emulator theory of representation: motor control, imagery and perception. Behavioral and Brain Sciences **27**, 377–442 (2004)
[11] Haikonen, P.: The Cognitive Approach to Conscious Machines. Imprint Academic, Exeter, UK (2003)
[12] Haikonen, P.: Robot Brains. John Wiley & Sons, Chichester, UK (2007)
[13] Hesslow, G.: Conscious thought as simulation of behaviour and perception. Trends in Cognitive Sciences **6**(6), 242–247 (2002)
[14] Hinton, G., McClelland, J., Rumelhart, D.: Distributed representations. In: D. Rumelhart, J. McClelland (eds.) Parallel Distributed Processing, vol. 1. MIT Press, Cambridge, MA (1986)
[15] Holland, O., Goodman, R.: Robots with internal models - a route to machine consciousness? Journal of Consciousness Studies **10**(4-5), 77–109 (2003)

[16] Holland, O., Knight, R., Newcombe, R.: A robot-based approach to machine consciousness. In: A. Chella, R. Manzotti (eds.) Artificial Consciousness, pp. 156–173. Imprint Academic, Exeter, UK (2007)
[17] Hopfield, J.: Neural networks and physical systems with emergent collective computational abilities. Proc. Nat. Acad. Sci. USA **79**, 2554–2558 (1982)
[18] Kleinfeld, D.: Sequential state generation by model neural networks. Proc. Nat. Acad. Sci. USA **83**, 9469–9473 (1986)
[19] Kleinfeld, D., Sompolinsky, H.: Associative network models for central pattern generators. In: C. Koch, I. Segev (eds.) Methods in Neuronal Modeling, Bradford Books, pp. 195–246. MIT Press, Cambridge, MA (1989)
[20] Kuipers, B.: Consciousness: Drinking from the firehose of experience. In: Proceedings of the Twentieh National Conference on Artificial Intelligence, pp. 1298–1305. AAAI Press, Menlo Park, CA (2005)
[21] Macaluso, I., Ardizzone, E., Chella, A., Cossentino, M., Gentile, A., Gradino, R., Infantino, I., Liotta, M., Rizzo, R., Scardino, G.: Experiences with CiceRobot, a museum guide cognitive robot. In: S. Bandini, S. Manzoni (eds.) AI*IA 2005, *Lecture Notes in Artificial Intelligence*, vol. 3673, pp. 474–482. Springer-Verlag, Berlin Heidelberg (2005)
[22] McCarthy, J.: Making robots conscious of their mental states. In: K. Furukawa, D. Michie, S. Muggleton (eds.) Machine Intelligence 15: Intelligent Agents, pp. 3–17. Oxford University Press, Oxford, UK (1999)
[23] McDermott, D.: Mind and Mechanism. MIT Press, Bradford Books, Cambridge, MA (2001)
[24] Minsky, M.: The Emotion Machine. Simon and Schuster, New York, NY (2006)
[25] Rao, R., Ballard, D.: Dynamic model of visual recognition predicts neural response properties in the visual cortex. Neural Computation **9**, 712–763 (1997)
[26] Rao, R., Ballard, D.: Predictive coding in the visual cortex: a functional interpretation of some extra-classical receptive-field effects. Nature Neuroscience **2**(1), 79–87 (1999)
[27] Rosenthal, D.: Consciousness and Mind. Oxford University Press, Oxford, UK (2005)
[28] Sanz, R., Lopez, I., Rodriguez, M., Hernandez, C.: Principles for consciousness in integrated cognitive control. Neural Networks **20**, 938–946 (2007)
[29] Sloman, A., Chrisley, R.: Virtual machines and consciousness. Journal of Consciousness Studies **10**(4-5), 133–172 (2003)
[30] Weyhrauch, R.: Prolegomena to a theory of mechanized formal reasoning. Artificial Intelligence **13**(1-2), 133–170 (1980)

The Consciousness Circuit – An Approach to the Hard Problem

Sulamita Frohlich and Carlos A. Franco

Abstract This paper is the continuation of a study about Consciousness as a resulting function between attention and luminosity, presented at the Neuroscience International Congress, which was held in the city of Natal, in Brazil, in 2006. There it was described how visual Consciousness is generated by the interaction of the cortical area activity through FFS (Feedforward Sweep Processes), RP (Recurrent Processes) and WSRP (Wide Spread Recurrent Processes), and its relationship with the luminosity that hits the eyes. We have applied the reciprocal interaction model, which says that the eye reacts to luminosity through the regulation of sleep-awake states through EEG wave synchronization; this, in turn, is regulated by the thalamic cortical neural activity from the brainstem monoaminergic and cholinergic nuclei.

Such an understanding has led us to construct a consciousness model which can be represented by an orthogonal graph. Through this model, we can represent all states of human consciousness (emotional consciousness, unconsciousness states, dreaming states, awareness states, pre-consciousness states and others) making it possible, in theory, to construct a Consciousness parameter which yields to the understanding of consciousness state observation without a subjective approach to experience.

We have also applied on this orthogonal graph the Quantum Orch OR Model. According to it, subjective, phenomenal conscious vision depends on quantum computation in microtubules where the quantum is the smallest quantity of radiant energy, in a scale in which matter and energy interact. This model helps us to build the vertical axe of the consciousness orthogonal graph, a cholinergic-aminergic scale, in which activation triggers the quantum mechanism relevant to the consciousness phenomenon.

Keywords Consciousness · Attention · Awareness · Self · Easy problem · Hard problem

C.A. Franco
Associate Professor, Computer Science Department, Mathematical Institute, Mental Health Area, Psychiatric Institute, UFRJ, BRAZIL
e-mail: carlosfranco@sensesac.org

S. Frohlich
Psychologist, member of the research group in Cognitive Neuroscience, Neuropsychology, Mental Health Area, Psychiatric Institute, UFRJ, BRAZIL
e-mail: sulafroh@arka2.com.br; arka2@arka2.com.br

A. Hussain et al. (eds.), *Brain Inspired Cognitive Systems 2008*, Advances in Experimental Medicine and Biology 657, DOI 10.1007/978-0-387-79100-5_16,

1 Introduction

1.1 The Consciousness Problem

The consciousness concept is hybrid, connoting a number of different concepts and phenomena. In order to clarify this issue, Chalmers separates the problems which are often clustered together under that name. For this purpose, he first distinguished between an "easy" and a "hard" problem of consciousness. The easy problems are, by no means, trivial - they are, actually, as challenging as most problems in psychology and biology - but it is within the hard problem that the central mystery lies. As he defines:

> "The ***easy problem*** of consciousness includes the following: *how can a human subject discriminate sensory stimuli and react to them appropriately? How does the brain integrate information from many different sources and use such information to control behavior? How can a person verbalize his or her internal states? Although all of these questions are associated with consciousness, they all concern the objective mechanisms of the cognitive system. Consequently, we have every reason to expect that continued work in cognitive psychology and neuroscience will answer them. The* **hard problem**, *in contrast, is the question of how physical processes in the brain give rise to subjective experience. This puzzle involves the inner aspect of thought and perception: the way things feel for the subject. When we see, for example, we experience visual sensations, such as that of vivid blue. Or think of the ineffable sound of a distant oboe, the agony of an intense pain, the sparkle of happiness or the meditative quality of a moment lost in thought. All are part of what I am calling consciousness. It is these phenomena that pose the real mystery of the mind.*"

As a working hypothesis, we have assumed that the hard and the easy problems are interrelated and, in order to organize such an issue, we have created an orthogonal graphic model which will help us to better organize the interaction between the organic pattern (the easy problem) and the subjective experiences (the hard problem).

To understand Consciousness, we have been studying visual consciousness as, in fact, it comprises some of the richest and most common aspects of the Consciousness processes. Attention focus can be provided by eye fixation, scanning eye movements, the non attentive look and many other visual states which connote different consciousness manifestations such as phenomenal consciousness, introspective consciousness, self consciousness etc.[1] Each eye movement leads to or expresses a state of the mind [2] and such a phenomenon gives the eyes the status of being the "windows of the soul".

The questions we intend to answer are: how does the experience of consciousness occur and what are the functions of so many related terminologies and manifestations? The main aspect of this study is to contribute to the understanding of the consciousness phenomenon as a subjective experience, in relationship with physiological processes.

This is a theoretical study and so far no related experiments have been planned. We are starting to understand the main aspects of such a rich subject.

1.2 About Terminologies

The different aspects of consciousness often create terminology confusion such as what awareness, attention, self-consciousness, sensorial consciousness, phenomenal consciousness, access consciousness, speech consciousness and action consciousness mean. This evidences that consciousness has different forms of manifestation.

Therefore, some questions arise: are consciousness and attention of the same nature? How are they related to each other, in fact? If we are paying attention to something, does this mean we are conscious of the subject? There is obviously a relationship between these two functions - but what actually is it?

Lamme define attention "*as a separate selection process, which is in principle independent from the conscious phenomenal experience and works as a limited capacity bottleneck-like process that allows stimuli to be processed deeper or faster, and which is necessary for storage in a durable working memory store or for a conscious report about stimuli*".[3]

What reaches visual consciousness is usually the result of an attentional step. In other words, consciousness and attention are intimately bound together. Note that although the results of attention are postulated to reach consciousness, the attentional mechanisms themselves are probably largely unconscious.[4]

It is more likely that Consciousness contain attention because the relationship between them is as follows: for the attentive state to occur, it is necessary to be conscious, but the contrary is not true. Consciousness is enriched by visual attention, though attention is not essential for visual consciousness to occur.[5]

Another important misunderstanding on the consciousness nomenclatures is about awareness, which is something independent from attention, such a term normally being used as synonymous to consciousness.

In order to help to grasp this entire concept better, we intend to give a more refined treatment to all of these terminologies, trying to understand the differences between them as well as their relationship properties. Therefore, we would like to

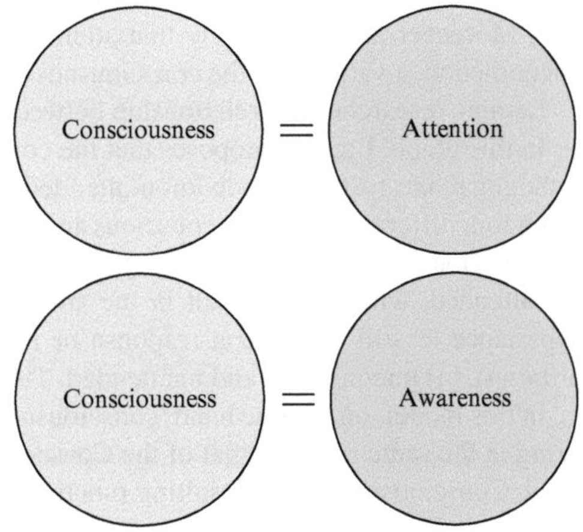

Fig. 1 "Consciousness" is often related to an attentive state, "awareness" being synonymous to it, or a phenomenon of the same level or quality. Thus, a question arises: what is the actual relationship between those two concepts?

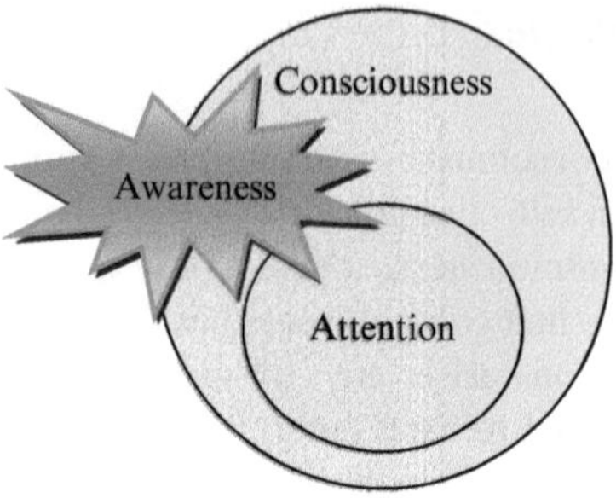

Fig. 2 The consciousness set contains the attention set. That can be explained by the fact that there are situations when we can be attentive to a stimulus and yet not be conscious of it, as it happens in some unconscious processes. Awareness, in turn, is a dynamic phenomenon of consciousness where one recognizes and vividly experiences his or her own conscious state. Consciousness, in turn, is the manifestation field of those phenomena

suggest the construction of a theoretical proposal of consciousness as an orthogonal graph. In this graph we intend to organize those concepts making them easy to be understood, and also to understand their dynamic processes.

1.3 Consciousness as an Orthogonal Graph

Based on computational theory for recurrent processing in the visual cortex, it is very likely that Consciousness can be represented as a two-parameter function. This idea is supported mainly by the fact that there is not a specific identified area for the Consciousness itself and that the Neurological Correlate of Consciousness (NCC) is not anatomically defined but just functionally described. This statement confirms the supporting idea that Consciousness is a resulting function and that just some neural activities lead to awaken consciousness.

As we have written above, there is reciprocity between consciousness and attention. Moreover, it is very likely that attention is an element of the Consciousness phenomenon, a variable of the consciousness process.

Lamme researched the relationship between attention and consciousness.

In this graph, Lamme proposes that the conscious/unconscious dichotomy is the orthogonal axe to the attended/non-attended dichotomy. Hence, visual inputs can reach four different states: (1) conscious and attended, resulting in access awareness, (2) conscious yet unattended, leaving only phenomenal awareness, (3) unconscious but attended, which may result in the stimuli for which there is no phenomenal experience to still generate a response or influence behaviour (as f.i. in masked priming), (4) unconscious and unattended. The fate of these inputs is uncertain.

In this model, on the one hand, consciousness and attention are in comparison as being in the same range as that of the Conscientious phenomenon and, on the other hand, Consciousness is the resulting function.

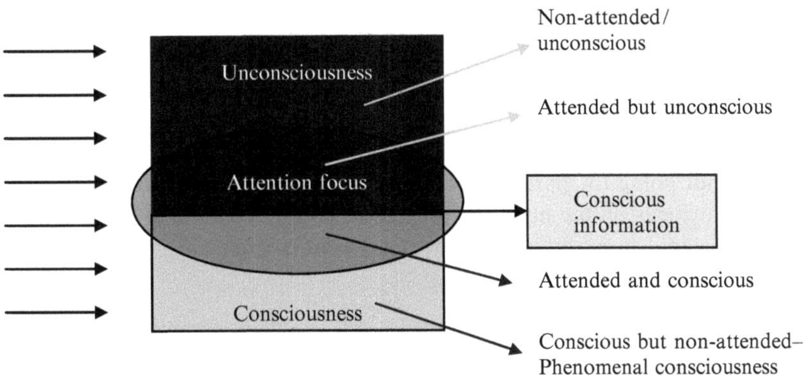

Fig. 3 Lamme's graphs on the relationship between attention and consciousness

A theoretical construction which states that a conscious state is not simultaneously conscious and attentive is illogical. In a logical perspective, that is a wrong assertion.

Such an imperfect idea is represented like this:

$$\boxed{\text{Consciousness (C)} = \text{consciousness (c)} \times \text{attention (s)}}$$

We agree with the idea that there exists a Consciousness function (C) in which one of its identified parameters is the attentional mode (S) but what we disagree about is that the second element of the orthogonal graph is consciousness (c).

But which element would be the second parameter for the Consciousness orthogonal graph?

One striking piece of evidence for a second element of the system comes from studies of Woolf-Hameroff [6]. According to the Orch OR model, subjective, phenomenal conscious vision depends on quantum computation in microtubules.

In this model, inputs from cortical arousal systems, in particular the cholinergic basal forebrain, select content for conscious attention via muscarin receptor activation. High levels of acetylcholine are correlated with increased attention and heightened conscious awareness.

They propose that quantum states supporting Orch OR are isolated in cytoplasmatic interiors of cortical pyramidal dendrites interconnected by gap junctions, forming a horizontal network or syncytium spanning visual cortical regions. Isolated quantum phases alternate with classical phases at approximately 40 Hz, the frequency that has been associated with synchronized cortical activity underlying conscious states.

We suggest here that the second element of the Consciousness function should be given by the Quantum computation in microtubules, which we are going to simplify as Quantum (Q).

Then the consciousness function variables are written as follows:

$$\boxed{\text{F Consciousness} = (\text{Q}) \times (\text{S})}$$

Where (Q) represents the Quantum and (S), Attentional modulation.

We are not going to approach the possibilities of measuring the event. We are just going to focus on the understanding of its dynamics through the cholinergic-aminergic balance in the arousal biochemistry phenomenon.

2 The Consciousness Parameters

In order to describe the consciousness function, we have studied the following two elements which are always present in visual consciousness: 1- attentional modulation (S) and 2- the quantum (Q).

1. Attentional modulation (S)

Attention is the cognitive process of selectively concentrating in one aspect of the environment while ignoring others. It implies a state of readiness for such attention, involving, especially, a selective narrowing or focusing of consciousness and receptivity. The problems in the recent literature on visual attention are focused mainly in the control of attention by top-down (or goal-directed) and bottom-up (or stimulus-driven) processes, which leads to the concept of direction.

We have also introduced the concept of modulation based on Lamme's proposal.

a) Attentional direction Theoretical accounts postulate that attention is controlled as an interaction between "bottom-up" (stimulus-driven) and "top down" (voluntary or cognitive) factors. Bottom-up control refers to the ability of a physically conspicuous object to attract attention automatically regardless of its task relevance. Top-down control refers to the ability of subjects to allocate attention according to a large class of behavioral influences, including spatial or temporal anticipation, statistical contingencies, or motor planning.[7]

Another focus is the attentional goal. It is widely accepted that exogenous and voluntary factors jointly determine the locus of attention[8].

Attentional modulation can be directed to an environmental stimulus (exogenous attention) or to inner feelings, memories, thoughts etc. stimulus (endogenous attention).

Exogenous attention leads the consciousness focus to environmental stimuli. With endogenous attention, the situation becomes more complex, but not fundamentally different. Now, an external event has to be translated as an inner image and then to an abstract cue. Parts of the brain that extract the meaning of the cue are able to relate this to current needs and goals must pre-activate or, otherwise, facilitate the appropriate sensory pathways, mostly via cortico-cortical feedback or sub cortical routes.[9]

This dynamics can be represented by a vector that has two opposite directions: an exogenous direction or an endogenous direction, meaning that exogenous attention leads consciousness to the environmental stimuli captured by the senses, to self expression and, also, to perform bioregulation.

Endogenous attention is related to thoughts, feelings and memory processes.

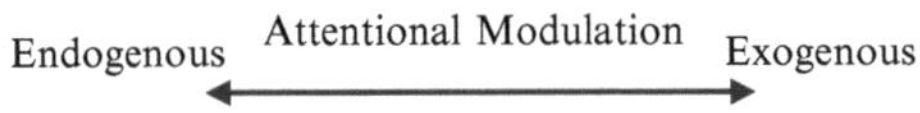

b) The attentional modulation stages: The neuronal processes are an important neurophysiologic cue for the understanding of attention. Lamme describes at least three fundamentally different types of processing that can be identified within this parallel flow.

- Feedforward processing (FFS)
- Recurrent processing (RI)
- Widespread recurrent processing (WRI)

The Feedforward sweep (FFS) is defined as the earliest activation of cells in successive areas of cortical hierarchy and is not sufficient for visual awareness. Recurrent interactions (RP) between areas are necessary for the visual features to be related to each other and are where binding and segregation may occur. Thus, perceptual organization is developed. Widespread recurrent interactions (WSRI) involve many regions of the brain, such as the frontal cortex, pre-frontal and temporal cortex, visual information being processed through the needs, goals and in the personal history of the person. It occurs when the stimulus has passed the intentional bottleneck between sensory and executive areas.

In short, visual attention can be processed in those three stages above described.

We can be (1) in attention (or in focus) to environment stimuli through a scanning visual process or through saccadic movements, activating the FFS processes; (2) examining, analyzing, forming and memorizing images through an RP process or (3) analyzing and processing visual information through the complex human soul system, comparing, programming, and valorizing contexts by the WSRI processes.

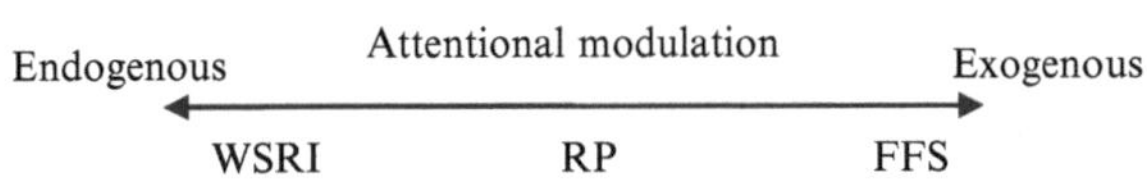

2. The quantum (Q)

Based on the Quantum Orch OR Model, the inputs from cortical arousal system, in particular, the cholinergic basal forebrain, select content for conscious attention via muscarin receptor activation. Selection may occur by direct cholinergic action on pyramidal dendrites as well as on GABAS interneuron.[10]

Recent evidence points to inhibitory GABAERGIC cortical interneurons as important mediators of 4O-Hz activity. The interneurons are themselves connected by gap junctions, and form 'dual' connections with each pyramidal dendrite: an inhibitory GAB chemical synapse and an electrotonic gap-junction connection.

On the Reciprocity Interaction Model, high levels of cortical acetylcholine are correlated with increased attention and heightened conscious awareness.[11]

The thalamocortical circuit and the aminergic-cholinergic projections are responsible for the desynchronization of the EEG during wakefulness. High aminergic activity during active wakefulness activates the thalamocortical circuits but is reduced during NREM sleep, and is absent during REM sleep.

Aminergic neurons are called wake-on-and-sleep-off cells. The cerebral cortex is aminergically demodulated during REM sleep due to the lack of hypocretin tone.

Normal sleep consists of alternation between REM and NREM stages. Characterised by the presence of synchronised waves in the electroencephalogram (EEG), NREM sleep can be subdivided into four phases (phases 3 and 4 correspond to slow wave sleep or delta sleep). The REM sleep stage is characterised by EEG desynchronization and low-amplitude waves. The synchronization-desynchronization of EEG waves during NREM-REM sleep and wakefulness is a consequence of neural activity in the thalamocortical circuits (reticular nuclei in the thalamus and cerebral cortex), derived from the interaction between monoaminergic and cholinergic nuclei in the brain stem.

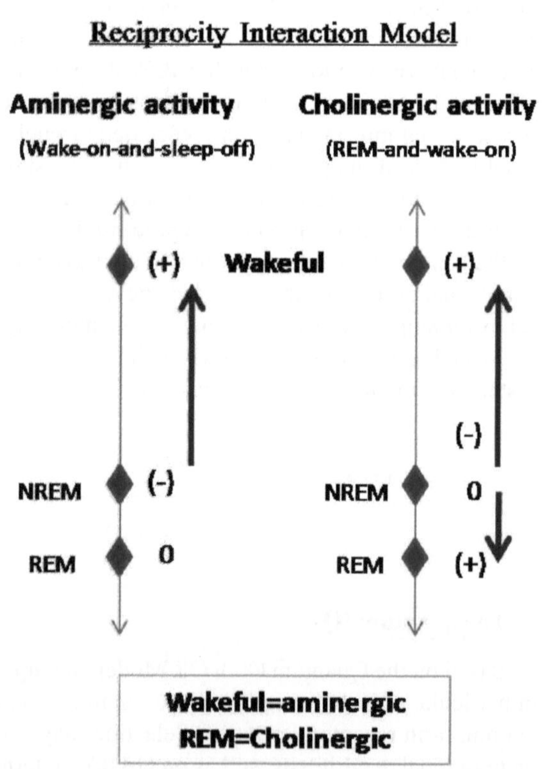

Fig. 4 The reciprocal interaction model is a functional model that establishes wakefulness as a predominantly aminergic state, REM sleep as a predominantly muscarinic cholinergic state and NREM sleep as an intermediate state

This idea is summed up in the reciprocal interaction model, which proposes two types of cell groups located in the reticular formation: the cholinergic REM-on cells and the serotoninergic-noradrenergic REM-off cells. During wakefulness, the aminergic REM-off system is tonically activated, generating EEG desynchronization, inhibiting the cholinergic REM-on system and suppressing REM sleep. On the other hand, during REM sleep aminergic REM-off cells as well as the cholinergic system are free from inhibitory influences and reach their peak. Therefore, REM sleep only occurs when the aminergic system suspends its inhibitory effect on cholinergic activity.

Thus, the vertical axe of the graph should be defined as a vector where the aminergic –cholinergic activities vary giving through this variation the consciousness tonus.

This subject is being very well studied by the Orch OR model, according to which it is defined that subjective, phenomenal conscious vision depends on quantum computation in microtubules.

Woolf and Hameroff proposed that three types of inputs or interconnections within the horizontal syncytium (i.e. Gap-junction network) could provide the basis of attention and modulation of Orch OR conscious events in visual cortex.[12]

1. Thalamus cortical inputs - along with excitatory local circuits cells, relay specific information along each vertical column of cortex. These provide non-conscious, neurophysiological information about the visual scene mapped in a point-to-point fashion by glutamatergic synapses.
2. High levels of acetylcholine are correlated with increased attention and heightened conscious awareness basal forebrain and select content for conscious attention via muscarin receptor activation.
3. Coherent cortical oscillations in EEG gamma frequency (30–70 Hz activity, also known in short as '40 Hz') appear to correlate with consciousness through gabaergic cortical interneurons. Selection can occur by direct cholinergic action on pyramidal dendrites, as well as on GABA interneurons.

We have decided to name the vertical axe as Quantum (Q) as we understand this field is the future for the development of this understanding.

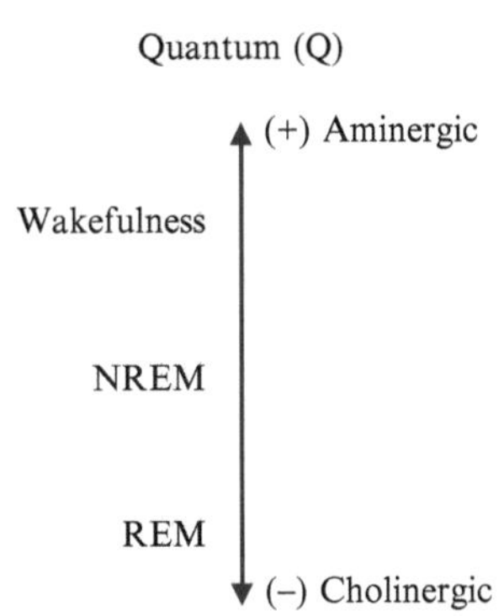

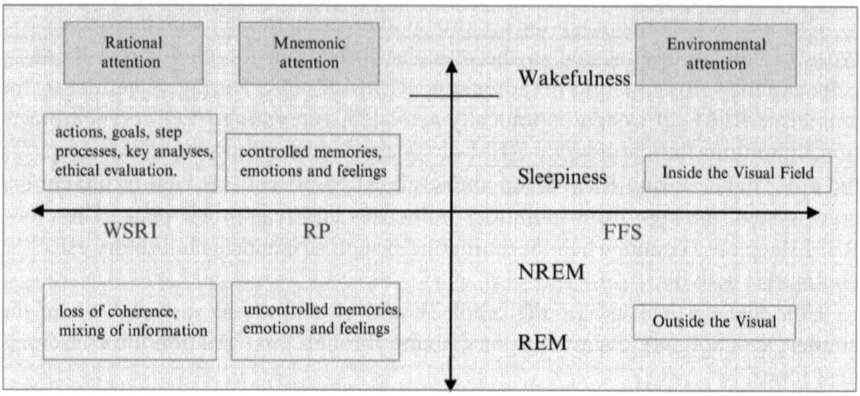

Fig. 5 The Consciousness graphic

This model helps us to build the vertical axe of the consciousness representative graph, where the more aminergic the circuit, the more Quantum is present in consciousness; the more cholinergic the circuit, the less Quantum is present in the consciousness processes. This will be better defined in future studies.

Our hypothesis is that attention, defined as the active neuronal area (S) and the quantum (Q), are measurable variables of the Consciousness function, establishing an orthogonal relationship between them.

$$\boxed{\boldsymbol{F}(\text{Consciousness}) = (\text{S}) \times (\text{Q})}$$

3 The Orthogonal Graphic

Attention and arousal are multi-dimensional psychological processes, which interact closely with one another. We propose here that the way they interact is thought as an orthogonal process that can be expressed graphically.

This orthogonal graph introduces the idea that consciousness is a macro neuronal circuit structure that alters itself in relation to parameter changes. Thus, rather than being something that can be found somewhere, consciousness itself is something fixed. Besides, it is not a biological phenomenon but something that activates biological processes. This leads to the idea that consciousness does not have a Neuronal Correlate (NCC) but is a result of separate and independent processes.

4 Analyzing the Graphic

There is ubiquitous confusion among researchers about consciousness and a major source of this confusion lies in the spectrum of different attempts to discriminate the concepts of attention, awareness, unconsciousness and consciousness.

In order to organize those concepts we will work on the Consciousness Orthogonal graphic model, proposing a new understanding of them based at the present moment on a symbolic understanding. Even if now it is just a symbolic reference, it is a first step to performing measurement procedures in the future.

4.1 Attention

We have hypothesized that the consciousness function produces attentional state.

A person can be attentive to an environmental stimulus and not attentive to his or her feelings, meaning that the consciousness circuit is activated by FFS modulation and not by RP.

If someone is looking at an environmental stimulus in an attentional state, this situation can be represented in our graph like this:

On the other hand, when someone is sleeping and dreaming, their eyes are closed and the light is hindered by their eyelids. We can conclude that their attention is then modeling an endogenous Recurrent Processing RP, developing illogical, confused and nonsensical images and thoughts as the cerebral cortex is aminergically demodulated. That should be represented like this:

The daydreaming situation should be represented as:

Attention to rational thoughts, preparation for action through strategy planning and aggressive regulation, and ethical valuation should be represented graphically as follows:

The graph above shows that Attention is a result of the Consciousness phenomenon. It is an abstract concept but it can be measured by physics means, just like the concepts of velocity (v) and acceleration (a).

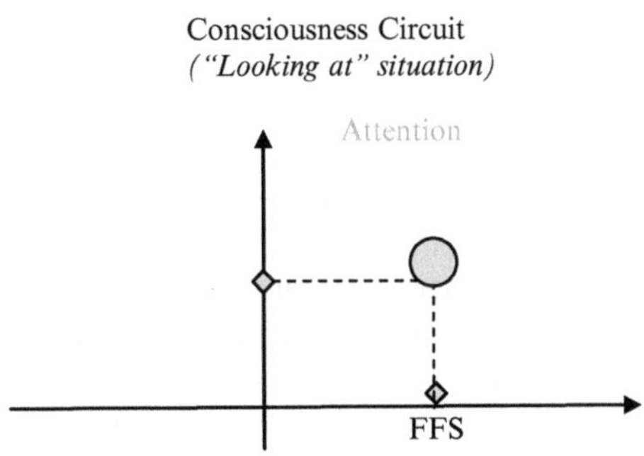

Fig. 6 Eyes open, focusing on the object

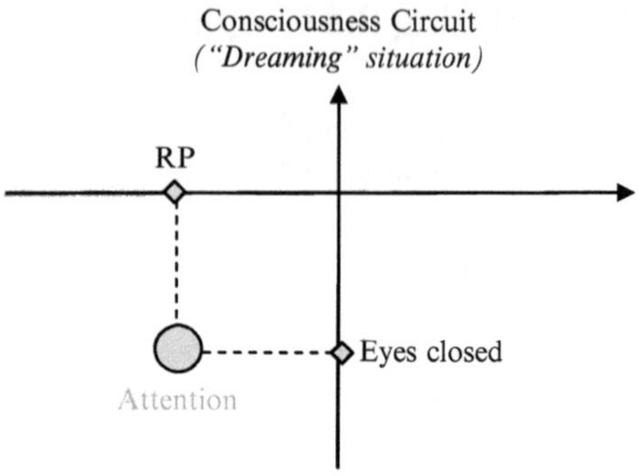

Fig. 7 Representation of a dreaming situation

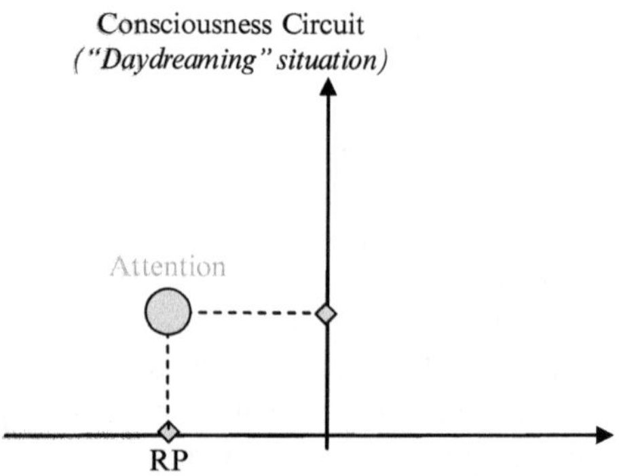

Fig. 8 Daydreaming situation: open eyes with objects out of focus (by rotation of the ocular globe)

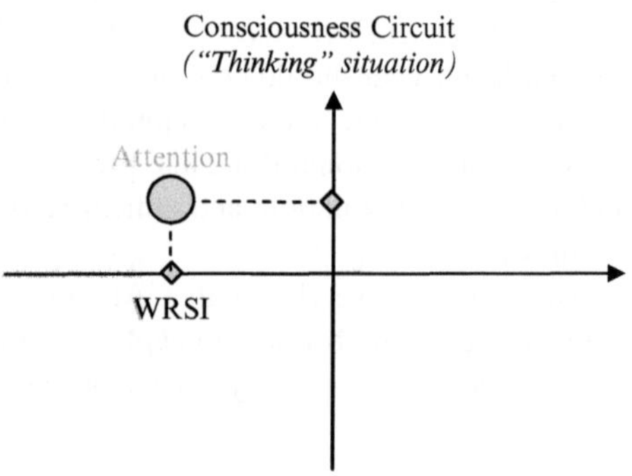

Fig. 9 Thinking situation: open eyes with objects out of focus (by rotation of the ocular globe)

We can also infer that there is no possibility of non attention states. Attention is always present, even in the so called unconsciousness states. In fact, the inattentive state means that the phenomenon is occurring in all graphic areas where the attentional focus is not allocated.

Here another term that is highly used in the Consciousness problem is "Unconsciousness". What does it mean? If Consciousness is a result of a dynamic process, how can there be antagonism?

Obviously the concept of unconsciousness has been used in a very confusing way. We will try to understand it better.

4.2 Unconsciousness

One of the applications of this concept is when it refers to all of the points of the graph where the awareness point is not on. In this case unconsciousness means "not attentive to". But in some cases a person may use current visual input to produce a relevant motor output without being able to say what was seen.[13]

The attention phenomenon exists anyway in living beings by the fact that the nervous system is always processing quantum. So there can not be a situation where attentional phenomena will not occur.

Another common understanding of the unconsciousness concept refers to the processes that are activated in the negative side of the vertical axe.

It is important to understand that even in the so called unconscious state there are attentive processes going on.

5 The Problem of Qualia

This concept leads us to another approach of Consciousness dynamics and here we can really establish a mark between the **Easy** and the **Hard problem** because we start to talk, not about the Consciousness physiological correspondences but about the ***QUALIA*** of the process, that is, how people can see the redness of the red, the nature of their feelings, etc. The problem here to be solved is: how can a person be aware of his or her own awareness state?

How can we understand and evaluate the possibility of awareness of the proper system regarding itself – "the awareness of the awareness state"? Here we can see the emergence of the concept of the Self, broadly referring to the cognitive representation of one's identity.

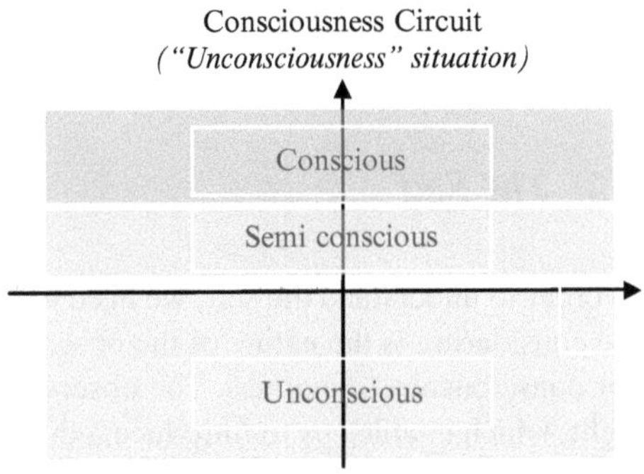

Fig. 10 Levels of consciousness are related to the variation of aminergic –cholinergic activities, which defines the tonus of consciousness

5.1 Awareness

"Awareness" is often used as synonymous to "consciousness" and is usually understood as being consciousness itself.

Awareness is a relative concept which may be focused on an internal state, such as a visceral feeling, or on external events, by means of sensory perception. Awareness provides the raw material from which one develops qualia, or subjective ideas about their experience and even self-awareness, which means that one is aware of one's own awareness state.

One characteristic from the awareness state is that subjective experiences stem from the first person's access to them. It is, first of all, a personal and subjective experience.[14]

Another one is that the awareness state is related to each personal development of cognitive capacity, that is, there exist "higher" forms of awareness, including self-awareness, which require cortical contributions, and "basic" forms of awareness, such as the ability to integrate sensations from the environment with one's immediate goals and feelings, in order to guide behavior. This springs from the brain stem, which human beings share with most vertebrates.

The awareness state has different degrees of perception. This is well described in the Indian tradition "scale of sentience", described in the consciousness processes as follows:

1. I'm aware of this.
2. I'm aware that this is so.
3. I'm aware that I am affected by this which is so.
4. I'm aware of that this is I who am affected by this which is so.

Each "stage" in this scale goes from mere experienced sensation to self-consciousness. As Merker puts it, "*reflective awareness is thus more akin to a luxury of consciousness on the part of certain big-brained species, and not its defining property; it is one of the many contents of consciousness available to creatures with sophisticated cognitive capacities.*"[**?**]

5.2 The Self

In order to understand the self, we need to enter a new order of references. What we are considering is the nature of the observer – "the one who is aware of" and his or her consciousness dynamics. The observer is the one who is able to see what is in sight, which enables us to introduce a third axe in our graph, leading us to a fourth dimension path: a quantum world perspective.

In 1916, Einstein abandoned the idea of space and time as something separate from the material content of the universe. The General Theory of Relativity becomes a theory of observables. He wrote then:

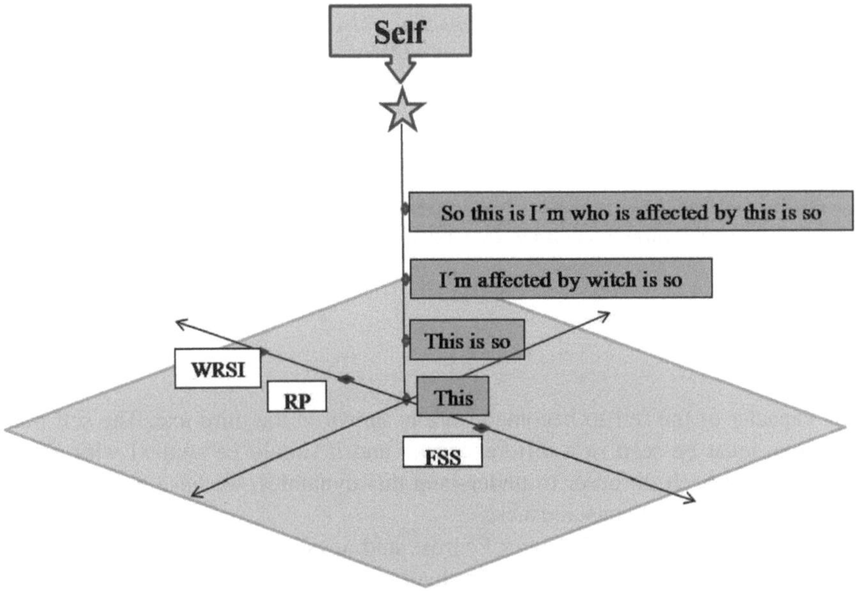

Fig. 11 The third axe of the graph opens a new reference order in the consciousness process and leads to another dimension outside time-space references. This is self dimension, which defines the uniccity of human existence outside the phenomenal world

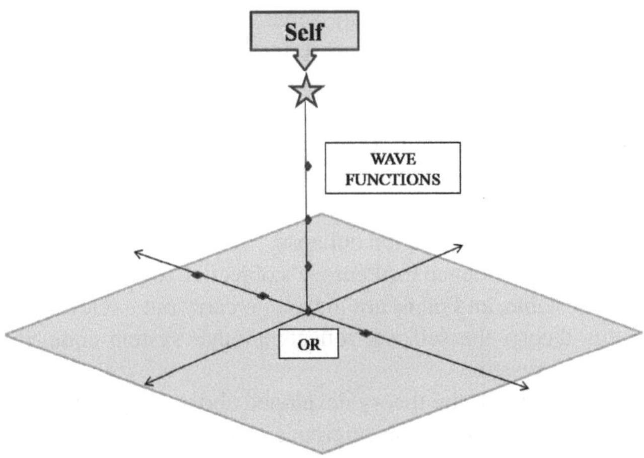

Fig. 12 The self dimension is the dimension of probabilities, where the person experiences the possibility of changing future actions, planning creative solutions and working out responses to feelings, in order to consolidate them into the matter through the nervous system

"All our space-time verifications invariably amount to a determination of space-time co-incidences. If, for example, events consisted merely in the motion of material points, then ultimately nothing would be observable but the meetings of two or more of these points. Moreover, the results of our measuring are nothing but verifications of such meetings of

the material points of our measuring instruments with other material points, coincidences between the hands of the a clock and points on the clock dial, and observed point-events happening at the same place at the same time. The introduction of a system of reference serves no other purpose than to facilitate the description of the totality of such coincidences". (Einstein 1916).

This idea is the main basis of our tridimensional graph, which includes the self as an observer of the attentional phenomenon as an awareness processor (the one who is aware of awareness) and with relative awareness capacity.

6 Conclusion

The capacity of the self to become aware is shown in the third axe. The self phenomenon must be seen in a different order and it should be studied with a new scientific approach. In order to understand this dynamics, we have been studying the Orch Or theory of consciousness.

Theoretical physicist Sir Roger Penrose and anesthesiologist Stuart Hameroff, [15] in there joint work the Orch Or theory of consciousness, assume that consciousness emerges from the brain, and the main focus is on complex computations that occur in synapses.

Penrose postulates that each quantum superposition (possible position of the particle) has its own piece of spacetime curvature. According to his theory, these different bits of spacetime curvature are separated from one another, and constitute a form of blister in spacetime. Penrose further proposes a limit to the size of this spacetime blister. This is the tiny Planck scale of (10–35 m). Above this size, Penrose suggests that spacetime can be viewed as continuous, and that gravity starts to exert its force on the spacetime blister. This is suggested to become unstable above the Planck scale, and to collapse so as to choose just one of the possible locations for the particle. Penrose calls this event "objective reduction (OR)", reduction being another word for wave function collapse.

There is no existing evidence for Penrose's objective reduction, but the theory is considered to be testable, and plans are in hand to carry out a relevant experiment.

Based on this theory, the self axe reflects another system code. It is a fourth dimension reference system.

Applying this concept to the theory developed above, the third axe nature could be the element that brings forth the reference of consciousness evolution and the possibility of understanding the easy problem of consciousness.

Applying the Orch OR model to our model, the awareness state could be an OR (objective reduction) point that has it own value measured by the amount of **photons** concentrated **in a given area**, in the specific moment, promoting a particular tuning of the brain:

- Feedforward processing (FFS)
- Recurrent processing (RI)
- Widespread recurrent processing (WRI)

We intend to continue this study on orthogonal graph and further develop the concept of awareness through wave function dynamics, that is, the third axe nature. From our point-of-view, that is the key to understanding the hard problem of consciousness.

References

1. Chun, Marvin M. and Wolfe Jeremy M., *Blackwell Handbook of Perception*, Chapter 9, Version of July 7, 2000
2. S. Frohlich and C.A.S. Franc *The neuropsychological function of the 12 cranials nerves*, http://www.nce.ufrj.br/ensino/posgraduacao/strictosensu/neurociencias/the_NEUROPSYCHOLOGICAL_FUNTION_OF_THE_CRANIAL_NERVES.pdf, 2007
3. V.A.F. Lamme, *Separate neural definitions of visual consciousness and visual attention; a case for phenomenal awareness*, Neural Networks 17 (2004) 861–872, 2004 Special Issue
4. Crick, Francis& Koch, Christof, *Towards a Neurobiological theory of consciousness*, seminars in The Neuroscience, Vol 2, 1990: pp 263–275, 1980
5. Jochen Braun, *It's Great But Not Necessarily About Attention*, http://psyche.cs.monash.edu.au/v7/psyche-7-06-braun.html PSYCHE, 7(06), March 2001
6. Woolf, Nancy and Hameroff, Stuart R. *A Quantum approach to visual consciousness.* Trends in Cognitive Science, vol 5 no. 11 November 2001
7. Howard E. Egeth and Steven Yantis, VISUAL ATTENTION: Control, Representation, and Time Course. Annu.Rev.Psychol.1997, 48:269–97
8. P.F Balan and Jacqueline Gottlieb, *Integration of Exogenous Input into a Dynamic Salience Map Revealed by Perturbing Attention* The Journal of Neuroscience, September 6, 2006 · 26(36):9239–9249 · 9239
9. V.A.F. Lamme, *Separate neural definitions of visual consciousness and visual attention; a case for phenomenal awareness*, Neural Networks 17 (2004) 861–872, 2004 Special Issue
10. Bentley, P. and Vuilleumier, P. and Thiel, C.M. and Driver, J. and Dolan, R.J. (2003) *Cholinergic enhancement modulates neural correlates of selective attention and emotional processing.* NeuroImage, 20 (1). pp. 58–70. ISSN 0049-9-X, pp. 165–170
11. Flávio Alóe; Alexandre Pinto de Azevedo; Rosa Hasan *Sleep-wake cycle mechanisms*, Rev. Bras. Psiquiatria. vol. 27 suppl. 1 São Paulo May 2005
12. Woolf, Nancy and Hameroff, Stuart R. *A Quantum approach to visual consciousness.* Trends in Cognitive Science, vol 5 no. 11 November 2001
13. Francis Crick and Christof Koch - *Consciousness and Neuroscience* -Cerebral Cortex, **8**:97–107, 1998
14. Chalmers, David - *First-Person Methods in the Science of Consciousness* - Arizona *Consciousness Bulletin*, 1999.
15. Bjorn Merker -*Behavioral and Brain Sciences* -Cambridge University Press 2006
16. Hameroff, Stuart. "*Quantum Computation in Brain Microtubules? The Penrose-Hameroff 'Orch OR' Model of Consciousness.*" Philosophical Transactions in the Royal Society of London 356 (1998): 1869–1896

Computational Consciousness: Building a Self-Preserving Organism

Allan Kardec Barros

Abstract Consciousness has been a subject of crescent interest among the neuroscience community. However, building machine models of it is quite challenging, as it involves many characteristics and properties of the human brain which are poorly defined or are very abstract. Here I propose to use information theory (IT) to give a mathematical framework to understand consciousness. For this reason, I used the term "computational". This work is grounded on some recent results on the use of IT to understand how the cortex codes information, where redundancy reduction plays a fundamental role. Basically, I propose a system, here called "organism", whose strategy is to extract the maximal amount of information from the environment in order to survive. To highlight the proposed framework, I show a simple organism composed of a single neuron which adapts itself to the outside dynamics by taking into account its internal state, whose perception is understood here to be related to "feelings".

1 Introduction

Although an old subject of psychological/philosophical discussion, consciousness has been attracting attention of the neuroscience community in recent years. Indeed, it has been controversial under different point of views to explain how consciousness arises or appears within the limits of the human brain. Even its definition is controversial. However, a number of worldwide known names have entered recently in the "quest for consciousness" [Koch, 2004].

Koch and Crick [Koch, 2004] suggested the idea of neural correlates of consciousness (NCC), the minimal brain mechanisms responsible for any one specific conscious percept, thought or memory. Some physiological phenomena such as masking, fading and attention are regarded as related to consciousness. Moreover,

A.K. Barros (✉)
Federal University of Maranhão, Brazil
e-mail: allan@ufma.br

A. Hussain et al. (eds.), *Brain Inspired Cognitive Systems 2008*, Advances in Experimental Medicine and Biology 657, DOI 10.1007/978-0-387-79100-5_17,

some researchers relate it to emotion. In this regard, Damasio, stated that feelings are also perceptions [Damasio 2000 and 2004].

In "does consciousness exist", William James [James, 1904] begins a discussion stating that rather than an entity, the word consciousness would stand for a function. As such, one can notice that still in the beginning of last century there was a concept being constructed where there would not exist any longer the idea of the homunculus, a little person in the brain, the "I", which looks out at the world and initiates the body actions to be taken.

In this regard, what happens if we try to mimic the human consciousness in a machine? Can it be a simple computer – a deterministic, pre-programmed one? Or does it need to be more elaborated – in this case, which kind of elaboration would be that one? Alexander and Dunmall [Alexander and Dunmall, 1904], for example, proposed a number of axioms that are useful for fully developing a conscious machine.

Here I focus on the idea that consciousness could be related to "the perception that an organism has of itself" [Damasio, 2004], aiming at self-preservation. Additionally, in the literature where consciousness is not directly defined, one can notice that some terms which are very much related to it appears constantly in the discussions. For example, the words "knowledge", "perception", "emotion" or "thought".

However, it is important to remember the advances made on the studies and modeling of how the cortex adapts to the environment. Indeed, a number of works have been published after Barlow relating perception to information theory [Barlow, 1989]. One concept which has grown consistently in recent years is that the brain would code information based on minimal mutual redundancy which is highly related to the idea of statistical independence [Hyvarinen, 2001]. This means that the brain copes with information processing by ignoring or reducing the redundant information (e.g. in a Shannon sense [Shannon, 1948]).

Information coding can be understood in the following way. If we imagine that, for example, the visual and auditory system evolved coupled to the external world stimuli, then we can infer that their organization evolved so that they extract the maximum amount of information from real world images or sounds.

Evidence for this reasoning are the works on the visual [Olshausen and Field, 1996] and auditory [Lewicki, 2002, Smith and Lewicki, 2006] cortices, where statistical independence was shown to be an efficient procedure to code information. Equivalently, by assuming that the autonomic responses evolved coupled to systemic demands in heart rate regulation, Barros and Principe [Barros and Principe, submitted] showed that it is possible that the autonomic interaction also exploits independence principles. Indeed, what happens if we think of the cortex as a computer implementing a sparse coding algorithm?

I propose here that consciousness arises when new information, which are important for the organism to self-preserve itself, modify the current state of the whole or of a part of the organism. Important information are those which a given internal cost function is based on. It is based on an information-theoretical framework,

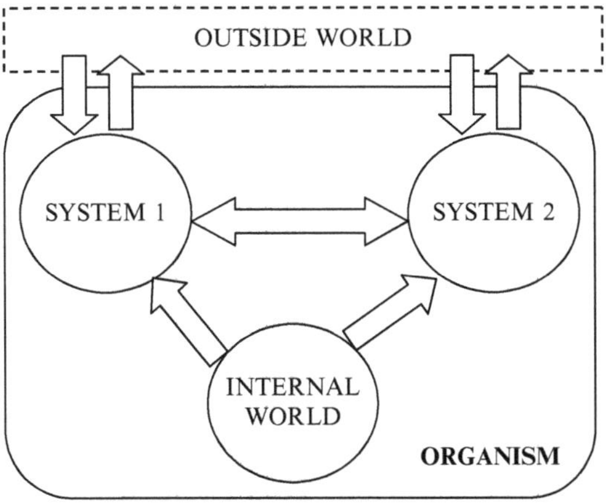

Fig. 1 Diagram for the self-preserving organism. It responds to outside stimuli taking into account the internal world. Between the outside and the internal world, there are some intermediate systems (1 and 2), all acting on each other. Every system consists of an automaton, which basically responds to stimuli that are already known, and a network which learns new stimuli (eq. 2). Obviously, the organism takes action on the outside world

such as Kullback-Lieber divergence, maximal likelihood or correlation. However, this learning only occurs when the current internal state of the organism yields so. In other words, emotion interferes in arousal of consciousness. It is important to stress here that I understand emotion as the perception of the internal state of the organism [Damasio, 2004]. To highlight this idea, I propose a simple cost function which can be easily implemented in a unicellular organism with multiple sensors.

2 Computational Consciousness

In order to avoid possible misuse of the word consciousness, which has no definition currently, I use "computational consciousness" to stress that it is related to information-theoretic concepts and can be implemented in a computer or any other programmable system. In general terms, the proposed system can be understood by the diagram as shown in Fig. 1. One can notice that there is both an outside and an inside world acting on two systems. There could be N systems, however, two are enough to highlight the idea here. Notice that they are interconnected and therefore act mutually on each other. Let us show then how a single system works.

enough to highlight the idea here. Notice that they are interconnected and therefore act mutually on each other. Let us show then how a single system works.

3 Sparse Codes

Information in the brain is represented by the pattern of activation of a large population of neurons. It is a multi-stage process of converting the sensory input into a "subjectively meaningful" experience. This process is composed of "cells" or stages. At each stage, the input information is processed, leading to an output which depends on an algorithm. This is carried out in a manner which is today understood as "neural code".

A sensory input, for example, an image, $I(x, y)$, can be coded as a linear superposition of some basis functions, $\varphi(x, y)$, which can be written as $I(x, y) = \sum_i a_i \varphi_i(x, y)$, where a are the coefficients which activate the corresponding basis function.

Thus, each image which "enters" the retina is processed in the early visual system is firstly coded as in (1). Results on the neuroscientific side suggests strongly that this is achieved through an overcomplete representation, where the number of neurons to represent an image is much larger than the dimensionality of the input pixels.

For example, in terms of cat primary visual cortex, there is an expansion rate of 25:1 in terms of axons projecting [Field, 1994, Olshausen and Field, 1996]. This suggests that there are few neurons active within a given time instant. This suggests that sparsificity is an strategy used in neural systems. Indeed, Olshausen and Field used an algorithm which enforced sparcificity among neurons and found that the coding of natural images yielded Gabor-like filters which resembled the receptive fields of V1 [Keffler, 1952, Olshausen and Field, 1996].

There are a number of tools to find the sparse codes. So far, all of them are based on higher order statistics, such as kurtosis [Hyvarinen et al., 2001], information maximization [Bell, 2003], maximum likelihood, etc. In figure 4 I show one example of Gabor wavelet-like basis functions generated by a code strategy based on kurtosis.

3.1 About a Single System

There is extensive literature on learning. However, how can one take into account the internal state of the system to guide learning and therefore consciousness? Usually in any learning system, there is some cost function which should be minimized or maximized. As an example, let us say that one system can be implemented by one single neuron (Extending this idea to more complex systems is straightforward).

Say that there is a vector with M inputs $\mathbf{x}(t)=[x_1(t), x_2(t), \ldots, x_M(t)]^T$, where T means transpose and t stands for time. One can thus think of an information-theoretic based function which depends on $\mathbf{x}$ and on a parameter vector $\mathbf{\Phi}(t)=[\mathbf{\Phi}_1(t), \mathbf{\Phi}_2(t), \ldots, \mathbf{\Phi}_M(t)]^T$. Let us define this function as $F(\mathbf{\Phi}, \mathbf{x})$, whose gradient is $f(\mathbf{\Phi}, \mathbf{x})$. Thus, one can think of a change in the parameter space from the current state t to another one $t+\tau$, $\mathbf{\Phi}(t)$ and $\mathbf{\Phi}(t+\tau)$, respectively. Let us also assume that there is a function which measures the current state of the internal world, $\psi(t, s)$, where s is an internal variables vector, which in this system can be understood as "emotion" – in the sense explained by Damasio, who relates emotion to the internal state of the body. This state can be for example the voltage level of a given device or the level of energy. Thus, I propose the following type of cost function,

$$[\textbf{next state}] = [\textbf{current state}] + [\textbf{internal state}][\textbf{gradient}], \tag{1}$$

or, in mathematical notation,

$$\mathbf{\Phi}(t) = \mathbf{\Phi}(t - \tau) + \psi(t, s)f(\mathbf{\Phi}, \mathbf{x}). \tag{2}$$

It is important to stress that the organism acts upon the outside world by the following function

$$y(t) = \mathbf{\Phi}(\mathbf{t})^{\mathrm{T}}\mathbf{x}. \tag{3}$$

3.2 An Example

I propose a simple organism which can be regarded as "conscious" in the framework proposed here. It is composed of a single neuron which can be "feeded" by a battery, i.e., it is sensitive to a voltage level of a battery. It is composed as well of M two dimensional sensors (which mimics eyes). Thus, besides the M sensors, the organism has an additional one, which "feeds" it (x_s). One can think of a photo-electric cell in this case, which loads the battery. A diagram of the system is shown in Fig. 2.

The learning can be verified by observing how the parameter vector changes according to a change in the environment, which obviously mean adaptation to this environment. The parameter vector in the case of images can be regarded as the receptive fields that are found in the primary visual cortex (V1) [Keffler, 1952, Olshausen and Field, 1996].

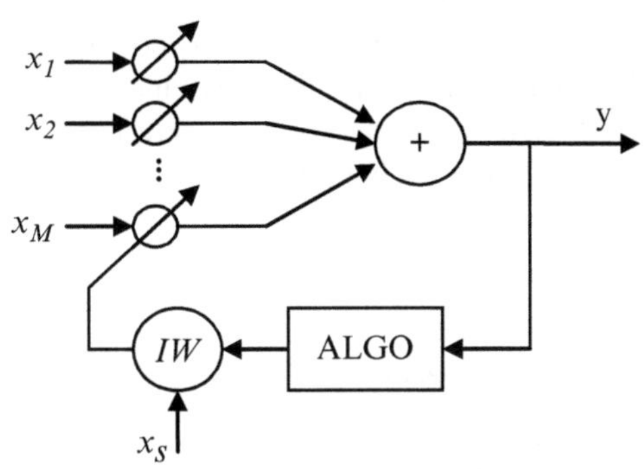

Fig. 2 Block diagram for a system with a single neuron. It has M inputs plus another one (xs) which controls the internal state (feeding). Its output acts upon the outside world (by making a movement towards food or to escape from a predator for example). An algorithm along with the internal world (IW) modifies the weights

3.3 *Learning Rule*

In order to show that the organism can be highly simple and still computationally conscious, I propose the following learning rule to be implemented in the system, which is a modified version of the algorithm proposed by Barros and Cichocki [Barros and Cichocki, 2001]. Firstly, let us define the following error function e(t)

$$e(t) = y(t) - y(t - t1), \tag{4}$$

where t1 is a previously chosen delay. Biologically speaking, one can think of a genetically pre-defined one.

Defining the cost function to be the expectation of the square of (4), i.e., the mean squared error, $E[e(t)^2]$ one can easily find that the weights $\Phi(t)$ can be updated by the following rule,

$$\Phi(t) = (1 - \psi(t, s))\Phi(t - \tau) + \psi(t, s)E[\mathbf{x}\mathbf{x}^T]^{-1}E[y(t - t1)\mathbf{x}], \tag{5}$$

where $\psi(t, s)$ can be either one or zero, as adaptation is required by the internal state the or not, respectively.

I have chosen this rule because one can prove that, for a given input statistics, it yields only one single independent component [Barros and Cichocki, 2001], although scaled. I.e., the mean squared error has only one single minimum. Other learning rules such as kurtosis have many ceiling poits (maximum and minimum) [Delfosse and Loubaton, 1995].

4 Simulation Results

In order to be as realistic as possible, I carried out some experiments where the inputs to the 2D sensors were two 512 × 512 natural images, as shown in Fig. 3. This was to simulate a change in the environment. The simulation was carried out by randomly taking a total of 20.000 16 × 16 images. I obtained the parameters for the image in Fig. 3(a), then, by changing the internal state $\psi(t, s)$ to 1, the other parameters regarding Fig. 3(b) were learnt. The outputs were delayed in space as in eq. (4), and for both images the parameters were found for the same delay. The results are shown in Fig. 4.

5 Discussions

Here I have proposed an artificial organism that can be regarded as "computationally conscious", based on how it adapts to changes in the environment by taking into account its internal state. The proposed organism is shown in Fig. 1. There are two key

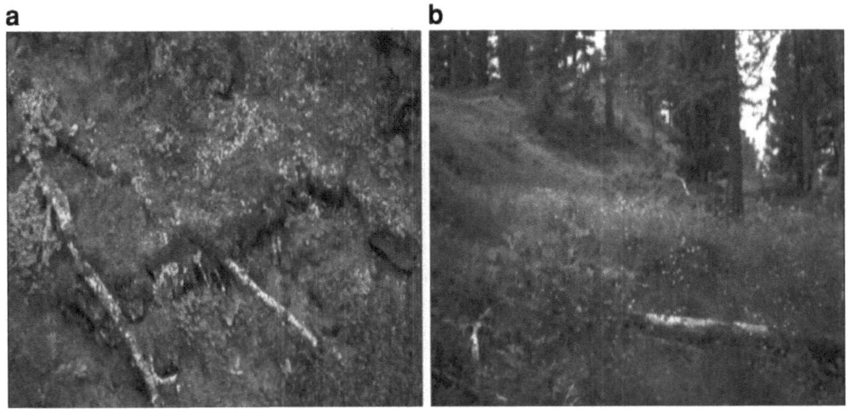

Fig. 3 Two original images used to simulate the learning of two different environments (**a**) and (**b**)

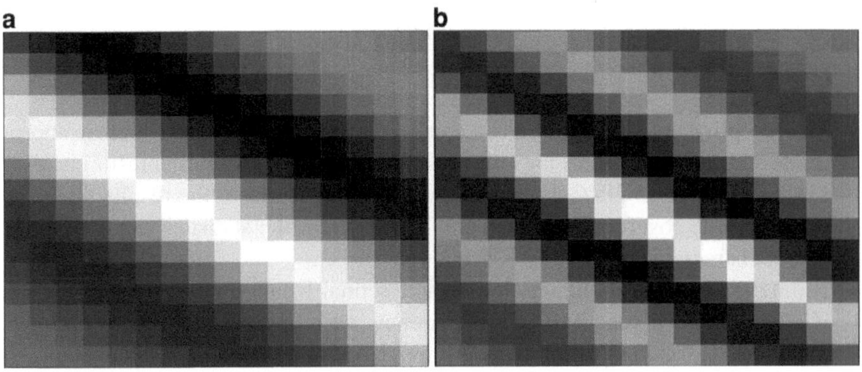

Fig. 4 Two set of parameters (equivalent to receptive fields) obtained using the proposed algorithm with one single delay. (**a**) Obtained with Fig. 3(a); (**b**) Obtained with Fig. 3(b). Notice the adaptation for each environment

features of this organism: 1) It is capable of learning and; 2) For that, it takes into account its internal state. The learning happens by using the concept of redundancy reduction. In this regard, it is important to stress that in this organism, counsciousness does not arise in a particular site, rather it is a function, just as thought by James [James, 1904]. Thus, the neural correlates of consciousness as called by Koch [Koch, 2004], can be found as the second term in the right side of eq. (1), standing therefore as a function, rather than a physical entity with an anatomical location.

As an example, I have proposed a very simple (unicellular) organism which uses a yet simpler leaning rule, based basically on the difference between pixels. By examining the found parameters in Fig. 4, one can clearly see a change in the parameters, although I was using the very same delay in the algorithm. This means that the algorithm could capture the changes in the environment, i.e., images. One can clearly see that while Fig. 3(a) is composed of features which do not change abruptly from one pixel to another, in Fig. 4(b) the parameter vector (or receptive field) learned the higher frequency content of Fig. 3(b). It is important to stress here that equivalent results were obtained by using other cost functions proposed in the literature [Hyvarinen, 2001].

However, more complex organisms can be thought using this simple model, and using different cost functions which carry out redundancy reduction strategies. In those complex organisms, one can think of, for example, that a given system (as shown in Fig. 1) can be active in a given time frame while other is active in another time, in a phenomenon called winning attention. Another feature that can be as well carried out is to use different cost functions to complex organism. One particularly elegant way to implement that is by using the concept of the co-information lattice suggested by Bell [Bell, 2001].

There are some other points that should be discussed in this work. The first one regards how plausible in physiological terms is the proposed organism. Moreover, one has to discuss the statistical grounds of it. Let us first discuss the physiology.

Physiological Plausibility

- The receptive field of a neuron is the region in the visual field in which an appropriate stimulus modulates the cell's response [Kuffler, 1952]. Interestingly, they can be modeled as 2D Gabor functions. One can see from the results that even a very simple cost function based only on second-order statistics can unfold this Gabor characteristics;
- Fading: Stabilizing an image to the same retinal location causes vision to fade [Coppola and Purves, 1996]. In the framework proposed here, it can be interpreted as information diminution. Although not implemented in the example, it can be easily added, as this fading property belongs to a sparse coding strategy acting upon time.
- Masking: A given stimulus, the mask, can interfere in the processing of another, the target. Again, I interpret it as a form of redundancy reduction coding, which responds only to one of the stimuli, whether they are close in time, space, statistics or other domain.
- Lower level internal state (i.e., sensory impulses from regions outside the brain). The thalamus is the intermediate relay point and processing center for most of sensory impulses, ascending to the cerebral cortex from the spinal cord, brainstem, cerebellum, basal ganglia, etc. The thalamus relays its output to the cerebral cortex. Reports on lesions in the thalamic intralaminar nuclei can completely knock out all awareness [Hunter and Jasper, 1949];
- Emotions (higher level internal state): Bilateral ablation of the amygdala has the effect of flattening emotion. Theoreticians relate consciousness to emotion. I propose here to use the idea of Damasio, who stated that feelings are also perceptions In this regard, Damasio, stated that feelings are also perceptions [Damasio 2000 and 2004]. Thus, by taking the previous item and this, we can say that, besides

the visual and auditory perception, feelings, in the context of computational consciousness, shall be interpreted as the perception of the current state of the system.
- Winning attention: In the case of two or more concomitant stimuli, attention conscious perception shall be of the winner one. This is again an example of redundancy reduction that can be easily implemented by a winner-take-all type of network.
- Although the nervous system is a single, unified communication network, it can be divided on a gross anatomical basis into the central nervous system (brain and spinal cord) and peripheral nervous system (cranial and spinal nerves, with afferent and efferent nerve cells). In the future, we can think of a model which makes use of afferent/efferent pathways to interact with the external world;

Following Barlow reasoning on perception, I believe that redundancy reduction plays a fundamental role on the arousal of consciousness. Indeed, Barlow suggested that a possible strategy for the brain to code information is by forcing redundancy reduction [Barlow, 1989]. Indeed, Olshausen and Field [Olshausen and Field, 1996] and Lewicki [Lewicki, 2001] showed that by applying algorithms which uses the concept of sparseness or statistical independence, the basis functions which were found resemble the receptive fields in primary visual cortex (see Fig. 3) or that the filters found using this strategy to sounds resembled colchlear wavelet-like filters. Similar strategy was used to learn the parts of objects [Lee and Seung, 1999]. It is also important to highlight that works relating complex systems to sparseness to information coding were also carried out and might be a good indication that redundancy reduction plays a fundamental role for the organism survival.

It is not difficult to think of a form to implement this organism as in a robot, where besides its main task, it should be capable of self-preservation. Therefore, it could survive in hostile environments. That is, it would be an automaton, in the sense that it would take immediate actions after a stimulus. In this regard, it would act as a zombie in the words of Koch [Koch, 2004]. For example, if a given visual or auditory stimuli is given and the system was already trained for that particular stimuli, an arm movement is made. This is basically the old-type robot style. However, it would be capable of self-preserving itself by sensing how hostile the new environment is. For example, if an aggression occurs, this fact has to be sensed and learned so that the organism would be capable of avoiding it in the future.

6 Remarks

REMARK 1: A certain line of thought in studying consciousness is to use determinist machines to mimic human beings. As determinist machines I mean those which have all the answers or codes previously fixed, and react to the external stimuli accordingly. One example are the so-called Turing Test. Turing [turing] described the following game. Suppose that we have a person, a machine, and an interrogator.

The interrogator is in a room separated from the other person and the machine. The task is: the interrogator should determine which of the other two is the person and which is the machine. The usual way to implement such test so far is by the use of determinist responses.

If we take the reasoning and claims of Koch, then definitely we should have some kind of learning in order for a given machine to be conscious. Indeed, a zombie behavior requires fast response and, contrary to conscious events, they can be regarded in general as automatic. We can conclude that zombies react to environmental inputs simply in a determinist fashion: it is a brain reflex to a given output. If we take as correct the reasoning and speculations of Koch, then definitely we should have some kind of learning in order for conscious to appear. Moreover, we can infer as well that consciousness is transient. Then we have the following remark:
REMARK 2: Consciousness is transient and requires learning. It is not difficult, therefore, to imagine a machine which fulfills the requirements of carrying out a sparse coding strategy to deal with information entering from the external world; that is both zombie and conscious – once at a time, of course.

References

Aleksander I, Dunmall B (2003) Axioms and tests for the presence of minimal consciousness in agents. J Conscious Stud 10(4–5):7–18.

Baars, B.J.: A Cognitive Theory of Consciousness (Cambridge Univ. Press, Cambridge, MA, 1988).

Barlow, H.B.: "Unsupervised learning". Neural Computation, 1:295–311, 1989.

Barros, A. and Principe, J: "A Model for Neural Regulation of Heart Rate Based on Statistical Independence". Submitted to neural computation.

Bell. A: "The co-information lattice". In Proceedings of the 4th Symposium on Independent component analysis and Blind signal separation (ICA2003), pp. 921–926. 2003.

Coppola, D. and Purves, D. "The extraordinary rapid disappearance of entoptic images", Proc. Natl. Acad. Sci. USA 93: 8001–8004, 1996.

Damasio, A.: Body and emotion in the making of consciousness. In Portuguese. Shchwarcz Editora. 2000.

Damasio, A.: Looking for Spinoza: joy, sorrow and the feeling brain. In Portuguese. Shchwarcz Editora 2004.

Dawson, G. D.: The central control of sensory inflow. Proc. Roy. Soc. Med., London 51 (5), 531–535 (1958).

Delfosse, N, Loubaton, P. "Adaptive blind separation of sources: a deflation approach", Signal Processing. 45: 59–83. 1995.

Edelman, G. M., W. Einar Gall, W. M. Cowan (eds.): Signal and Sense. Local and Global Order in Perceptual Maps. Wiley, New York 1990.

Field, D. J.: "What is the goal of sensory coding?" Neural Computation, 6:559–601. 1994.

Hagbarth, K. E., D. J. B. Kerr: Central influences on spinal afferent conduction. J. Neurophysiol. 17 (3), 295–297 (1954).

Hassler, R.: Interaction of reticular activating system for vigilance and the corticothalamic and pallidal systems for directing awareness and attention under striatal control. In: Buser et al. (eds.) 1978.

Hunter, J. and Jasper, H.H. "Effects of thalamic stimulation in anaesthetized cats," EEG Clin. Neurophysiol. 1: 305–315. 1949.

Hyvarinen, A, Karhunen, J, Oja, E. Independent Component Analysis. John Wiley and Sons. 2001.
James, W: Does 'consciousness' exist? Reprinted in: G. N. A. Vesey (ed) Body and mind: readings in philosophy. London: George Allen & Unwin, 1970, pp. 202–208. 1904.
Koch, C: The quest for consciousness. A neurobiological approach. Roberts and Company Publishers. 2004.
Kuffler, S.W: "Neurons in the retina: Organization, inhibition and excitatory problems", Cold Spring Harbor Symp. Quant. Biol. 17: pp. 281–292. 1952.
LeDoux, J. E.: Emotional networks in the brain. In: Lewis, M., J. M. Haviland (eds.): Handbook of Emotions. Guildford Press, New York 1993.
Lee D. D. and Seung, S. "Learning the parts of objects by non-negative matrix factorization". Nature, Vol 401. pp. 788–791, 1999.
Lewicki, M. S. "Efficient coding of natural sounds". Nature Neuroscience, vol. 5, 356–363, 2002.
Mangun, G. E., S. A. Hillyard, in: Scheibel, A. B., A. F. Wechsler (eds.): Neurobiology of Higher Cognitive Function. Guildford Press, New York 1990.
Meric, C., L. Collet: Attention and otoacoustic emissions. Neuroscience and Behavioral Reviews 18 (2), 215–222 (1994).
Newman, J., B. J. Baars: A neural attentional model for access to consciousness: a global workspace perspective. Conceptions in Neuroscience 4 (2) 255–290 (1993).
Olshausen, B. A. Field, D. J. "Emergence of simple-cell receptive field properties by learning a sparse code for natural images". Nature Vol. 381, 607–609, 1996.
Scheibel, A. B.: The brain stem reticular core and sensory function. In: Handbook of Physiology. The Nervous System, Vol. III, 1. American Physiological Society, Bethesda 1984.
Scheibel, A. B., A. F. Wechsler (eds.): Neurobiology of Higher Cognitive Function. Guildford Press, New York 1990.
Shannon, CE, "A mathematical theory of communication", Bell System Technical Journal, Vol. 27, pp. 379–423, 1948.
Valdes-Sosa, P., Sanchez-Bornot, JM., Lage-Castellanos A, Vega-Hernandez M, Bosch-Bayard J, Melie-Garcıa L and E Canales-Rodrıguez L, "Estimating brain functional connectivity with sparse multivariate autoregression" Philosophical Transactions of the Royal Society B. Theme Issue on Multimodal Brain Connectivity. (Eds) P. Valdes-Sosa, R. Kotter, K. Friston. in press.
Zeki, S.: Functional specialization in the visual cortex: the generalisation of separate constructs and their multistage integration. In: Edelman, G. M., et al. 1990, pp. 85–130.

The Hippocampal System as the Cortical Resource Manager: A Model Connecting Psychology, Anatomy and Physiology

L. Andrew Coward

Abstract A model is described in which the hippocampal system functions as resource manager for the neocortex. This model is developed from an architectural concept for the brain as a whole within which the receptive fields of neocortical columns can gradually expand but with some limited exceptions tend not to contract. The definition process for receptive fields is constrained so that they overlap as little as possible, and change as little as possible, but at least a minimum number of columns detect their fields within every sensory input state. Below this minimum, the receptive fields of some columns are expanded slightly until the minimum level is reached. The columns in which this expansion occurs are selected by a competitive process in the hippocampal system that identifies those in which only a relatively small expansion is required, and sends signals to those columns that trigger the expansion. These expansions in receptive fields are the information record that forms the declarative memory of the input state. Episodic memory activates a set of columns in which receptive fields expanded simultaneously at some point in the past, and the hippocampal system is therefore the appropriate source for information guiding access to such memories. Semantic memory associates columns that are often active (with or without expansions in receptive fields) simultaneously. Initially, the hippocampus can guide access to such memories on the basis of initial information recording, but to avoid corruption of the information needed for ongoing resource management, access control shifts to other parts of the neocortex. The roles of the mammillary bodies, amygdala and anterior thalamic nucleus can be understood as modulating information recording in accordance with various behavioral priorities. During sleep, provisional physical connectivity is created that supports receptive field expansions in the subsequent wake period, but previously created memories are not affected. This model matches a wide range of neuropsychological observation better than alternative hippocampal models. The information mechanisms required by the model are consistent with known brain anatomy and neuron physiology.

Keywords Pyramidal neuron · Cortical column · Receptive field · Hippocampus · Semantic memory · Episodic memory

---

L.A. Coward (✉)
Department of Computer Science, Australian National University, Canberra, ACT 0200, Australia
e-mail: andrew.coward@anu.edu.au

A. Hussain et al. (eds.), *Brain Inspired Cognitive Systems 2008*, Advances in Experimental Medicine and Biology 657, DOI 10.1007/978-0-387-79100-5_18,

1 Introduction

Since the observations of the combination of memory deficits observed in patients after surgical removal of parts of their hippocampal system [Scoville and Milner, 1957], there has been strong interest in the role of this structure in memory.

However, these and subsequent observations demonstrated three dissociations which have presented challenges to understanding the actual role of the hippocampal system. One is that although there can be severe anterograde amnesia for both semantic and episodic memory, retrograde amnesia is stronger for episodic memory. The second is that speech capabilities, general intelligence, and previously acquired skills are unaffected, despite the memory deficits. The third is that although the ability to create new declarative (i.e. semantic and episodic) memories is strongly affected, a significant ability to learn sensorimotor skills and to perform repetition priming is retained.

Furthermore, lesions to diencephalic structures such as the mammillary bodies and the anterior thalamic nuclei can generate similar combinations of deficits in the absence of damage to the hippocampal system. Thus damage to the mammillary bodies of the hypothalamus can result in anterograde memory deficits [Tanaka et al. 1997], damage to the anterior thalamic nuclei can result in both anterograde and retrograde amnesia [Caulo et al. 2005], but again in such cases all other cognitive capabilities are unaffected. It has also been observed that the amygdala plays a role in enhancing the memory of emotional events [Phelps 2006]. However, there is developing evidence that these structures have these effects through their action on the hippocampal system [Caulo et al. 2005; Dolcos et al. 2004], and should therefore be regarded as an integral part of that system.

A wide range of functional roles has been proposed for the hippocampal system to account for the observed combination of deficits. Many of these models propose two component systems to account for the combination of global anterograde semantic and episodic amnesia with stronger retrograde episodic amnesia. Typically these models have a component supporting stimulus memory and a component supporting episodic retrieval [e.g. Gluck et al. 2003], and argue that detailed stimulus information is initially registered in the hippocampal system and gradually transferred to long-term storage in the neocortex. The models in general have issues in providing an account for the full range of experimental observations [Cohen et al. 1999], and do not provide any functional reason for the roles of the mammillary nuclei, anterior thalamic nuclei and amygdala other than speculation about possible redundancy [e.g. Graff-Radford 1990].

An alternative concept [Coward 1990; 2000; 2005a] is that the primary role of the hippocampal system is management of the information recording resources of the cortex. A major part of this role is determining at each point in time where information about current sensory inputs will be recorded in the neocortex, performing this function by managing a competition between all cortex areas to determine the most appropriate combination of locations. A side effect of this function is that the hippocampal system acquires information about which cortex locations record information at the same time, information critical for episodic memory retrieval and navigation. Retrieval of semantic memory, on the basis of associations between cortex locations that are frequently active at the same time (not necessarily with information recording), can become independent of the hippocampal system. The role of sleep includes configuration of some neocortex resources to be as appropriate as possible for recording information in the immediate future, using past experience (with a bias in favor of the most recent) as the best available estimate for future experience.

As described in this paper, the resource management concept can be developed into a detailed model that provides an intuitively simple reason for the existence of the hippocampal system, eliminates the need for complex information transfers back and forth between neocortex and hippocampus, and provides straightforward reasons for the existence of the various dissociations. An integrated account can be provided for the roles of different parts of the hippocampal system, the anterior thalamic nuclei, the mammillary bodies and the amygdala in memory. High level functional processes can be mapped into known or plausible neuron processes, for example functional learning into long term potentiation (LTP). Finally, it includes a memory related role for sleep including dream sleep that is more consistent with experiment that the alternative memory consolidation models.

Furthermore, there are system architectural arguments [Coward 2001] indicating that any system that must learn a complex combination of behaviours will tend to be constrained into an architectural form with separations between some specific subsystems. The constraints include a separation between one subsystem that records stimulus information, another that records response information, and a third that manages the resources of the stimulus recording subsystem, determining where new information will be recorded in response to a novel stimulus. The constraints require a new stimulus to be learned instantaneously (and relatively permanently) by slight expansions to the receptive fields of a small set of modules within the subsystem, but responses to those stimuli are learned gradually by continuous variation of behavioural weights assigned to the outputs from stimuli detection modules. With the neocortex corresponding with the stimulus recording subsystem, cortical columns corresponding with stimuli detection modules, the thalamus and basal ganglia corresponding with the response recording subsystem, and the hippocampal system with the resource manager, the dissociations in damage following hippocampal system damage can be understood, because although no new stimuli can be learned, the behavioral associations of existing stimuli are preserved and can continue to change. Knowledge acquired by the resource manager about which columns recorded information at the same time can be used to reconstruct episodic memories, but not memories based on frequent simultaneous presence of different stimuli (i.e. semantic memories such as word meanings).

This paper presents a detailed model of hippocampal functions based on the resource management concept. Firstly, a range of previous models for the role of the hippocampus are described. Then the architectural constraints on complex learning systems (as described fully in Coward [2001]) and how they apply to the brain are outlined. The differences between the information models for semantic, episodic, priming and procedural memory that follow from the architectural constraints are described. Next, an overview of the relevant physiology of the hippocampal system is presented. The proposed resource management model is then described, including the roles of each structure in the hippocampal system, the roles of the participating hypothalamic, thalamic and amygdala nuclei, and the role of sleep including REM sleep. The way in which memory processing occurs as a sequence of detailed physiological steps is then described. The resource management model is used to provide an account for a range of experimental observations in human beings and various animal models, and comparisons are made with various alternative models.

2 Models of Hippocampal System Function

In 1957, Scoville and Milner reported on a striking combination of memory deficits that had followed surgical resection of the medial temporal lobes, including substantial hippocampal damage, in three patients DC, HM and MB. In each case the patients appeared to have lost the ability to remember any events subsequent to their surgery, and memory of events for a period of time prior to surgery was also affected. However, general intelligence, conversation skills, perception and reasoning ability appeared unaffected. The details of the deficit in patient HM have been extensively investigated in the half century since than [Corkin 2002].

The deficit in HM included global anterograde amnesia for declarative type memories: an inability to learn any new events, facts or words. There also appeared to be a retrograde deficit for memory of events. In the 1957 paper it was remarked that in conversation with HM a couple of years after his operation, he reverted constantly to boyhood events and appeared to have a partial retrograde deficit "inasmuch as he did not remember the death of a favorite uncle three years previously. . .yet could recall some trivial events that had occurred just before his admission to the hospital". Later, HM was tested more formally using Crovitz's test, in which subjects are asked to relate a personally experienced event incorporating each of ten nouns. HM's memories were only of events earlier than 11 years prior to his operation, in striking contrast with normal controls [Sagar, Cohen, Corkin and Growden 1985]. This apparent 11 year retrograde deficit in episodic memory contrasts with his retained semantic memory for word meanings learned in the same 11 year period [Kensinger, Ullman, and Corkin, 2001].

However, HM retained a significant ability to learn sensorimotor skills [Corkin, 1968] and his repetition priming capability was normal [Milner, Corkin and Teuber, 1968].

Given the association of the hippocampus with navigation [e.g. Maguire et al. 2000], it is of interest that immediately after surgery, HM "could no longer . . . find his way to the bathroom" and after moving to a new house a few blocks away on the same street he could not be trusted to find his way home alone [Scoville and Milner 1957]. His ability to recall spatial location was severely impaired [Smith 1988]. However, after living in a new home (an 860 square foot bungalow) for 8 years he was able to draw an accurate map of the location of the rooms and he retained that capability three years after leaving that home [Corkin 2002].

The combination of (1) global anterograde amnesia for semantic and episodic memory, (2) the continued ability to learn simple motor skills and repetition priming, (3) retrograde amnesia for episodic memory covering limited time periods, but not for semantic memory or complex skills learned prior to onset of amnesia and (4) unaffected general intelligence etc. appears fairly typical of damage to the hippocampal region. There is, however, considerable variation in detail that will be discussed in later sections. This combination of deficits has presented a major problem for modelling the function of this region. A wide range of models has been proposed, but encounter difficulties in accounting for the exact combination of deficits that are exhibited.

Tulving et al. [1996] proposed that the role of the hippocampal system is determination of the novelty of a stimulus, and encoding of current incoming information by frontal lobe cortical areas depends on the novelty of that information. This model does not address the apparent dissociation between semantic and episodic information in retrograde amnesia following hippocampal system damage. As pointed out in a review of models by Cohen et al. [1999], issues with the model include observations in some cases of greater hippocampal activity for old vs. new items; other observations of differences in hippocampal activation when there was no difference in novelty; and yet other observations in which systematic variation in the degree of novelty produces no change in hippocampal activation.

The simple consolidation model [e.g. Squire and Alvarez, 1995] was developed to explain the combination of global anterograde amnesia with retrograde amnesia covering a limited period of time. In this model, information is initially registered in the hippocampal system, and gradually transferred to long term storage in the neocortex. McClelland, McNaughton and O'Reilly [1995] argued that one value of gradual transfer is that it could reduce the interference between prior and later learning.

This simple consolidation model was criticized by Nadel and Moscovitch [1997] for a number of reasons, including (1) that it does not account for the differences in retrograde amnesia between episodic memories (generally the most severe amnesia), personal semantic memories and semantic memories of public events and persons (less severe) and general semantic memory (least affected), and (2) that the retrograde amnesia period observed in some cases for autobiographic memories implies that consolidation require the entire lifetime.

Teyler and DiScenna [1986] proposed that the role of the hippocampus is to form an index of neocortical areas activated by each experienced event. Only the location and temporal sequencing of activated cortical modules is encoded, there is no coding of any neuronal transformation of the event itself in the hippocampus. Reactivation of the indexed neocortical modules in the appropriate spatio-temporal sequence simulates the original experience. If a new event activates only a fraction of the index of some past event, and the fraction exceeds a threshold, then the remainder of the index for the past event is activated, and this activates all the neocortical modules active during the event. Teyler and DiScenna suggested that the operation of the index could be regarded as a pattern matching function: the hippocampus continually and automatically tests each pattern of cortical activation to see if it matches previously stored patterns.

Nadel and Moscovitch [1997] extended the memory indexing theory to provide an account for the dissociations between episodic and semantic memory. In their multiple trace theory, as in the memory indexing theory, the hippocampal complex rapidly encodes all information that is attended or consciously apprehended. The hippocampal record acts as a pointer to neocortical neurons that represent the information, and binds them into a coherent memory trace. The memory trace for an episode is thus the entire hippocampal-neocortical ensemble. The difference in the multiple trace theory is that reactivation of a trace results in creation of new traces, and for semantic information some traces can become independent of the hippocampus. Damage to the hippocampal system will affect a memory to a degree dependent on the proportion of the traces for that memory that are damaged. Multiple traces within the hippocampus and traces outside the hippocampus will reduce the effect of the damage. Another approach is to argue that incremental learning and storage and retrieval of episodic memories are performed by separate subsystems of the hippocampal system. Gluck et al. [2003] have claimed a synthesis of this type, with incremental learning supported by representational transformations in the input regions to the hippocampus (especially the entorhinal cortex), and the storage and recall of previously processed representations supported by the CA3 and CA1 regions.

Another two component model is that of Eichenbaum et al. [1994]. These authors argue that the hippocampal system performs two sequential functions corresponding with anatomically separate structures. First, the hippocampal system can fully represent current sensory items (without relationships between them) in a memory buffer that can hold information for at least several minutes. Second, while these representations are held in the buffer, the hippocampal system compares and relates them to other memory representations, creating relational representations between the items and linking with any previously created relations involving the items. They propose that the temporary storage of sensory representations occurs in the entorhinal, perirhinal and parahippocampal cortices, and the relational processing in the CA fields, dentate gyrus and subiculum.

Lisman [1999] emphasized that a key characteristic of episodic memory is that sequences are recollected, and introduced a model in which memory sequences are recalled by a combined dentate-CA3 circuit. In this model, CA3 is a recurrent network that contains heteroassociative information making it possible for one item in a memory sequence to recall the next item. Long chains of such recalls are liable to become increasingly noisy and error prone. To correct such a buildup of errors, the dentate gyrus is a second recurrent network that contains autoassociative information making it possible for one item in a memory sequence to recall itself with lower noise. In this model, the role of CA1 is to convert the CA3 output to a cortical representation, and to compare CA3 predictions of the next items in a sequence with actual sensory input derived more directly from the neocortex. As a rat moves through the place field of a hippocampal place cell, the cell fires with progressively earlier phase in successive cycles of theta activity [Skaggs et al. 1996]. Lisman [1999] suggested that this phase advance is the means by which memories of successive points in time are associated. This mechanism allows association of events separated by less that about 100 milliseconds, and a buffering function (supported by neuron activity for many seconds after the stimulus is removed) allows association of events with wider separations. Lisman's model is an ambitious attempt to link processes at molecular, cellular, network and behavioural levels, but as Lisman points out, does not address how the information stored in the hippocampus is utilized by other brain networks.

2.1 Hippocampal Subregional Information Models

As discussed in the previous section, Lisman [1999] argued that the dentate gyrus operates as an autoassociator, CA3 as a heteroassociator, and CA1 as a comparator. These information functions together operate to retrieve episodic memory sequences.

Kesner et al. [2004] extended the pattern matching paradigm introduced by Teyler and DiScenna [1986] to attempt to understand the roles of different hippocampal regions within the framework of consolidation type theories. They identified different types of information processing, including pattern separation, pattern association, pattern completion, novelty detection, and memory (short, intermediate and long term). Pattern separation is the mechanism for separating partially overlapping patterns of activation so that one pattern can be retrieved separately from other patterns. Pattern association links patterns that are discontiguous in space or time [Wallenstein et al., 1998]. Pattern completion retrieves previously stored patterns on the basis of partial inputs.

Kesner et al. [2004] review a range of experimental evidence to argue that different subregions within the hippocampal formation support different information processes. In particular, the dentate gyrus acts as a competitive network to reduce the redundancy of sensory inputs and produce sparse, orthogonal outputs. These outputs are used by CA3 to perform spatial pattern separation. CA3 supports spatial pattern association, spatial pattern completion, short term memory and novelty detection. Outputs from CA3 are used to support temporal pattern separation in CA1. CA1 performs temporal pattern association, temporal pattern completion and intermediate term memory.

However, the way in which these information processes combine to support the memory phenomena as described by the higher level models has not been made fully clear.

3 Architectural Model of the Brain

As pointed out by Lisman [1999], a full theory of the hippocampus must link processes at molecular, cellular, network and behavioural levels. A critical element in such a theory is some concept of what information processes the hippocampus contributes to the rest of the brain and to the neocortex in particular. Such a concept requires a system architecture of the brain as a whole.

A general system architectural model of the brain called the recommendation architecture has been proposed by Coward [1990; 2001; 2005a], including theoretical arguments that any system which must learn a complex combination of behaviors with limited information handling resources will tend to be constrained into the forms of this model by a number of practical considerations. These considerations include the need to limit resources, the need to learn without interference with past learning, the need to recover from damage and failure, the need to construct the system itself without errors, the need for synchronicity or maintenance of associations between the results of processing the same input state (i.e. information derived from the environment or other sources at the same time) by different parts of the system, and the need to use the same resources to simultaneously process input states from different times. These practical considerations generate interacting and conflicting pressures on system architecture, and some remarkably specific architectural requirements result from the need to find an adequate compromise that satisfies these conflicting pressures.

The general architectural form of the model is illustrated in figure 1. For a complex learning system, the greater the ratio of behaviors to resources, the more tightly the system will be confined within this architectural form [Coward 2001]. As illustrated in figure 1, there are a number of separations between subsystems which perform different types of information processes, and evidence from physiological structure, dissociations between different cognitive processes, and the deficits resulting from local damage has been offered [Coward 1990; 2000; 2005a] to support the view that there is a correspondence between these subsystems and the physiological structures of the mammal brain identified in the figure.

The model makes it possible to create a hierarchy of causal descriptions of the same phenomenon at a number of different levels of detail, from physiological to psychological, in such a way that descriptions on one level can be mapped into descriptions on other levels. Such a description hierarchy is essential for understanding a complex phenomenon [Coward and Sun 2007].

In the architectural form illustrated in figure 1, there is a primary separation between a modular hierarchy (called clustering) and a component hierarchy (called competition). The difference between a module and a component is that component inputs and outputs can only have simple behavioral meanings (i.e. recommending performance of one behavior, or against performance of anything except one behavior). Module inputs and outputs can have complex behavioral meanings (i.e. recommending performance of many different behaviors, generally with different recommendation strengths) [Coward 2001; 2005a]. In the mammal brain the cortex corresponds with clustering and the thalamus and basal ganglia correspond with competition.

Clustering defines and detects information conditions within the information available to the system, each module being programmed to detect a set of fairly similar conditions which define the receptive field of the module. An output from a module indicates the detection of a significant subset of its programmed conditions,

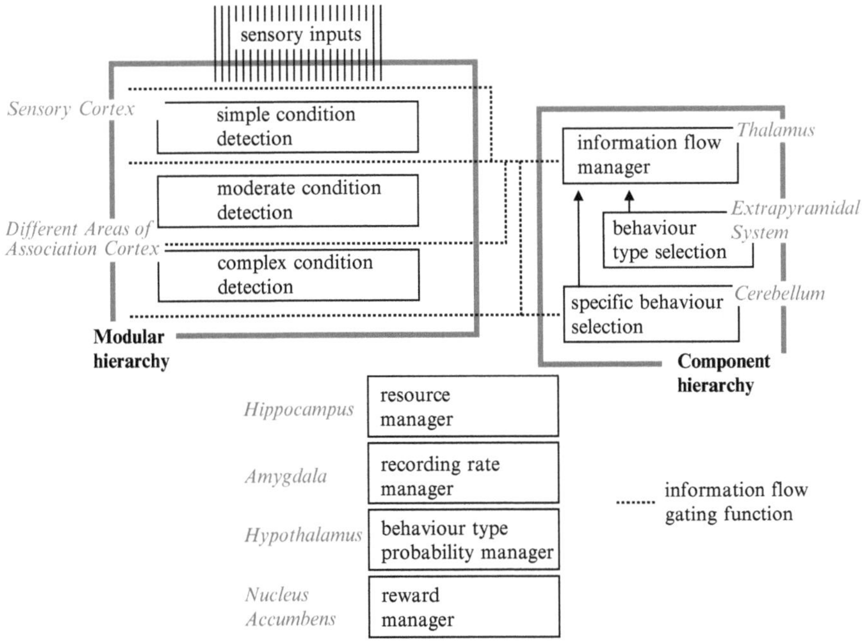

Fig. 1 The architectural form into which a system will tend to be constrained if it must learn a large number of different behaviors with limited resources is illustrated. The modular hierarchy defines and detects conditions within the information available to the system. Many condition detections flow to the component hierarchy where they are interpreted as recommendations in favor of a wide range of behaviors. Conditions are detected on different levels of complexity, with simpler conditions being generally (but not exclusively) more useful for recommending information flow behaviors, moderate complexity conditions for recommending selections of general types of behavior, and high complexity conditions for recommending specific (e.g. motor) behaviors. The component hierarchy selects and implements the most strongly recommended behaviors, including behaviors that are information releases from one part of the modular hierarchy to another. Selection of, for example, a general type of behavior is effectively a selection of one information flow within the modular hierarchy over another. For efficiency reasons, most modular hierarchy information flow decisions are implemented through a single subsystem within the component hierarchy (the information flow manager), relevant selections by other parts of the component hierarchy are funneled through the information flow manager as illustrated. Reward feedback is managed by a separate subsystem. Such reward feedback acts upon the component hierarchy to change recommendation weights but cannot change condition definitions in the modular hierarchy. Decisions on where to record conditions in the modular hierarchy at each point in time are made by the resource manager on the basis of inputs from the modular hierarchy. Selection of a general type of behavior can be influenced by general circumstances via the behavior type probability manager. Such selections include influencing the rate of condition recording. Special circumstances can also result in elevation of the rate of condition recording by the recording rate manager. The human brain structures corresponding with these subsystems are indicated [Coward 2005a and this paper]

i.e. the detection of its receptive field. The term "feature" could perhaps be used instead of "information condition" or "module receptive field", but the implication of the word feature is that the information condition detected and resultant module output corresponds exactly with some simple cognitive feature. As discussed below, and in detail in Coward [1990; 2001], an information condition or module receptive field is a circumstance that is in general detected within many different cognitive features, the indicator of the difference between two cognitive features is the different (but partially overlapping) populations of conditions or receptive fields detected.

The system information within which conditions are detected includes raw information about the state of the external environment and about the internal state of the system itself. Competition receives inputs indicating detections of various receptive fields (i.e. module outputs) and interprets each such input as a recommendation in favour of many different behaviours, each with an individual weight. The total recommendation weights of each behaviour is determined, and competition drives the implementation of the behaviour with the largest current weight. Reward feedback modifies recommendation weights in competition but does not change condition definitions in clustering.

3.1 The Modular Hierarchy

A condition is ultimately defined by a set of raw system inputs and an associated state specified for each input in the set, and the condition occurs if a high proportion of its set of inputs is in the state specified for the condition. Conditions are defined on different levels of complexity, where the complexity of a condition is the number of raw inputs (including duplicates) that contribute to the condition, either directly or via intermediate conditions. Conditions on different levels of complexity (and the receptive fields which contain the conditions) may be more appropriate for recommending different types of behavior [Coward 2005a].

In physiological terms, a condition is a group of inputs to a pyramidal neuron that are integrated as a group before contributing to the potential injected into the soma. Figure 2 illustrates this staged integration process. A condition programmed on an arm of the dendrite is present if enough action potentials arrive at the different synapses defining the condition within a short enough period of time to exceed the threshold for injection of potential deeper into the dendrite. One neuron will be programmed with many such conditions, and will produce an output if a significant proportion of its conditions are present. The receptive field of the neuron is specified by the group of conditions that it detects, and its output indicates the detection of that receptive field.

Pyramidal neurons in sensory areas in the neocortex detect relatively simple receptive fields, with complexity increasing in later sensory areas. Pyramidal neurons in association areas detect even more complex conditions that are combinations of receptive fields detected in the areas from which they derive their inputs. Cortical

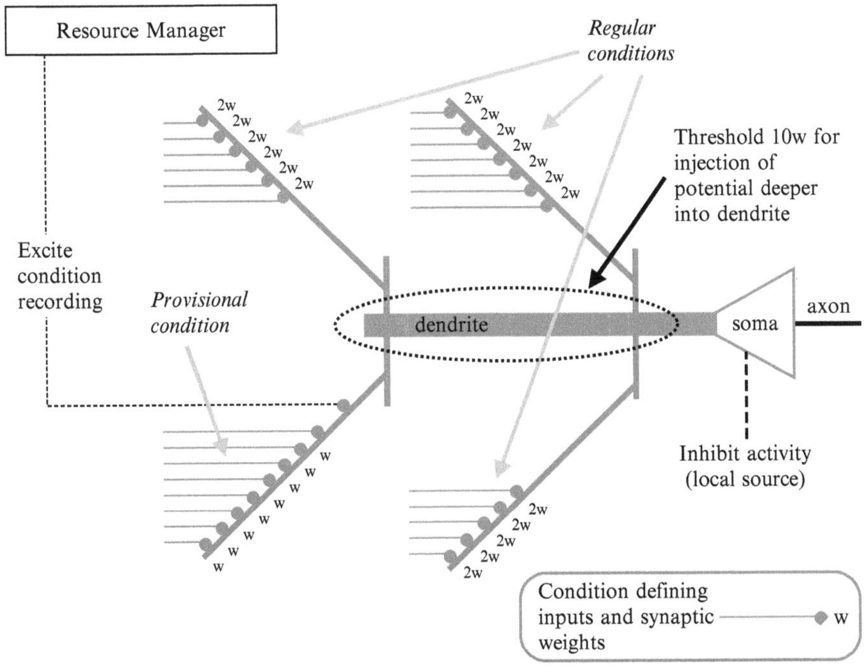

Fig. 2 Conceptual diagram of a pyramidal neuron. There is separate integration of postsynaptic potentials on different arms of its dendrite. If the integration product within an arm exceeds a threshold, potential is injected further into the dendrite, and such injection from enough arms causes the soma to produce an action potential. The lower left arm does not have enough synaptic strengths in its condition defining inputs to inject potential. However, if its input exciting condition recording is active, the total potential in the arm can exceed the threshold. If shortly afterwards the neuron produces an action potential and there is also an action potential that backpropagates into arms that have recently injected potential, the postsynaptic strengths will increase by the LTP mechanism. This increase will tend to enable the arm to inject potential in the future without activity from the input exciting condition recording, in other words, a condition has effectively been recorded. Inputs exciting condition recording will in general come from the hippocampal system

columns detect receptive fields defined by the receptive fields of their constituent pyramidal neurons. Cortical arrays detect at least a minimum number of columnar receptive fields on one level of complexity in every sensory input state. Cortical areas are made up of sequences of one or more arrays, with each array detecting receptive fields on a different level of complexity but within the same input space.

An important aspect of the evolution of receptive fields is that because one receptive field has many different behavioural meanings, changes to receptive fields must be tightly controlled to minimize interference with existing such meanings. Coward [2001; 2005a] has argued that a good first approximation is that receptive fields can expand (by addition of conditions) but cannot change or eliminate conditions once they have been added.

Two important qualifiers to this approximation are that changes to receptive fields can be reversed on very short and very long time frames with limited behavioural risk. If a condition is added but does not occur again within a relatively short period of time, it could be eliminated. In other words, the receptive field of a pyramidal neuron could expand slightly, and if the expansion is detected again soon afterwards becomes permanent, otherwise relaxing again to its prior state. If a long period of time elapses during which a condition or group of conditions do not occur, then again the behavioural risk of eliminating the condition may be low. An extreme example is if there is a major change in sensory inputs which results in some inputs no longer occurring, such as the significant changes to receptive fields observed when the surface skin of two digits of an owl monkey are connected [Clark et al. 1988]. A rather different qualification is that in the early experience of the brain, condition detections may not yet be associated with behaviors. If these associations are not being created, then there can be greater freedom to change and eliminate conditions. This greater freedom may be important to heuristically establishing an effective set of modules in early learning of infants [Coward 2005a].

An implication of this restriction on change is that in general modular receptive fields cannot be evolved to correspond with features or categories of sensory objects. This is consistent with the observations of Tanaka [1993] that even in the inferotemporal cortex functionally associated with object recognition, pyramidal neuron and column receptive fields correspond with ambiguous shapes and not clear visual features.

There are a number of considerations that can be used to define the circumstances in which receptive field expansion can occur. The first is that a reasonable range of recommendations is required in response to any input state in order to achieve a high integrity behavior. Because behavioral recommendations are also module outputs, this implies that at least a minimum number of modules (i.e. cortical columns) must be producing outputs in response to every input state. If the number of columns producing outputs to the component hierarchy is low, expansion of module receptive fields by recording of additional conditions will generally be required. An implication is that the degree of condition recording will be higher during novel experiences. A second consideration is that the conditions within one module must be similar, both for resource economy reasons and to ensure that the behavioral meaning of module outputs is not excessively diluted. A third consideration is that adequate discrimination must be achieved between input states with different behavioral implications: the populations of conditions detected with such input states must be sufficiently different to adequately guide behavior. Although reward feedback cannot be used directly, there is a contradictory reward feedback mechanism that can be used to drive increases in resolution [Coward 2005a]. In this mechanism, if a similar group of columns generate outputs on a number of different occasions, these outputs result in the same behaviour being implemented, but the reward feedback following the behaviour is sometimes positive and sometimes negative, the implication is that the columns are not providing adequate discrimination. In this situation, additional discrimination could be provided (for instance, by causing two columnar output pyramidals with very similar receptive fields to diverge by forcing change to one of them). A fourth consideration is that the overall use of resources (i.e. neurons and connectivity) must not be excessive.

One approach to guiding change is to expand the receptive field of a module if the module was not producing an output, provided that the current input state was fairly similar to past input states that resulted in a module output. Another approach is to limit expansions to modules with the greatest similarity between the current input state and past state which caused them to produce outputs. This second approach means that the resultant module receptive fields are more orthogonal and therefore better able to discriminate between behaviourally different circumstances, but requires extensive communication between modules. As discussed later, both of these approaches are important.

There are some conceptual similarities between column arrays and components in independent components analysis [Hyvärinen et al. 1999], in the sense that arrays decompose a sequence of input states into partially statistically independent "features" in an unsupervised manner. The critical difference is that independent components are calculated prior to use with a preselected set of sensory inputs, and do not change in response to actual sensory inputs. Columns constantly evolve by addition of new conditions. This evolution means that columns are less rigorously statistically independent, but new types of input states can be decomposed using an existing column array.

3.2 *The Pyramidal Neuron Model*

The proposed leaky integrator model for a cortical pyramidal neuron is illustrated in figure 2. In this model, action potentials arriving at synapses inject potential into local regions (arms) of the dendrite. These postsynaptic potentials follow a rapid increase and slower decay cycle, and add together locally within each arm. If the total arm potential at some point in time exceeds a threshold, potential is injected deeper into the dendrite. This potential also follows a rapid increase and slower decay cycle, and further integration within the dendritic tree and within the soma of the neuron determine whether the neuron will produce an output action potential. This type of staged integration across a dendritic tree appears to be physiologically plausible [Hausser and Mel 2003].

A condition is defined by inputs to one arm of the dendritic tree from pyramidal neurons with simpler receptive fields (the condition defining inputs in figure 2) and their associated synaptic weights. The receptive field of the neuron is defined by the set of conditions programmed on its arms, and the neuron will detect its receptive field if a significant proportion of its conditions are detected (i.e. inject potential deeper into the dendrite) within the integration time for the soma. The rate of action potential generation by the neuron indicates the degree of presence in the current input state of its receptive field.

Expansion of the receptive field of a neuron can take place in two ways. One is by increases to some of the input synaptic strengths of an existing condition, the other is addition of a new condition with a somewhat different set of inputs from existing conditions. It would be impractical to create the connectivity needed to define a new condition at the instant the condition was required. Configuration of provisional conditions in advance is therefore required, and condition recording occurs by increases in some of the input synaptic strengths of the new condition.

The mechanism by which synaptic strengths are increased is the LTP mechanism described by Bi and Poo [1998]. If, within the integration time for the arm, action potentials arrive at a significant proportion of the condition defining inputs, the total postsynaptic potential could exceed the threshold for the arm. If the arm injects potential deeper into the dendrite, and if shortly afterwards the soma generates an action potential, a backpropagating action potential that only enters arms which have recently injected potential into the dendrite increases the weights of recently active synapses on such arms. This mechanism will cause the receptive field of the neuron to increase whenever the neuron produces an output.

Such an unmanaged increase in receptive fields would reduce their behavioural value. There are several ways in which the increases can be managed more effectively. Firstly, by limits to the weights of individual synapses to ensure that individual synapses do not dominate a condition or individual conditions dominate a receptive field. Secondly, by requiring that an increase will be reversed unless several increases to the same synapse occur within a short period of time, so that rarely occurring conditions are not recorded. Thirdly, by eliminating any synapses on an arm which are close to their initial value when most other synapses have reached their maximum. Fourthly, by providing condition recording management inputs as illustrated in figure 2, particularly to provisional conditions.

For the provisional condition illustrated in figure 2, the total weight of the condition defining synapses is in general too small to result in injection of potential deeper into the dendrite for further integration. However, if action potentials arrive at a significant proportion of the provisional condition defining inputs within the integration time and in addition arrive at the condition recording management inputs, the total postsynaptic potential could exceed the threshold. If the arm injects potential deeper into the dendrite, and if shortly afterwards the soma generates an action potential, the LTP mechanism increases the weights of recently active synapses on the arm. This increase means that the total synaptic strengths of those synapses would in future be enough to result in injection of potential from the arm into the dendrite independent of the state of the management inputs. In information terms, a new condition has been recorded on the neuron.

A fifth way to manage receptive field expansion uses inhibitory interneurons. If activity within a column is already at a high level, general internal inhibitory connectivity (derived from devices that received inputs from a wide range of pyramidal neurons within one layer of the column and generated inhibitory outputs) would prevent any further increase, effectively preventing recording. Such inhibitory connectivity would be directed to the pyramidal somas and perhaps to individual provisional conditions.

The critical issue of what could be an appropriate source for condition recording management inputs is discussed in the next section.

3.3 *Columns, Arrays and Management of Receptive Field Expansions*

To understand the condition recording management process in more detail, consider the operation of the simple cortical column model illustrated in figure 3. Such columns are arranged in arrays, where the array is managed so that it detects at least a minimum number of columnar receptive fields at one level of condition complexity in all input states from a given input domain.

Receptive fields of pyramidal neurons in the top layer are defined as combinations of receptive fields of columns in the array providing inputs to the column. Receptive fields of pyramidal neurons in the middle layer are combinations of receptive fields detected by pyramidal neurons in the top layer, and bottom layer receptive fields are combinations of middle layer fields. There is therefore a gradual increase in receptive field complexity from top to bottom, with bottom layer pyramidals providing column outputs to the next array.

This increase in receptive field complexity means that pyramidal neurons in the top or middle layer could detect their receptive fields even if there were no such detections in the bottom output layer. However, the input state within which the middle layer detections occurred would have to have significant similarity to past input states which actually generated detections in the bottom layer. Hence if the receptive fields of some columns within an array must be expanded to reach the minimum required level of columns generating outputs, a high degree of pyramidal activity in the middle layer of one of the columns is a reasonable indicator that the needed expansion in its receptive field would be small.

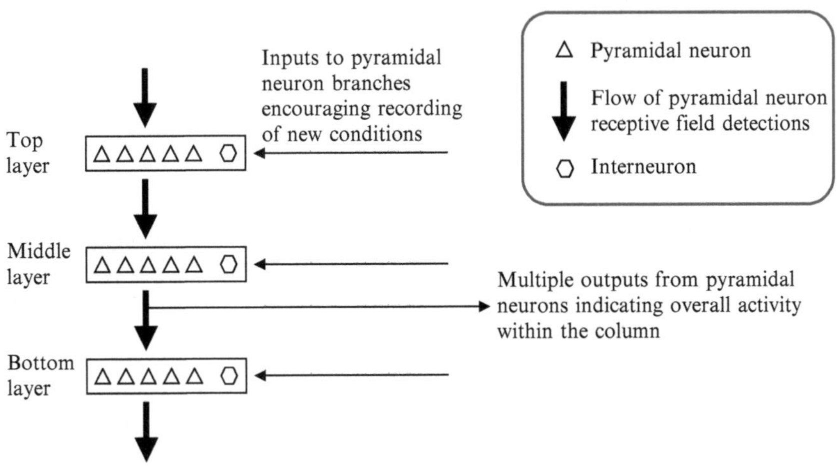

Fig. 3 Column structure which could manage the condition recording process. The column records conditions in the current input state if that state is adequately similar to past input states which generated column outputs (see text). There is a competition to determine the most appropriate columns to record information in response to each input state

To provide the condition recording management inputs discussed in the previous section, there is therefore a requirement for a process that can identify the columns with the greatest middle layer activity, and provide the pyramidals in those columns with the management signals. Such condition recording management connectivity could in principle be derived locally: excitatory connectivity from within the column, inhibitory from all peer columns. This local approach requires considerable (all-to-all) inter-column connectivity resources. Condition recording management that is equivalent in an information model sense could be performed with much less connectivity resource if every column was reciprocally connected to a global resource manager. This manager would perform a competition function to select columns to record information and generate outputs to devices in those columns to excite such recording. Local inhibitory connectivity would be required to limit internal column activity. Short range inhibitory connectivity could also implement local competition between columns with similar receptive fields to improve discrimination.

Such a global resource manager could also make use of information on which columns had expanded their receptive fields at similar times in the past to improve the selection of current columns. As discussed below, this global management function is proposed in this paper as the primary role of the hippocampal system.

At the start of sensory experience, there is a need to bootstrap receptive fields. If there were no column activity at this initial point, there would be no hippocampal activity and therefore no condition recording management inputs. Initially, therefore, receptive fields must be defined randomly (with some genetically imposed bases) and expand only on the basis of internal similarity until enough column activity is generated for the management process. Provisional conditions in the very early stages must therefore have enough condition defining input weight to lead to initial neuron activation without condition recording management inputs.

3.4 Configuration of Provisional Conditions

Provisional conditions could be defined by random selection of their constituent conditions. However, placing a statistical bias on the selection process can increase the probable usefulness of provisional conditions. This bias is in favor of constituent conditions that have often been present in the past at the same time as each other and at a time when the target neuron has been detecting other conditions.

Such a bias is using past experience to estimate the type of conditions that are most likely to be needed in the future. The recent past is a particularly relevant guide. A simple way to achieve such a bias is to take the system off-line, and perform a rapid rerun of a sample of past activity. Provisional condition connectivity would then be created between axons indicating the presence of possible constituent conditions and target neurons being programmed with provisional conditions if input axons and target neurons were often active at the same time. Coward [1990] suggested that a primary role of REM sleep was to perform this partial rerun process. A central resource manager has a natural access to the type of information needed to guide this rerun process.

3.5 Behavioral Specialization of Areas

For maximum resource economy, a cortical array at a given level of complexity would support all possible types of behavior. However, if there were two important types of behavior for which the receptive fields at the same level of complexity that provided the best discrimination were different, the behavioral advantages of separate parallel arrays might outweigh the resource costs. In general this would occur for a limited range of receptive field complexities and for a limited range of behavioral types. Columns in one array would tend to recommend only one type of behavior, in a parallel array only a second type of behavior and so on. For example, Coward [1990] suggested that separate arrays on some levels of complexity might exist for major behavioral types such as aggressive, fearful, and food seeking. Given such separation, the arrays could be optimized for their different behavior types [Coward 2005a].

One side effect of this behavioral parallelism is that detection of general conditions indicating the appropriateness of a type of behavior could be used to bias the brain towards that behavioral type, for example low blood sugar could favor arrays generating food seeking behaviors. Such favoring could include preferential condition recording within the arrays targeted at the behavior type.

3.6 The Component Hierarchy

As described in Coward [2001; 2005a], the component hierarchy (thalamus and basal ganglia) is made up of components corresponding with different behaviours and types of behaviour. It receives indications of the presence of the conditions currently being detected by the modules in the modular hierarchy (cortical columns) and interprets each such indication as a set of recommendations in favor of a range of different behaviors, each recommendation having a specific weight. These recommendations are instantiated by the connection weights of column outputs into the appropriate components.

The component hierarchy determines the behavior with the largest total weight across all currently detected conditions using inhibitive connectivity between components and implements that behavior.

Components in the hierarchy correspond with individual behaviours or types of behaviours. The hierarchy must determine that one and only one behaviour (or perhaps a consistent set of behaviours) is selected in response to each input state, and that reward feedback is applied appropriately.

Coward [1990, 2000] suggested that there must be a competition between components corresponding with all behaviours, modulated so that there is one and only one "winner" producing an output, and offered evidence that the basal ganglia was the primary site for these processes. In this model, as illustrated in figure 4,

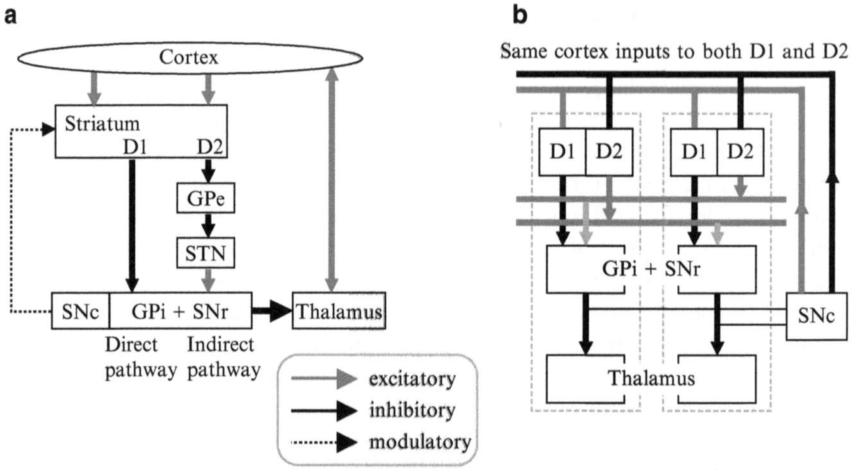

Fig. 4 Basal ganglia and thalamus structures [Albin et al. 1989, Alexander et al. 1986] interpretated in terms of the information role of selecting behaviours most strongly recommended by the cortex. As illustrated in **a**, spiny neurons in the striatum receive excitatory inputs from the cortex. There are two populations of spiny neurons in the striatum. One population (D1) directly inhibits two similar structures, the globus pallidus internal segment (GPi) and the substantia nigra pars reticula (SNr). The other population (D2) indirectly excites the same two structures via intermediate structures, the globus pallidus external segment (GPe) and the subthalamic nucleus (STN). GPi and SNr generate tonic inhibitive outputs to the thalamus, the direct path reduces thalamic inhibition, the indirect path increases it. The thalamus receives strong excitatory input from the neocortex, and returns excitatory outputs to the same cortical areas from which the inputs were received. The thalamus provides these outputs to the cortex only if the tonic GPi and SNr inhibitive outputs are reduced by striatal D1 activity. A further path goes from the substantia nigra pars compacta (SNc) back to the striatum. SNc is closely associated with GPi and SNr. The return path has different effects on the D1 and D2 populations. These structures are interpreted in terms of the information model in **b**. Two components corresponding with different behaviours are illustrated. The observed patch and matrix structure in the striatum [Goldman-Rakic 1982] may reflect this component structure. Each component has segments in the striatum, in GPi and SNr, and in the thalamus. The same cortical inputs are available to both the D1 and D2 of one component. These inputs have weights that can be interpreted as recommendations in favour of the behaviour corresponding with the component. Outputs from D1 within a component can also be interpreted as recommendations in favour of the component behaviour, while outputs from D2 target other components and can be interpreted as recommendations against any behaviour other than the component behaviour. A behaviour is implemented by outputs from the thalamus that result in release of internal cortical activity either to other cortical regions or to the cerebellum, brain stem and spinal cord to drive motor behaviour. If many behaviours are strongly recommended, the SNc detects a high level of activity in GPi and SNr across all components, and increases D2 activity relative to D1 in all components until only one component has strong GPi and SNr activity. If GPi and SNr overall activity is too low, the implication is that no behaviour is currently recommended, and SNc increases D1 activity relative to D2 until activity in GPi and SNr supports a behaviour implementation

different components of the striatum correspond with different behaviours. Each component has a D1 and D2 section. The direct pathway from a striatum component D1 section to the GPi and SNr excite the behaviour corresponding with the component. If the resultant degree of excitation is either too high or too low (i.e. either multiple or no behaviour being selected), the substantia nigra pars compacta indirect pathway modulates the D2 section of the component to increase or decrease the general excitation in GPi and SNr until one and only one behaviour is selected. This selection is communicated through the thalamus back to the cortex, resulting in release of cortical outputs to drive the selected behaviour.

Reward feedback following a behavior affects the weights of recently active connections within the components corresponding with recently implemented behaviors. If the reward is positive, excitatory weights in those components are increased and inhibitory weights decreased, and vice versa if the reward is negative. The effect of a reward is therefore to modulate the probability of the same behaviors being selected in similar circumstances in the future. As illustrated in figure 5, information indicating that a reward is appropriate could come from

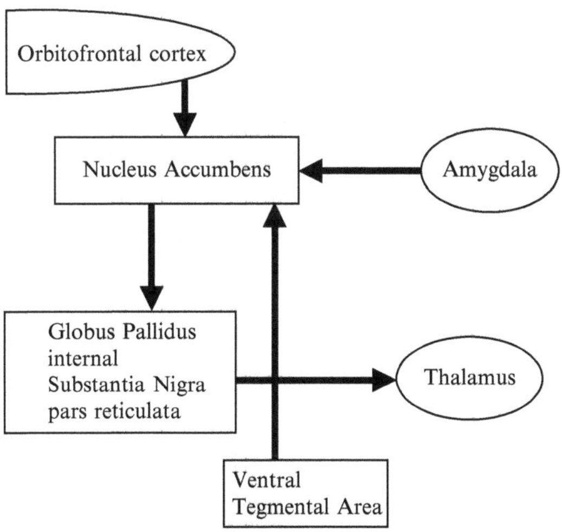

Fig. 5 Interpretation of the nucleus accumbens and associated structures in terms of the information role of modulating recommendation weights of recently selected behaviours. The nucleus accumbens has been associated with positive rewards [Kelley 1999] and negative rewards [Schoenbaum 2003]. The nucleus accumbens receives strong inputs from the orbitofrontal cortex and the basolateral amygdala [McDonald 1991; Haber et al. 1995], and from the ventral tegmental area [Sesack and Carr 2002]. The nucleus accumbens projects strongly to the globus pallidus and substantia nigra [Nauta et al. 1978]. In terms of the information model, the orbitofrontal cortex provides inputs correlating with general circumstances in which rewards are appropriate (e.g. detection of social signals). The amygdala provides inputs correlating with somewhat more specific circumstances, and the ventral tegmental area provides inputs correlating with very specific circumstances (e.g. perception of pain). The nucleus accumbens integrates these inputs and provides signals that modulate the connection weights of inputs to GPi and SNr that have recently resulted in the selection of a behaviour

different sources corresponding with different types of situation: general (e.g. social) circumstances from the orbitofrontal cortex, emotional from the amygdala, and pain from the ventral tegmental area. These sources are integrated in the nucleus accumbens, which then drives changes to recommendation weights in the GPi and SNr.

In this competition system model, the thalamus gates the flow of information within the cortex, and the basal ganglia act by influencing the thalamus. Coward [2005a] has argued that the thalamus manages internal flows of information within the cortex, mainly on the basis of the recommendation strengths of cortical inputs directly to the thalamus, but modulated by basal ganglia inputs. For example, the basal ganglia could determine the general type of information flow (e.g. which cortical areas send outputs to which other areas), while the thalamus would determine the exact set of columns that provide those outputs.

Components in the basal ganglia correspond with basic motor movements and internal cortical activation actions. However, there is also a requirement to learn and perform effectively many complex sequences of such basic movements and actions. Examples of sequences of motor movements include the sequences of motor movements required to utter words or phrases, or to perform skilled finger manipulations for typing or playing the piano. Examples of sequences of internal cortical activation actions include speech generation and generation of episodic memories.

There is therefore a requirement for a subsystem which can learn to perform such sequences. Such a subsystem makes it possible to recommend, select and reward such behavioural sequences as a whole, making it less likely that their performance will be inappropriately interrupted, and Coward [2005a] has proposed that the cerebellum performs this function in the mammal brain. If invoked, such a sequence component biases each individual behavioral component (located in the basal ganglia or thalamus) in turn. The effect of biasing the first component is that the weights of currently active cortical columns in favour of the corresponding behaviour are favoured, and if there is a reasonable level of such weight the behaviour will tend to be implemented. Once the first behaviour has been performed, the component corresponding with the second behaviour is biased and so on. The behaviour sequence will therefore proceed only if there is enough total recommendation weight of the appropriate type in the active column population, but there will be a tendency to hold off from other behaviours until such a total is present. Damage to such a subsystem will not remove the ability to perform behaviours, because the individual behavioural components are driven directly by cortical outputs. However, such damage would result in problems with detailed coordination. This type of problem is a typical result of damage to the cerebellum [Gilman et al. 1981].

Indirect activation behaviours will also benefit from this coordination function. One example is the sequence of activation behaviours for activating episodic memories as described in the next section. Many such indirect activation sequences will be needed to support different types of cognition, and as expected for this model, the cerebellum appears to have a role in language and cognitive functions [Liener et al. 1993].

3.7 Memory and Indirect Activation of Information

The relatively permanent recording of information in cortex columns makes a number of information activation mechanisms behaviorally useful. Suppose that a set of columns is producing outputs because they are detecting conditions within current sensory inputs. There may be other columns that are currently inactive but which may have recommendation strengths relevant to the current circumstances. For example, if a column is inactive but has recently been active at the same time as many currently active columns, or has often been active in the past at the same time as many currently active columns, or has expanded its receptive field (i.e. recorded conditions) in the past at the same time as many currently active columns, it may have relevant recommendation strengths.

A column may therefore have recommendation strengths in favor of activation of other columns on the basis of temporally correlated past activity. Such a capability makes it possible to expand the information available to guide behavior beyond that present within current sensory inputs. As discussed in Coward [2005a; 2005b], indirect activation on the basis of frequent past simultaneous activity makes semantic memory possible, activation on the basis of simultaneous past recording supports episodic memory, and activation on the basis of recent activity supports priming.

In the case of semantic memory, hearing a particular word (e.g. "bird") activates a fairly consistent set of auditory columns, while seeing different instances of a category of objects (e.g. different types of bird) activates different sets of visual columns with a fair amount of overlap on some levels of condition complexity because of visual similarity. The auditory columns will therefore often be active at the same time as a frequently occurring subset of the visual columns. Hearing the word "bird" will therefore (on the basis of frequent past simultaneous activity) generate a visual activation at some levels of condition complexity as if an "average" bird were being seen. Conditions close to visual inputs will not be active, so there will not be a visual hallucination.

In the case of episodic memory, during an experience there will be some degree of condition recording in a range of columns that may be otherwise relatively unrelated. The degree of condition recording will be particularly high during an experience with a significant degree of novelty. Subsequently, activation on the basis of past temporally correlated condition recording could partially reconstruct the pattern of column activation during the experience, resulting in an episodic recollection. As an example, consider episodic memory of watching news of the first Bali Bombing (as discussed in Coward [2005a]). At the time of the original experience, there would have been considerable novelty in the sensory experience, and therefore considerable information recording in the columns active at the time. If later the words "Bali" and "bombing" were heard, they would generate activity on the basis of frequent past simultaneous activity of the auditory and visual columns. There might be some overlap with the population active during the earlier experience, but probably not enough to support verbal recall. However, if this active population were evolved on the basis of past simultaneous condition recording, it would tend to move towards an approximation to the population active at the time of the original experience, in other words an episodic recollection. Evolution on the basis of recording shortly after the current population makes it possible to move through the experience.

Viewed from this perspective, navigation is closely related to episodic memory, since it depends on learning what visual experiences come at the same time and after other visual experiences, based on prior visual experience.

A resource manager function that selected the columns to record conditions at each point in time would have a natural access to the information required to activate columns in the future on the basis of temporally correlated condition recording, but no such natural access to information required to activate on the basis of frequent simultaneous past activity. Such a resource manager can therefore be expected to play a role in creation of both episodic and semantic memories, but long term only in access to episodic memories. For example, when a category name is first learned, the link between the word and the visual information that defines its meaning would be on the basis of simultaneous auditory and visual information recording. Once the word has been used a number of times, the basis of the link will shift to frequent past simultaneous activity of that information.

In the case of procedural memory, information is recorded in the receptive fields of columns and in the recommendation weights of those columns into the thalamus and basal ganglia. Access to previously learned skills requires activation of the relevant columns and interpretation of column outputs as recommendations in the thalamus and basal ganglia. Because skills are learned by a process of repetition, there may be a requirement for indirect activation on the basis of frequent past simultaneous activity, but not for indirect activation on the basis of past simultaneous recording. A new simple skill in a familiar motor domain could possibly be learned by changes to column recommendation weights alone, but learning a more complex skill would also require changes to column receptive fields.

3.8 Resource Management Function

The brain model proposed on system theoretical grounds thus has a requirement for a resource management function with a number of primary roles. One role is to assign cortex condition recording resources, including additional provisional conditions to devices, additional device resources to columns and additional columns to arrays etc. A second role is to manage the configuration of the resources by ensuring that assigned resources have connectivity to and from the appropriate targets and input sources. A third role is to determine which columns will record conditions at each point in time. This third role includes biasing condition recording towards arrays generating recommendations in favor of currently selected general behavior types and to increase the degree of condition recording in circumstances determined to be critical.

In this resource management model, the hippocampal system receives inputs indicating the degree of internal activity within each cortical column. However, there is another source of information which could be used to improve the selection of the most appropriate columns. Suppose that there is some set of columns that have often expanded their receptive fields at the same time in the past. Suppose further that a large proportion of the set is indicating that receptive field expansion is appropriate, In such a situation, expansion of the receptive fields of the other columns in the set could be appropriate, even if their degree of internal activity is a little smaller. An approach making use of both current activity and averaged past activity can therefore improve recording management.

The resource manager has a natural access to information needed to perform a number of secondary roles. One is the support of episodic memory, but not semantic memory. A second is the detection of the degree of novelty in a situation (indicated by the overall demand for condition recording.

4 Anatomy of the Hippocampal System

For the purposes of this paper, the hippocampal system will be defined to include the CA fields (CA1, CA2 and CA3), the dentate gyrus, the subicular complex (subiculum, presubiculum, and parasubiculum), and the entorhinal, perirhinal and parahippocampal cortices. Another widely used term for this hippocampal system is the medial temporal lobe. The CA fields plus the dentate gyrus will be labeled the hippocampal formation. The discussion in this section will draw largely on work with monkeys, but work on rats and rabbits will be cited in some cases. There appear to be strong anatomical similarities between hippocampal system structures in different mammals. The CA fields contain a single layer of pyramidal cells, in contrast with the neocortex which contains four or five major layers of pyramidal cells. The subiculum, located physically between the CA fields and the entorhinal cortex and the CA fields, is a transition zone in terms of numbers of layers. The primary cell types in the dentate gyrus are granule and mossy cells, both generally excitatory. Inhibitory interneurons are present in all the structures.

As illustrated in figure 6, the extensive connectivity from the neocortex into the hippocampal system is organized hierarchically [Lavenex and Amaral 2000]. Inputs from unimodal areas are directed to the perirhinal and parahippocampal cortices. Outputs from these cortices target the entorhinal cortex, along with outputs from a number of other polymodal cortical areas. Outputs from the entorhinal cortex target the hippocampal formation. There is a high degree of reciprocity in this connectivity. For example, projections from the parahippocampal and perirhinal cortices largely reciprocate the projections from the neocortex to the parahippocampal and entorhinal cortices. There is some lower degree of connectivity (not illustrated in figure 6) from the perirhinal and parahippocampal cortices and from some polymodal areas directly into CA1 [Suzuki and Amaral, 1990]. There is connectivity from CA1 directly back to those cortices. There is also some lower degree of connectivity (again, not illustrated in figure 6) from similar cortical areas directly into the subicular complex [e.g. Naber, Witter and Lopes da Silva 1999].

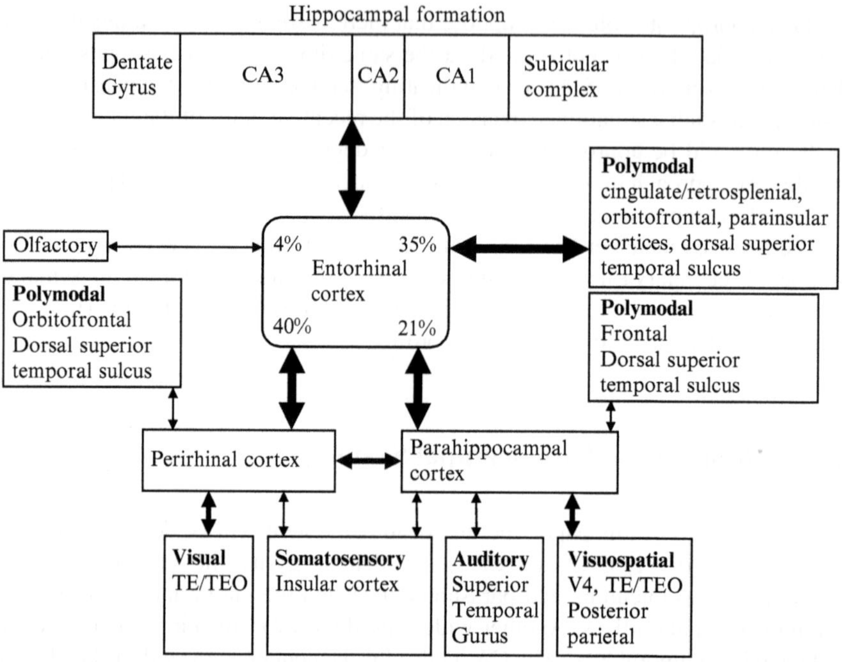

Fig. 6 Neocortical sources of afferent projections to the hippocampal system. Based on Lavenex, Suzuki and Amaral 2004; Insausti and Amaral 2004; Brown and Aggleton 2001; Lavenex and Amaral 2000; Suzuki and Eichenbaum 2000; and Suzuki 1996. Percentages indicate relative volume of different sources of input to the entorhinal cortex [Suzuki 1996]. Weaker connectivity from the perirhinal and parahippocampal cortices and from some polymodal areas directly into CA1 [Suzuki and Amaral, 1990] and also into the subicular complex [e.g. Naber, Witter and Lopes da Silva 1999] is not illustrated

The major flows of information into and out of the hippocampal formation are illustrated in figure 7. Inputs from the wide range of cortical areas illustrated in figure 6 enter pyramidal layer II of the entorhinal cortex. Outputs from layer II target the dentate gyrus and the CA fields via the perforant path. There is also flow of connectivity from layer III of the entorhinal cortex to CA1. Most of the output from the hippocampal formation to the cortex goes from the subicular complex to layers V and VI of the entorhinal cortex, and from there to the cortical areas that provided input to the hippocampal system.

Information flows within the hippocampal formation and between that formation and the entorhinal cortex, the amygdala, the thalamus, and the mammillary bodies and surrounding supra-mammillary area of the hypothalamus are also illustrated in figure 7. The three subcortical structures have been selected because, as discussed earlier, damage to these structures has been associated with memory deficits in human patients.

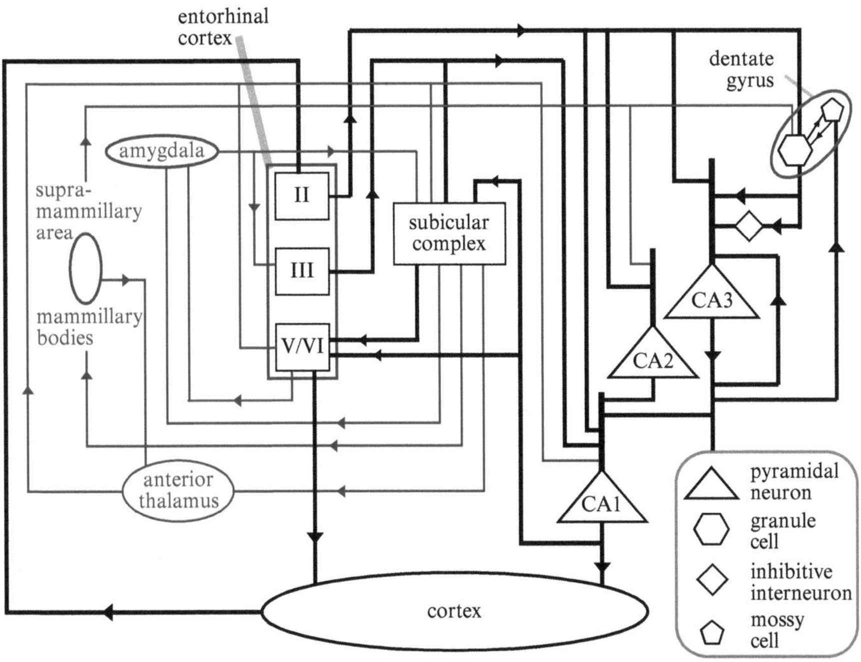

Fig. 7 Major connectivity within the hippocampal formation and between the hippocampal formation and the entorhinal cortex. Also shown is the connectivity between the hippocampal formation and subcortical structures that have a role in declarative memory as demonstrated by the effects of damage to the structures in human patients: the amygdala, anterior thalamic nuclei, and mammillary bodies of the hypothalamus. Cortical information reaches all of the substructures of the hippocampal formation via the entorhinal cortex. Most return information to the neocortex proceeds via the entorhinal cortex, although there is a small proportion directly from CA1. The anterior thalamus receives outputs from the subicular complex and target the entorhinal cortex, the subicular complex and CA1. The amygdala receives inputs from the subicular complex and the entorhinal cortex and targets the same structures (different layers in the case of the entorhinal cortex). The mammillary bodies receive inputs from the subicular complex and target the anterior thalamus, and also target (via the supramammillary area) both CA2 and the dentate gyrus

There is strong connectivity from CA3 pyramidals to CA1 pyramidals, from CA1 to the subicular complex, and from CA1 and the subicular complex to the deep layers of the entorhinal cortex [Lavenex and Amaral 2000].

From the deep layers of the entorhinal cortex information flows back to the wide range of cortical areas that provided input. There is a much smaller flow of information directly from CA1 to external cortical areas as discussed earlier.

There is very prominent internal feedback in CA3. A typical CA3 pyramidal cell in the rat has about 3,600 excitatory inputs from the perforant path, but 12,000 excitatory inputs from other CA3 pyramidals [Amaral, Ishizuka and Claiborne 1990]. It also has about 50 excitatory inputs directly from the dentate gyrus, but strong inhibitory inputs from interneurons that are much more heavily targeted by dentate gyrus outputs [Acsady et al. 1998].

Outputs from layer II of the entorhinal cortex target the dentate gyrus and all the CA fields, and outputs from the somewhat deeper entorhinal layer III target only CA1 and the subicular complex [Insausti and Amaral 2004]. There is a pronounced flow of information from the (excitatory) granule cells of the dentate gyrus to both CA3 pyramidal neurons and inhibitory interneurons, with a typical granule cell innervating 11–15 CA3 pyramidal cells (in the rat hippocampus) and a much larger number of CA3 inhibitory interneurons [Acsady et al. 1998]. There is some evidence for direct inhibitory (GABA) effects by granule cells on CA3 pyramidal neurons [Walker et al. 2001]. Overall, it is possible to interpret the effect of a dentate gyrus granule cell as "communicating with a handful of CA3 pyramidal cells while silencing most others" [Mody, 2002].

CA3 pyramidal neurons extensively target CA1 pyramidals. A single CA3 pyramidal cell may contact CA1 cells throughout 75% of the length of the hippocampus [Lavenex and Amaral 2000], and CA2 pyramidals target CA1 pyramidals [Tamamaki et al. 1988]. However, CA1 does not have the extensive internal feedback observed in CA3.

There is also very strong internal feedback within the dentate gyrus [Buckmaster and Schwartzkroin, 1994]. Granule cells excite mossy cells in the hilar region, and the mossy cells make excitatory connections back on to granule cells. As indicated earlier, granule cells have two types of target within CA3 [Acsady et al., 1998]. Firstly, they make excitatory contacts on to the apical dendrites of a relatively small number of pyramidal cells. These contacts are to specific structures known as thorny excrescences. Secondly, they make excitatory contacts on to a wider range of CA3 interneurons, which in turn make inhibitory contacts on to CA3 pyramidals. Overall, the net general effect of the dentate gyrus on CA3 is inhibitory [Bragin et al. 1995]. CA3 pyramidal cells have axon branches that produce excitatory feedback to the dentate network, exciting mossy cells in the hilus, which in turn excite granule cells. [Ishizuka et al. 1990; Muller and Misgeld, 1991; Penttonen et al. 1997].

There are therefore two excitatory recurrent circuits, one in CA3 and the other in the dentate gyrus, with reciprocal connectivity between them [Lisman, 1999].

The amygdala provides inputs to the subicular complex and to layer III of the entorhinal cortex. The subicular complex and layers V/VI of the entorhinal cortex provide connectivity back to the amygdala. [Insausti and Amaral 2004].

The anterior thalamus receives inputs from the subicular complex via the fornix, and projects back to CA1 [Wyss et al. 1979; Bayat et al. 2005], and to the subicular complex and layer V/VI of the entorhinal cortex [Shibata 1993].

The mammillary bodies receive a strong connectivity from the subicular complex [Allen and Hopkins, 1989], and project strongly to the anterior thalamic nuclei over the mammillothalamic tract. The associated supramammillary area has a substantial projection to CA2 pyramidals and dentate gyrus granule cells [Veazey, Amaral, and Cowan, 1982].

5 Hippocampal System Model

The primary role of the hippocampal system in the resource management model is to select the cortical columns that will record information at each point in time. It performs this role by (i) collecting information on the degree of internal activity in each cortical column, (ii) processing this information to determine the relative activity of different (partially overlapping) groups of columns that have tended to record information at similar times in the past, (iii) performing a competition between the groups to determine the appropriate locations for recording, and (iv) generating outputs to pyramidal neurons in appropriate columns that drive the recording.

Various structures including the thalamus, hypothalamus and amygdala act upon the hippocampal system to modulate the selection of the appropriate cortical columns and the overall degree of condition recording. The amygdala increases the degree of condition recording above the base level in strongly emotional circumstances. In information terms, this reflects the probability that such circumstances may be more useful than average for guiding future behaviour, justifying extra information recording. The hypothalamus biases information recording in favour of cortical areas that tend to generate recommendations in favour of different general types of behaviour (aggressive, food seeking etc.). The bias is one way in which the probability of selection of a behaviour of the type is increased. The thalamus receives inputs from cortical columns that are interpreted as recommendations in favour of condition recording, with the weights depending upon reward feedback following such recording in different past experiences, and modulates the outputs from the hippocampus on this basis. All of these biases are implemented by changing the activity of pyramidal neurons generating condition recording management signals at some appropriate point within the hippocampal system.

The primary hippocampal role results in the collection of information that makes the hippocampal system useful for a number of secondary roles. One such secondary role is providing information about groups of columns that have recorded information at the same time in the past, to permit indirect activation of columns on this basis. This type of indirect activation is the mechanism for accessing episodic memories.

Another secondary role is supporting navigation. For every physical location, visual similarity means that a particular group of neocortical columns will have recorded conditions the first time the location was visited and will tend to record additional conditions whenever there is a novel experience while at the location. There will therefore tend to be cells in all parts of the hippocampus that correspond with different specific locations. These cells can contribute to solving navigation problems.

A further secondary role of the hippocampal system could be providing an indication of the novelty of an experience, on the basis of the overall demand for condition recording.

The hippocampal system also has natural access to information about which columns have recently recorded information, and therefore where additional resources could be required. Hence it plays a role in the assignment of such resources.

Its information about which columns have recorded information at the same time as other columns is relevant to configuring those resources by creating appropriate provisional conditions.

As discussed earlier, a column can expand its receptive field if conditions are present within current cortical information that are similar enough to previously recorded conditions to avoid excessive dilution of the behavioral meanings of column outputs. "Similar enough" means that the column has produced outputs in the past in response to moderately similar cortical information. A significant level of internal column activity indicates such moderate similarity. Condition recording will be discouraged if there is already significant output from a column. Such discouragement is on the basis of activity within the same column, and could be implemented by local inhibitive connectivity.

To minimize dilution of behavioral meanings, condition recording should occur only in enough columns to meet the minimum required total column activity (i.e. the level at which there is an adequate range of behavioural recommendations to support a high integrity behavioural selection). These columns should generally be those with no outputs but high levels of internal activity. An all-to-all competition is required to determine the identity of those columns, which in connectivity terms is most efficiently performed by a central resource manager as discussed earlier. However, if this simple competition is biased in favor of groups of columns that have recorded conditions at the same time in the past, the chance of identifying a consistent group with minimum dilution of behavioral meanings is improved.

The requirement is therefore to perform a competition between different groups of cortical columns on the basis of current activity within the group and the degree to which the group has tended to record conditions at the same time in the past. Any one column could of course appear in multiple such groups. The competition determines the identity of groups of columns to record conditions at each point in time, and a translation back into condition recording management signals directed to the pyramidal neurons in individual columns that appear in many of the selected groups is then required.

Computer simulations have been performed using three layer columns, a very simple pyramidal neuron model with binary inputs and binary outputs, and a learning algorithm in which receptive fields expand but do not contract. These simulations demonstrate that a set of columns employing a competitive process based on middle layer activity can self organize to discriminate between input states with behaviourally different implications [Gedeon et al. 1999; Coward 2001; Ratnayake et al. 2003]. Such self organized columns can be used in combination with a simple basal ganglia model to learn high integrity behaviour selections that are resistant to interference between prior and later learning [Coward et al. 2004].

5.1 Physiological Operations Supporting Memory Functions

This resource manager model can be understood in more detail by consideration of the major physiological connectivity routes as illustrated in figures 6 and 7. In the model, as in figure 6, information on the internal activity of cortical columns (i.e. activity of pyramidal neurons in an appropriate middle layer) is communicated to the parahippocampal and perirhinal cortices. Conditions programmed on pyramidal neurons within these cortices are inputs from sets of cortical columns that recorded conditions at the same time in the past, and the receptive fields of columns in these parahippocampal and perirhinal cortices are therefore groups of cortical columns that have often recorded information in the past at the same time. These groups will generally be limited to columns within one cortical area. Outputs from these parahippocampal and perirhinal columns target the cortical columns providing their inputs, and constitute the management inputs that excite condition recording to those columns. These outputs are generally not activated without inputs derived from the hippocampus via the entorhinal cortex.

Outputs from another layer of parahippocampal and perirhinal columns target the entorhinal cortex. In the entorhinal cortex, columnar receptive fields resulting from these parahippocampal and perirhinal inputs are groups of groups of cortical columns that have recorded information at the same time in the past. These groups of groups will generally include columns in multiple cortical areas. Entorhinal columnar outputs from one layer target the parahippocampal and perirhinal columns providing their inputs, and constitute the management inputs that excite condition recording in those columns. These outputs are not activated without inputs derived ultimately from the hippocampal formation. Entorhinal outputs from another layer are provided to all of the components of the hippocampal formation.

A high level of entorhinal input to the hippocampal formation thus reveals a strong activation of cortical columns in response to the current input state without any changes to receptive fields, indicating that the state is relatively familiar and a low level of condition recording is appropriate. Conversely, a low level of entorhinal input indicates a novel situation, requiring a high level of condition recording in order to activate enough columns to achieve an adequate range of behavioural recommendations. Note that for a given sensory input state there could be both familiar and novel aspects: for example individual sensory objects could be familiar but their spatial arrangement could be novel. The activity of the entorhinal cortex would carry all this information. A competition occurs within the hippocampal formation to determine the groups of columns most appropriate for recording information, and the hippocampal output structure (the subicular complex) begins the conversion of the outputs of this competitive process into signals that can drive recording. This conversion process continues back through the cortices associated with the hippocampus to the sensory, association and motor cortices.

The competition process can be understood by consideration of the physiological connectivity illustrated in figure 7. Input from the entorhinal cortex comes into granule cells in the DG. These cells detect conditions that indicate activity of groups of groups of groups of cortical columns. Dentate gyrus granule cells do not have condition recording management inputs, and their receptive fields therefore develop without the benefit of the management process. As a result, these receptive fields will be relatively poorly focussed on groups of entorhinal cortex columns that tend to be active at similar times.

DG granule cells have two types of target in area CA3. Firstly, they directly excite specific structures (thorny excrescences) on the dendrites of those CA3 pyramidals that have similar receptive fields to the source granule cells. These structures define provisional conditions on the CA3 pyramidals that are combinations of entorhinal inputs, and the DG inputs are functionally the inputs that excite condition recording. Secondly, they excite CA3 interneurons that in turn inhibit a wider range of CA3 pyramidals that have different receptive fields from the source granule cells. CA3 pyramidal neurons also have large numbers of excitatory inputs from other CA3 pyramidals. CA3 pyramidal outputs target granule cells in the DG, and also CA1 pyramidal neurons.

If there is strong input from the entorhinal cortex, the implication is that the input situation is familiar and little information recording is required. In this situation, granule cells will be strongly excited, generating strong CA3 interneuron activity, which will prevent significant CA3 pyramidal activity. If entorhinal cortex input is weak, there will be relatively weak activity by granule cells, and weak activity of CA3 interneurons. Initial CA3 pyramidal activity is driven by inputs from the entorhinal cortex and indicates detection of the activity of groups of groups of groups of cortical columns. Direct input from granule cells triggers recording of additional conditions. Feedback from other CA3 pyramidals biases activity in favour of groups of groups of groups that recorded information at the same time in the past. The effect is to activate a population of CA3 pyramidals corresponding with a set of groups of groups of groups of cortical columns that have all tended to record information at the same time in the past. Condition recording on the CA3 pyramidals will slightly expand their receptive fields to include groups about to record information at the same time. As CA3 pyramidal activity increases as a result of condition recording, feedback to DG granule cells via mossy cells increases, and the resultant increased activity of the granule cells increases the inhibition back into CA3 and limits the buildup of CA3 activity. The larger the input from the entorhinal cortex, the smaller the total CA3 activity. In other words, CA3 activity will be proportional to the degree of novelty in the current input state.

CA1 pyramidals receive inputs from the entorhinal cortex and detect conditions that indicate activity of groups of groups of groups of cortical columns. The use of condition recording management inputs derived from DG granule cells means that CA3 pyramidal neurons will have receptive fields more sharply focussed than DG granule cells on groups of columns that tend to have recorded information at similar times in the past. Outputs from CA3 pyramidals target CA1 pyramidals with similar receptive fields (i.e. that have often been active in the past at the same time), and both directly excite those pyramidals and form their condition recording management inputs. CA1 pyramidals thus take the results of the CA3-DG competitive process and generate stable outputs that drive receptive field expansions throughout the cortex. Because CA1 pyramidals have condition recording management inputs derived from CA3 pyramidals with more focussed receptive fields that DG granule cells, the receptive fields of CA1 pyramidals will be even more sharply focussed on groups of columns that recorded information at the same time in the past than the CA3 pyramidals.

Thus the staged use of management signals means that neuron receptive fields become more and more sharply focussed on groups of columns that have tended to record information at similar times in the past, going from the very weakly focussed DG granule cells to the somewhat more sharply focussed CA3 pyramidals and then to the sharply focussed CA1 pyramidals. The CA1 pyramidals are therefore the most appropriate for driving current condition recording.

CA1 outputs will not occur without appropriate inputs from the anterior thalamus, and these inputs from the anterior thalamus will be triggered by inputs to the anterior thalamus from CA3 and perhaps other hippocampal structures indicating the completion of the competition.

The first step in this condition recording process is that CA1 outputs target the columns in the entorhinal cortex from which they derive their inputs. The entorhinal columns that occur most frequently in the inputs to the active CA1 pyramidals therefore receive strong inputs encouraging condition recording. These strong inputs trigger both condition recording and the generation of outputs from the entorhinal columns. The outputs are targetted on the perirhinal and parahippocampal columns that occur most frequently in the inputs to the active entorhinal cortex columns. A similar process results in condition recording in, and output generation from, the most heavily targetted perirhinal and parahippocampal columns. In turn, condition recording and output occurs in the cortical columns that occur most frequently in the inputs to the active perirhinal and parahippocampal columns.

This process results in columns in the parahippocampal and perirhinal cortices expanding their receptive fields to include the new groups of cortical columns about to record information at the same time. Columns in the entorhinal cortex add new groups of groups of columns. Pyramidal cells in CA3 add conditions that are groups of groups of groups of columns that record information at the same time, but these conditions also include information on other groups of groups of groups that recorded information at the same time in the past and are also currently appropriate for such recording. CA1 pyramidals add conditions that are groups of groups of groups of columns that record information at the same time.

DG granule cell outputs to encourage condition recording target CA3 pyramidals with similar receptive fields, and CA3 pyramidals target CA1 pyramidals with similar receptive fields. This targetting and receptive field similarity can be achieved to an adequate degree of approximation by biasing the creation of the condition recording connectivity in favour of connectivity between cells that are very frequently active at the same time. A group of entorhinal inputs that forms a provisional condition will be made up of inputs from neurons that have often recorded information at the same time in the past.

The entire information recording process occurs over a period during which the neocortical columns must be receiving a consistent set of sensory inputs (i.e. derived from a stable sensory input defined by the attention function). Time is required for flow of information to the hippocampus and back to the cortex, and time must be available to achieve the several repetitions of conditions required to create a long term LTP-supported condition recording. The implication is that memories cannot be created for visual experiences lasting less than about 100 milliseconds. The memory is fully defined in information terms at the end of the few hundred millisecond period. There may be chemical processes required to consolidate the memory [Tronson and Taylor, 2007], but the information content of the memory is not significantly changed or relocated.

Management of action potential timing is an important factor for the operation of the hippocampus. Coward [2004; 2005a] has proposed that action potentials generated by sensory inputs derived from the focus of attention are temporally bunched around peaks in a modulation signal corresponding with the gamma band in the EEG. This bunching means that conditions will be detected in sensory inputs derived from the attention focus but not in sensory inputs not derived from that focus and therefore not bunched. In order for the LTP mechanism to operate effectively, signals from the hippocampal system must arrive at neocortical columns in phase with the modulation signal. Lisman [2001] has proposed a model in which interactions between gamma and theta band oscillations can link memories of a sequence of events, and some version of this approach is required in the resource manager model.

Inputs from the neocortex arrive at pyramidals in layer II of the entorhinal cortex. Outputs to the hippocampal formation are derived primarily from the same layer. The primary outputs from the hippocampal formation are derived from CA1 and pass through the subicular complex to pyramidals in layer V/VI of entorhinal columns. These layer V/VI pyramidals generate the outputs to the neocortex.

The amygdala generates outputs indicating that a higher level of condition recording is justified because of "emotional" circumstances. These outputs target the subicular complex and layer III of the entorhinal cortex, and have the effect of increasing the level of output to neocortical columns selected for condition recording. This increase is limited to such columns within specific cortical areas recording complex associative conditions more likely to be useful in the future that simpler sensory conditions.

The hypothalamus acts on the hippocampus to bias the competition in favour of certain cortical areas that tend to result in behavioural recommendations of particular types. Outputs from the hypothalamus (supramammillary area) therefore influence the actual competition process by targetting the dentate gyrus, and increase the degree of recording in the target areas by targetting CA1 pyramidals. The role of the CA2 field is to receive inputs from the supramammillary area reflecting a selection of a general behavioral type to receive priority. CA2 then biases the CA1 selections in favor of cortical areas that tend to generate recommendations of the selected behavioral type. Feedback on the degree of current output from the subicular complex to the mammillary bodies regulates the degree of total output. It is also important to note that one source of information that the hypothalamus can use to select appropriate behavioural priorities is the relative degree of activity in different cortical areas. This information is therefore provided to the mammillary bodies over the fornix.

The thalamus in general gates the flow of information between different cortical areas. These information flows ultimately result in behaviours, and the thalamus uses reward feedback following behaviours to modulate the probability of similar information flows in the future. The anterior thalamus therefore excites the structures generating condition recording outputs (CA1, the subicular complex and layer V/VI of the entorhinal cortex) to different cortical areas. The type of behaviour currently favoured is relevant to the selection of information flows, and there is therefore connectivity from the mammillary bodies to the anterior thalamus.

An input state that is sufficiently novel at the sensory level will require information recording in the relevant primary sensory cortex, and recording in other areas is not an alternative (although it may well occur in addition). Hence decisions on information recording in primary sensory areas could be made locally without reference to the central resource manager, because connectivity to support such local decisions would not be excessive. This would account for the exclusion of the primary sensory areas from the cortical regions providing input into the hippocampal system (figure 7).

5.2 Creation of Connectivity

At a number of points in the description of the resource manager, the existence of appropriate connectivity has been implicitly assumed. One key requirement is for connectivity to support provisional conditions. Recorded conditions must be similar to other conditions already recorded on the same neuron, where "similar" means that they share inputs with and/or occur at the same time as previously recorded conditions. Inputs to provisional conditions could be selected randomly from the same sources as existing conditions, but this approach risks wasting a high proportion of connectivity and including irrelevant inputs that happened to be active at the time of recording but are rarely active at the same time as the other inputs. However, information about past temporally correlated activity can be used to improve the probability of creating useful conditions [Coward 1990; 2000].

For example, suppose that the cortex and hippocampal system were submitted to an internally generated activation that was a weighted average of past condition recording activity with a bias in favour of more recent activity. Then suppose that new dendrite arms accepted inputs from groups of axons in their neighbourhood, provided that those axons tended to be active at the same time. Such provisional conditions would have a significantly higher probability of generating useful regular conditions. Simulations have demonstrated that this type of approach reduces the required connectivity in cortical models using simple binary input and output neurons by about 20% [Coward 2000]. Coward [1990] proposed that one role of sleep is to support the creation of provisional connectivity, with REM sleep providing the weighted average rerun of past activity required to identify the most appropriate connectivity. In this proposal, unlike consolidation models [Squire and Alvarez, 1995], there are no changes to past memories. The effect of REM sleep is to create connectivity that is as appropriate as possible for recording information in the future, using a weighted average of past experience as the best available estimate for future experience. The averaged rerun will be required both in the hippocampal system and throughout the neocortex to support configuration of provisional conditions.

Similarity of receptive fields of two neurons means that the two neurons will have a high tendency to be active at the same time over an average of past experience. The rerun can therefore be used to support creation of the required appropriate connectivity between granule cells and CA3 pyramidals, between CA3 pyramidals and granule cells, and between CA3 pyramidals and CA1 pyramidals on the basis of receptive field similarity. Furthermore, the creation of appropriate connectivity between CA3 interneurons and CA3 pyramidals can be managed on the basis of low tendency to simultaneous activity of the CA3 pyramidal and the granule cells providing inputs to the interneuron.

Because the receptive fields of hippocampal neurons contain information on groups of columns that have recorded information at the same time in the past, the system is the appropriate place to drive the rerun activity both within the hippocampal system and throughout the neocortex

5.3 Access to Memories: Episodic Memory

As discussed earlier, episodic memory depends upon indirect activation of neocortical columns on the basis of simultaneous past receptive field expansions. Activation on this basis generates an overall active column population that approximates to the one active during the experience that is being recalled. The hippocampal system preserves information on such past simultaneous information recording, and it therefore contains the information needed to drive access to episodic memories.

Pyramidal neurons in the parahippocampal, perirhinal and entorhinal cortices, and in CA1 and CA3 all contain information on columns that recorded information at the same time in the past, as do granule cells in the dentate gyrus. However, CA1 pyramidal cells are the primary driving force for information recording, and use of these cells for access to episodic memories risks recording of irrelevant information.

The mechanism for accessing an episodic memory is a sequence of steps. Firstly, an initial neocortical column population is activated by, for example, hearing trigger words. The columns in this population have recommendation strengths in favour of indirect activation of other columns on the basis of simultaneous past condition recording. These recommendation strengths are instantiated by connection weights of column outputs into components of the thalamus and basal ganglia that correspond with the behaviour type. Such recommendation weights are set relatively high at the moment that simultaneous recording occurs, and decay with time. Use of the recommendation strengths, particularly if followed by positive reward feedback, blocks the decay and even increases the weights.

The same columns have many other types of recommendation strengths, and the second step is a competition within the thalamus and basal ganglia between the different types of behaviour. If indirect activation on the basis of past simultaneous information recording is the selected behaviour, the third step is that outputs from this neocortical population are released by the thalamus to drive activation of pyramidal neurons in the parahippocampal, perirhinal and entorhinal cortices and CA3, and granule cells in the dentate gyrus. All of the activated neurons in these structures contain significant numbers of conditions made up inputs from different currently active neocortical columns. The activated neurons therefore correspond with groups and groups of groups of neocortical columns that recorded information at the same time as the currently active group of columns. Fourthly, activity in CA1 does not reach a level resulting in strong output to drive information recording because of lack of input from the thalamus. This lack of input is the result of reward feedback in similar past circumstances in which episodic memory has been encouraged. However, there could be a small degree of output, which would result in the ability to remember recalling the memory. Fifthly, hippocampal system activity drives activation of neocortical columns by feedback connectivity that targets pyramidal neurons in the columns from which the hippocampal system structures received input. This release will again be a behaviour managed by the thalamus. However, this feedback connectivity does not target provisional conditions, but the soma, basal dendrites or proximal apical dendrite of target pyramidal neurons, in the output layer of the neocortical columns. Sixthly, neocortical columns receiving substantial hippocampal system input of this type produce outputs. This secondary population of neocortical columns will tend to be made up of columns that all recorded information in the past at the same time, seeded by the original trigger words. This activation approximates to the activation during the past experience. Seventhly, a second cycle of indirect activation through the hippocampal system could increase the consistency of the population on this simultaneous recording basis, and therefore the degree to which the population corresponds with that during the original experience.

The receptive fields of pyramidal neurons in CA3 correspond with complex groups of groups of groups of neocortical columns from a wide range of cortical areas. Receptive fields in the entorhinal cortex correspond with somewhat less complex groups of groups, and columns in the perirhinal and parahippocampal cortices with even less complex groups from just one or a few cortical areas. The more complex the episodic memory to be accessed, the higher in the hippocampal hierarchy will be the regions that must participate. For example, retrieving an episodic memory of a complex event but including detailed sensory imagery would be expected to require participation of CA3.

Recall of a memory sequence requires activations on the basis of information recording slightly after past information recording in a currently active column population. Management of such activations requires dynamical processes similar to those described in Lisman [2001].

5.4 Access to Memories: Semantic Memory

As discussed earlier, semantic memory requires indirect activation of neocortical columns on the basis of frequent past simultaneous activity, without any requirement for information recording. Thus the ability to access the word "Paris" from an activation generated in response to "What is the capital of France?" is based upon neocortical columns activated in response to hearing the question activating columns with recommendation strengths in favour of speaking the answer. These recommendation strengths are based upon frequent past simultaneous activity. When the statement "Paris is the capital of France is heard for the first time, there will be condition recording, and initially access to the response would be on the basis of simultaneous past recording. The degree of information recording will be much less in subsequent exposures to the information. Retrieval of the response on the basis of simultaneous past condition recording will be much less efficient that retrieval on the basis of past simultaneous activity. However, if hippocampal system pyramidal neurons were also required to store information on past simultaneous activation without condition recording, the additional information would make identification of the locations for new recording less effective.

Study of the deficits following local cortical damage indicates that the anterior temporal cortex is important for semantic memory [Rogers et al. 2006], but functional imaging of semantic memory tasks results in activation of a very wide range of different cortical areas [e.g. Thompson-Schill, 2003]. The anterior temporal cortex can therefore be interpreted as the location within which information on frequent past simultaneous activity of different cortical columns is stored. On this model, columns in the anterior temporal cortex will develop receptive fields corresponding with groups of columns in other areas that are often active at the same time, with connectivity back to the same columns that can activate those columns without condition recording (i.e. targetting somas, basal dendrites or proximal apical dendrites rather than provisional conditions in the distal apical dendrites). Release of outputs from the anterior temporal cortex to other cortical areas will of course be gated by the thalamus. Control of access to frequently retrieved episodic type memories could of course be shifted to the anterior temporal cortex.

5.5 Creation of Imaginary Events

Consider the process for imagining an event that has not actually taken place, such as a party with Albert Einstein as a guest. Trigger words such as party, guest, Albert Einstein result in a pseudovisual activation on the basis of frequent past activity at the same time as the auditory columns directly activated by hearing the words. The effect is creation of a population of columns that would be activated if a party or if Albert Einstein were seen, although as discussed earlier the column activity close to sensory input would be weak and there would not be experience of a visual hallucination. However, although this population would be fairly strong at the object level of receptive field complexity (because parties and pictures of Albert Einstein have been seen in the past) it will also be weak at some of the more complex receptive field complexities level (because an event combining a party and Albert Einstein has not been seen in the past). A strong activation at this more complex level would correspond with a memory. CA1 then generates outputs resulting in condition recording that brings the weak neocortical activation up to the minimum level.

The process can be viewed as generation of an active cortical column population as though the imaginary scene was being perceived, using the same resources as would be used for recalling an actual memory, with slight expansions of receptive fields as required. This model predicts that the cortical resources used to imagine events are the same as those used for remembering similar events, as observed by Addis et al. [2007]. It also predicts that, because of the loss of the ability to record new conditions, patients with hippocampal amnesia will not be able to imagine new experiences, as observed by Hassabis et al. [2007]. The resource manager model predicts that area CA1 should be more active during imagination of events than during recollection of past events.

5.6 *Novelty Detection*

The degree of novelty in a visual experience will be indicated by the magnitude of input to and output from the part of the hippocampal system that manages information flows with the higher visual cortices. Hence activity in that interface will indicate the degree of novelty in a visual experience.

6 Evidence for Hippocampal System Model

6.1 *Cognitive Deficits Following Hippocampal System Damage*

Physical damage to the resource manager function as described would result in loss of the ability to select the neocortical columns in which new information is recorded and to drive that condition recording. Because all existing columns and their associated recommendation strengths are preserved, there is minimal disruption to most other cognitive capabilities. However, because the hippocampal system acquires information identifying groups of columns that recorded conditions at the same time in the past in the course of its resource management role, and is therefore the natural source for information to guide indirect activations of cortical columns on that basis, there will therefore some loss of episodic memory through inability to access the information. Information to identify groups of columns on the basis of frequent past simultaneous activity or recent simultaneous activity must be collected to support semantic memory and priming. Use of the resource manager to collect such information would interfere with its primary role. Other cortical structures must therefore collect such information, and damage to the resource manager will not affect these types of memory. All existing neocortical columns and their associated recommendation weights into the thalamus and basal ganglia are unaffected by damage to the hippocampal system. Hence all previously learned physical and cognitive skills are unaffected. Recommendation weights of existing columns could still be changed by reward feedback. Hence if a skill could be acquired without changes to column receptive skills, learning of such a simple skill could proceed despite hippocampal damage. Even relatively complex skills might be acquired using a population of columns with fixed receptive fields, if past experience had accidentally provided even a suboptimal discrimination and enough repetition of the tasks occurred to establish appropriate recommendation weights. The model therefore provides a straightforward account for the striking combination of cognitive symptoms associated with damage to the hippocampal system.

6.2 Correspondence with Observed Physical Connectivity

As demonstrated by the description of the detailed model, the resource manager provides a functional account for all of the major connectivity paths observed within the hippocampal system and between the hippocampal system and other brain structures. In particular, it provides functional reasons for the memory deficits observed as a result of amygdala, hypothalamus and thalamus damage.

6.3 Differential Effects of Damage to Subregions

Each pyramidal neuron in CA3, CA1, and the associated cortices preserves information identifying groups of cortical columns that have recorded conditions at the same time. The number of columns in the groups decreases from CA3 and CA1 to the entorhinal cortices and decreases further in the parahippocampal and perirhinal cortices. This information is needed for indirect activations in support of episodic memory. However, because outputs from CA1 drive condition recording, the use of CA1 for such a purpose could result in inappropriate condition recording.

Consistent with this understanding, in human subjects damage to CA1 alone generates anterograde amnesia but little if any retrograde amnesia, and no signs of significant cognitive impairment other than this loss of memory (e.g. patients RB [Zola-Morgan et al. 1986] and GD [Rempel-Clower et al. 1996]). When damage extends to other hippocampal formation structures, retrograde amnesia becomes significant in addition to anterograde amnesia (e.g. patients LM and WH [Rempel-Clower et al. 1996], patient HM [Corkin et al. 1997]).

In the resource management model, information derived from sensory experiences is recorded immediately in columns in the neocortex, and subsequent damage to the hippocampal system will not affect the information. The only effect of such damage will be on the capability to access such information on the basis of past temporally correlated information recording. If the basis for activation shifts over time towards temporally correlated activity, access will become more and more independent of the hippocampal system. For example, when a word is first learned, there will be information recording in auditory columns activated in response to hearing the word, and in visual columns activated in response to the object or concept of the word. The capability to understand the word depends upon the recorded information and in the short term, actual understanding of the word exist because the auditory columns indirectly activate the visual columns on the basis of simultaneous past information recording, utilizing information from the hippocampal system. After a number of occasions on which the word has been understood in this way, the auditory columns will acquire the ability to activate the visual columns on the basis of frequent past simultaneous activity, supported by connectivity paths within the neocortex and independent of the hippocampal system. To the degree to which this has occurred, access to the information would be expected to result in the most severe retrograde amnesia for episodic memories, less for personal semantic memories and semantic memories of public events and persons, and least for general semantic memory. This graduation is consistent with the observed amnesias [Nadel and Moscovitch, 1997].

An autobiographical memory is the record of complex, unique events. The link between the information active at the time of the event will therefore be temporally correlated information recording across a complex population of cortical columns, and there is no reason for these columns to be frequently active at the same time. At the other extreme, a new word is the record of a relatively simple link between auditory columns and visual columns and the columns are active at the same time each time the word is used. A shift to information access on the basis of frequent past simultaneous activity is likely to be rapid. The association between the names and faces of public individuals, and personal semantic facts represent an intermediate state. The observed graduation in retrograde amnesia with hippocampal system damage from most severe for autobiographic to negligible for word knowledge is as expected by the model. An exception could be if a particular memory were very frequently described. In such a case, frequent repetition could result in some ability to access the memory independent of the hippocampal system on the basis of frequent past simultaneous activity.

Regular autobiographical memory does not become independent of the hippocampal system. In the experiments of Rekkas and Todd Constable [2005], the activation of the hippocampal formation was observed during retrieval of both recent and remote autobiographical memories, and activity was greater for remote memories. In these experiments, the research design stressed depth of recall and encouraged visualization of details. Thus subjects were asked "Can you recall a specific high school teacher?" or "Can you recall the school yard of your elementary school?" but it was stressed to participants that the questions were meant to cue an actual episode, such as "The time the English teacher brought in a recording of Hamlet and made us listen" rather than a series of facts (like the name of the teacher). Follow-up questions like "Do you recall a time when you were playing in a specific area of the school yard?" This design aimed to exclude facts recall and recall of highly salient emotional events (e.g. weddings, graduations, loss of a pet) that may be more common in autobiographic self reports.

6.4 Hippocampal System and Navigation

In the resource management model, CA1 pyramidal neurons correspond with groups of neocortex columns that have frequently recorded information at the same time in the past. CA3 pyramidals also correspond with such groups, but have interconnectivity with many other CA3 pyramidals corresponding with groups that have recorded information at the same time in the past, but somewhat less frequently. The entorhinal cortex brings together the inputs from such groups and provides inputs indicating the activity in the groups to the hippocampal formation. The entorhinal cortex also receives outputs from the hippocampal formation and translates them back into outputs directed to the individual columns to drive information recording. Hence a mapping between groups of cortical columns that have recorded information at the same time and individual pyramidal neurons can be expected in CA1, CA3, and the entorhinal cortex.

One situation in which recording of information at the same time can be expected is in navigation, where simultaneous recording across a specific population of cortical columns can be expected when in the same location. Pyramidal neuron "place cells" which are active when a rat is in a specific location have been observed in CA1 and CA3 fields [Leutgeb et al. 2004] and in the entorhinal cortex [Fyhn et al. 2004], but not in the more peripheral areas of the hippocampal system [Fyhn et al. 2004]. This distribution of place fields is as expected by the model.

Furthermore, given that the role of CA3 is to focus CA1 on an optimal group of columns to record information at the same time, place fields developed in the absence of CA3 would be expected to be less sharp, as observed by Brun et al. [2002]. Changes to the environment would be expected to result in greater changes to CA3 than CA1 place fields, as observed by Leutgeb et al. [2005].

6.5 Damage to Hypothalamus and Amygdala

In the model, the role of the hypothalamus is to influence current information recording in favor of current general behavioral priorities. Loss of this function would be expected to affect the ability to record information in the future, but to have no effect on access to past information on the basis of temporally correlated past recording. Consistent with this interpretation, damage strictly limited to the mammillary bodies (bilaterally) results in anterograde amnesia but minimal retrograde amnesia [Tanaka et al. 1997]. The thalamus influences the level of hippocampal system outputs in general, and damage to the anterior thalamic nucleus can therefore result in both anterograde and retrograde amnesia as observed [Graff-Radford et al. 1990].

In the resource management model, the role of the amygdala in memory is to adjust the degree of hippocampal system driven information recording during emotional events in favor of cortical areas where the increased recording is likely to be useful in determining behavior in the future. Consistent with this role, it is found that emotional arousal biases the memory of the event in favor of the gist and reduces the memory for visual details, and that bilateral damage to the amygdala eliminates the bias [Canli et al. 2000; Adolphs et al. 2001; 2005]. Furthermore, it is the interaction between the amygdala and the hippocampal system that correlates with the enhanced emotional memory [Dolcos et al., 2004].

It is also relevant that, as expected by the model, the effect of lesions to the mammillary bodies and anterior thalamic nuclei is to depress the operation of the hippocampal system. Thus the hippocampal system activity observed during memory encoding and retrieval tasks in normal subjects was not observed during attempted performance of the same tasks in a subject with such diencephalic lesions but no damage to the hippocampal system [Caulo et al. 2005]. Similarly in the case of emotional memory it is a correlation of activity between the amygdala and hippocampus that occurs during creation of stronger memories in emotional situations [Dolcos et al. 2004].

6.6 Role of Sleep

In the resource management model the role of sleep including REM sleep in declarative memory is radically different from that proposed in the consolidation model. In the latter model, the role is to rerun recent sensory experiences to expedite the long-term registration of the memories of the experiences in the neocortex. The rerun must presumably be fairly precise to achieve this function. One view is that rerun of very recent experience could occur during slow wave sleep, and rerun of more remote experience during REM sleep [Hoffman and McNaughton 2002].

In the resource management model, sleep including REM sleep addresses the problem that it would not be practical to create the physical connections required to support the recording of new information at the instant such recording was needed. Rather, provisional connectivity is created in advance, and information is recorded as the most appropriate subsets of this provisional connectivity. Sleep is the period in which provisional connectivity between pyramidal neurons that could be used to record new information in the next wake period is created. Such provisional connectivity would be established between the appropriate levels of condition complexity, but if it was otherwise completely random, a high degree of connectivity would be required to ensure that subsets existed that corresponded with useful information. Much of this connectivity would be wasted and could result in significant behaviorally confusing noise. If the connectivity can be biased in favor of connectivity somewhat more likely to be useful, significant resource advantages would follow. Past experience provides the only available guide to probable usefulness. Any past experience could be relevant, but recent experience will on average be a somewhat better guide than remote experience. The proposed role of sleep is to establish provisional connectivity, but with a bias in favor of connectivity between neurons that have often been active in the past at similar times, especially the recent past. Activating neurons in a manner that reflected past temporal correlations could impose such a bias, and the presence of information on such correlations in the hippocampal system would make that structure appropriate to drive such an activation pattern. This activation pattern would be experienced as a partial, approximate rerun of past experience, with a bias in favor of the most recent [Coward 1990]. Simulations indicate that such a biasing process reduces the information recording resources required to achieve a given level of performance for a relatively simple behavioral problem by about 20% [Coward 2001].

In the resource management model, as in the case of consolidation models, there would be a rerun of recent experience (perhaps in slow wave sleep) and more remote experience (in REM sleep). However, there would be no need for an exact rerun of past experience, only a reactivation that provides a reasonable approximation to the temporal correlations between pyramidal neuron activities in the past. Elimination of REM sleep would not prevent learning, rather it would increase the resources required for learning by some degree.

Correlations between neuronal activity during waking and during the subsequent sleep period have been observed [Skaggs and McNaughton, 1996], and it can be argued that slow wave sleep reflects activity in recent waking experience and REM sleep more remote experience [Hoffman and McNaughton 2002]. However, although dreams include clearly recognizable waking elements, they do not reproduce real-life events [Fosse et al., 2003] as required by the consolidation models. This situation is, however, fully consistent with the resource management model. Consolidation models also have the problem that REM sleep deprivation appears to have relatively little effect on memory capabilities, and REM sleep can be substantially or completely suppressed (by various antidepressant drugs, or by bilateral damage to the pons) without apparent effect [Vertes and Eastman 2000]. In the resource management model, deprivation of REM sleep would be expected to increase the resources required for memory support to some degree, but would not qualitatively interfere with memory creation.

In the resource management model, the averaged rerun of past experience would be required to configure appropriate provisional conditions on pyramidal neurons in the hippocampal system and throughout the neocortex, including the prefrontal cortex as observed by Euston et al. [2007].

7 Comparisons with Other Models

The central place occupied by the driving of neocortical receptive field expansions in the resource manager model makes the model qualitatively different from the alternative models discussed earlier, and conceptually much simpler. However, there are some similarities between certain functions of the resource management model and parts of other models. Indexing theory [Teyler and DiScenna 1986] suggests that the hippocampal system maintains an index of the neocortical areas activated by each experienced event. In the resource manager model, records of past simultaneous information recording by different groups of columns are maintained, but not as an explicit index for each event. However, using the records for different groups of columns as described earlier can access memories of past events. Indexing theory does not make a clear distinction between events in which there is strong recording and events in which there is just simultaneous activity. Hence the support of semantic memory outside the hippocampal system is not explained.

The multiple trace model [Nadel and Moscovitch 1997], like the resource manager model, proposes that additional memory traces are created in subsequent, similar experiences, but provides no reasons why memory traces should initially be within the hippocampal system and subsequently in a separate structure such as the anterior temporal cortex. The role of REM sleep is not explained in indexing theories.

Consolidation models [Squire and Alvarez 1995] propose that memories are initially registered in the hippocampal system, and over a period of time transferred to other neocortical structures, perhaps to avoid interference between prior and later memories [McClelland, McNaughton and O'Reilly 1995]. Such transfers of complex information constructs between different physiological structures would be physiologically costly and implausible, and are not required by the resource manager model, in which neocortical information records are immediately created in their long term storage locations.

Two component models attempt to account for the complex combination of memory deficits following hippocampal damage by arguing that the hippocampal system is actually two relatively independent subsystems. In the proposal of Gluck et al. [2003], incremental learning is supported by representational transformations in the input regions to the hippocampus (especially the entorhinal cortex), and the storage and recall of previously processed representations supported by the CA3 and CA1 regions. It is not clear how this synthesis can be compatible with the observation that cell loss strictly limited to the CA1 field within the hippocampal system could result in the observed combination of severe anterograde amnesia and little if any retrograde amnesia [Zola-Morgan et al. 1986]. In this specific case, even the perforant path connectivity from the entorhinal cortex through CA1 to CA3 and the dentate gyrus was intact. In the resource manager model, CA1 is the driving force for information recording, but plays a minimal role in retrieval, consistent with the Zola-Morgan et al. observations.

Eichenbaum et al. [1994] propose a two component model in which the hippocampal system contributes only to declarative memories, and plays two orthogonal roles in such memories: the temporary maintenance of memories by the parahippocampal region; and the processing of a particular type of relational memory representations by the hippocampus itself. In this model, sensory information generates a representation in the neocortex that is very sensitive to replacement by subsequent sensory inputs, and can only be maintained as long as the level of intervening interference is low. Activation of the parahippocampal region does not preserve the actual representation but creates a trace of the neocortical activation that is less sensitive to subsequent processing. Activation of the hippocampal formation processes comparisons between current and previous sensory input states. These comparisons require changes to connectivity within cortical areas, but the nature of these changes is not described in the model. Interactions between intermediate term storage in the parahippocampal region and relational processing in the hippocampus are fed back and forth for a significant period of time, contributing to the long-term consolidation of memories.

The Zola-Morgan et al. [1986] observations of severe anterograde amnesia for all types of declarative memory including simple recognition memory following CA1 damage is also inconsistent with the Eichenbaum et al. [1994] two component model in which the role of the hippocampal formation is limited to declarative memories with a strong relational content. The suggestion that the parahippocampal region plays a role in maintaining neocortical activations that are otherwise very sensitive to replacement by subsequent sensory inputs does not appear to be consistent with observations that short term memory of up to several minutes does not appear to be affected by hippocampal damage [e.g. Scoville and Milner 1957], but the dorsolateral prefrontal cortex plays an important role in working memory, appearing to direct attention to internal representations of sensory stimuli and motor plans that are stored in more posterior regions [Curtis and D'Esposito, 2003].

Wallenstein and Eichenbaum [1998] have suggested that the primary role of the hippocampus is supporting creation of memories associating items that are discontiguous in terms of their temporal and/or spatial positioning. In the resource manager model, the primary role is driving receptive field expansion throughout the neocortex, which also results in support for memories of discontiguous events. The observations that object recognition memory is also affected by hippocampal damage [Broadbent et al. 2004] is consistent with the resource manager model but demonstrates a hippocampal role beyond memory for discontiguous events.

In addition to the functions of encoding, retrieval and consolidation proposed in various alternative models, some models emphasize subsidiary information processes. For example, Kesner and Hopkins [2006] argue that pattern separation, pattern association and pattern completion are important hippocampal functions, where pattern separation is defined as the ability to encode and separate events in time and space, pattern association as the ability to form arbitrary associations between events and items, and pattern completion as the ability to retrieve complete information on the basis of partial or incomplete inputs.

In the resource manager model, the hippocampal system selects an appropriate group of cortical columns to record information in response to a sensory input state, records the identities of different subgroups of the selected group that are (therefore) recording information at the same time, and uses these records both to better determine appropriate groups for future experiences and to activate groups that all recorded information at the same time in the past (i.e. to access episodic memories).

Different aspects of the resource manager model can be interpreted as functions like those proposed by Kesner and Hopkins [2006]. For example, because the visual environment is relatively stable at a particular point in space, there is a group of columns that will tend to be activated when the subject is at that point. Novel events occurring at that point in space (including the first visit to the point) will tend to result in recording of information at the same time in a group of columns that includes the group corresponding with the point. Hippocampal neurons have receptive fields corresponding with groups of neocortical columns that recorded information at the same time in the past, and some such neurons will tend to correspond with points in space. These neurons can be interpreted as performing pattern separation.

Because hippocampal neurons in the entorhinal cortex and hippocampal formation correspond with large, heterogeneous groups of neocortical columns, connected only because they have recorded information at the same time in the past, these neurons effectively perform pattern association across arbitrary associations (such as object location and object identity). The activity of these neurons can also be interpreted as pattern completion, because they can identify the larger group of neocortical columns that recorded information in the past at the same time as all of a smaller group.

Although different components of the resource manager model could be interpreted as performing the various suggested information processes, the model as a whole has the advantages that it provides an integrated account of the hippocampal system performing a single critical brain process (i.e. neocortical resource management) and a secondary process which can use the information already present in the resource manager without interfering with that information (i.e. episodic memory) implemented by detailed connectivity paths and physiological processes consistent with those observed in the hippocampal system.

8 Conclusions

The proposed model of the hippocampal system as resource manager for the neocortex has a number of advantages over alternative models. Firstly it is conceptually simpler, with one primary functional role accounting for all the subsystems and observed behavioural functions. Secondly, it provides a straightforward account for the specific combination of deficits observed in patients with hippocampal damage. Thirdly, it provides an integrated account for the memory related deficits observed with damage to mammillary bodies, anterior thalamus, and amygdala. Fourthly, the major connectivity paths observed within the hippocampal system and between the hippocampal system and other brain structures are as expected for the model. Fifthly, it indicates how the LTP mechanism supports the information processes required for memory. Sixthly, it provides a role for sleep in memory that has no issues with experiment. Seventhly, its role is consistent with theoretical constraints on the architectures of complex learning systems.

In general, a major advantage of the model is that it makes it possible to describe the same memory phenomenon on four different levels of detail (neuron physiology; cortical column activity; interactions between major anatomical structures; and psychology), with mapping between different levels. This mapping between different levels of description is critical for scientific understanding [Coward and Sun 2007].

References

Acsady, L., Kamondi, A., Sık, A., Freund, T. and Buzsaki G. (1998). GABAergic Cells Are the Major Postsynaptic Targets of Mossy Fibers in the Rat Hippocampus. *Journal of Neuroscience* 18, 3386–3403.

Addis, D.A., Wong, A.T. and Schacter, D.L. (2007). Remembering the past and imagining the future: Common and distinct neural substrates during event construction and elaboration. *Neuropsychologia* 45, 1363–1377.

Adolphs, R., Tranel, D. and Buchanan T.W. (2005). Amygdala damage impairs emotional memory for gist but not details of complex stimuli. *Nature Neuroscience* 8, 512–518.

Adolphs, R., Denburg, N.L. and Tranel, D. (2001). The Amygdala's Role in Long-Term Declarative Memory for Gist and Detail. *Behavioral Neuroscience* 115, 983–992.

Albin, R. L., Young, A. B. and Penney, J. B. (1989). The functional anatomy of basal ganglia disorders. Trends in Neuroscience 12, 366–375.

Alexander, G. E., DeLong, M. R. and Strick, P. L. (1986). Parallel organization of functionally segregrated circuits linking basal ganglia and cortex. Annual Reviews of Neuroscience 9, 357–81.

Allen, G.V. and Hopkins, D.A. (1989). Mamillary body in the rat: topography and synaptology of projections from the subicular complex, prefrontal cortex, and midbrain tegmentum. *Journal of Comparative Neurology* 286, 311–336.

Amaral, D.G., Ishizuka, N. and Claiborne, B. (1990). Neurons, numbers and the hippocampal network. *Progress in Brain Research* 83, 1–11.

Bayat, M., Hasandeh, G.R., Barzroodipour, M. and Javadi, M. (2005).The effect of low protein diet on thalamic projections of hippocampus in rat. *Neuroanatomy* 4, 43–48.

Bi, G-q. and Poo, M-m. (1998). Synaptic modifications in cultured hippocampal neurons: Dependence on spike timing, synaptic strength, and postsynaptic cell type. *Journal of Neuroscience* 18, 10464–10472.

Bragin, A., Jando, G., Nadasdy, Z., van Landeghem, M. and Buzsaki, G. (1995). Dentate EEG spikes and associated interneuronal population bursts in the hippocampal hilar region of the rat. *Journal of Neurophysiology* 73, 1691–1705.

Broadbent, N.J., Squire, L.R. and Clark, R.E. (2004). Spatial memory, recognition memory, and the hippocampus. *Proceedings of the National Academy of Sciences (USA)* 101, 14515–14520.

Brown, M.W. and Aggleton, J.P. (2001). Recognition memory: what are the roles of the perirhinal cortex and hippocampus. *Nature Reviews Neuroscience* 2, 51–61.

Brun, V.H., Otnaess, M.K., Molden, S., Steffenach, H-A., Witter, M.P., Moser, M-B. and Moser, E.I. (2002). Place Cells and Place Recognition Maintained by Direct Entorhinal-Hippocampal Circuitry. *Science* 296, 2243–2246.

Buckmaster, P.S. and Schwartzkroin, P.A. (1994). Hippocampal mossy cell function: a speculative view. *Hippocampus* 4, 393–402.

Canli, T., Zhao, Z., Brewer, J., Gabrieli, J.D.E. and Cahill, L. (2000). Event-Related Activation in the Human Amygdala Associates with Later Memory for Individual Emotional Experience. *Journal of Neuroscience* 20(RC99), 1–5.

Caulo, M., Van Hecke, J., Toma, L., Ferretti, A., Tartaro, A., Colosimo, C., Romani, G.L. and Uncini, A. (2005). Functional MRI study of diencephalic amnesia in Wernicke-Korsakoff syndrome. *Brain* 128, 1584–1594.

Clark, S.A., Allard, T., Jenkins, W.M. and Merzenich, M.M. (1988). Receptive fields in the body-surface map in adult cortex defined by temporally correlated inputs. *Nature* 332, 444–445.

Cohen, N.J., Ryan, J., Hunt, C., Romine, L., Wszalek, T. and Nash, C. (1999). Hippocampal System and Declarative (Relational) Memory: Summarizing the Data From Functional Neuroimaging Studies. *Hippocampus* 9, 83–98.

Corkin, S. (1968). Acquisition of motor skill after bilateral medial temporal-lobe excision. *Neuropsychologia* 6, 225–264.

Corkin, S, Amaral, D.G., Gonzalez, R.G., Johnson, K.A. and Hyman, B.T. (1997). H. M.'s Medial Temporal Lobe Lesion: Findings from Magnetic Resonance Imaging. *Journal of Neuroscience* 17, 3964–3979.

Corkin, S. (2002). What's new with the amnesic patient H.M.? *Nature Reviews Neuroscience* 3, 153–160.

Coward, L.A. (1990). *Pattern Thinking*, New York: Praeger.

Coward, L.A. (2000). A Functional Architecture Approach to Neural Systems. *International Journal of Systems Research and Information Systems*, 9, 69–120.

Coward, L.A. (2001). The Recommendation Architecture: lessons from the design of large scale electronic systems for cognitive science. *Journal of Cognitive Systems Research* 2(2), 111–156.

Coward, L.A. (2004). Simulation of a Proposed Binding Model. *Brain Inspired Cognitive Systems*, L. S. Smith, A. Hussain and I. Aleksander, (editors), University of Stirling: Stirling.

Coward, L.A. (2005a). *A System Architecture Approach to the Brain: from Neurons to Consciousness*. New York: Nova Science Publishers.

Coward, L.A. (2005b). Accounting for episodic, semantic and procedural memory in the recommendation architecture cognitive model. *Proceedings of the Ninth Neural Computation and Psychology Workshop: Modelling Language, Cognition, and Action.*

Coward, L. A. and Gedeon, T.D. (2008). Implications of resource limitations for a conscious machine. Neurocomputing in press.

Coward, L.A. and Gedeon, T.D. and Ratanayake, U. (2004). Managing Interference between Prior and Later learning. *ICONIP 2004*, Calcutta.

Coward, L. A. and Sun, R. (2007). Hierarchical Approaches to Understanding Consciousness. *Neural Networks* 20(9), 947–954.

Curtis, C.E. and D'Esposito, M. (2003). Persistent activity in the prefrontal cortex during working memory. *Trends in Cognitive Sciences* 7, 415–423.

Dolcos, F., LaBar, K.S. and Cabeza, R. (2004). Interaction between the Amygdala and the Medial Temporal Lobe Memory System Predicts Better Memory for Emotional Events. *Neuron* 42, 855–863.

Eichenbaum, H., Otto, T. and Cohen, N.J. (1994). Two component functions of the hippocampal memory system. *Behavioral and Brain Sciences* 17, 449–517.

Euston, D.R., Tatsuno, M. and McNaughton, B.L. (2007). Fast-Forward Playback of Recent Memory Sequences in Prefrontal Cortex During Sleep. *Science* 318, 1147–1150.

Fosse, M.J., Fosse, R., Hobson, J.A. and Stickgold, R.J. (2003). Dreaming and Episodic Memory: A Functional Dissociation? *Journal of Cognitive Neuroscience* 15(1), 1–9.

Fyhn, M., Molden, S., Witter, M.P., Moser, E.I. and Moser, M-B. (2004). Spatial Representation in the Entorhinal Cortex. *Science* 305, 1258–1264.

Gedeon, T.D., Coward, L.A. and Bailing, Z. (1999). Results of Simulations of a System with the Recommendation Architecture, *Proceedings of the 6th International Conference on Neural Information Processing*, Volume I, 78–84.

Gilman, S., Bloedel J. R., and Lechtenberg, R. (1981). *Disorders of the Cerebellum*. Philadelphia, PA: FA Davis.

Gluck, M.A., Meeter, M. and Myers, C.E. (2003). Computational models of the hippocampal region: linking incremental learning and episodic memory. *Trends in Cognitive Sciences* 7(6), 269–276.

Goldman-Rakic, P. S. (1982). Cytoarchitectonic heterogeneity of the primate neostriatum: subdivision into island and matrix cellular compartments. Journal of Comparative Neurology 205, 398–413.

Graff-Radford, N.R., Tranel, D., Van Hoesen, G.W. and Brandt, J.P. (1990). Diencephalic Amnesia. *Brain* 113, 1–25.

Hassabis, D., Kumaran, D., Vann, S.D. and Maguire, E.A. (2007). Patients with hippocampal amnesia cannot imagine new experiences. *Proceedings of the National Academy of Sciences (USA)* 104(5), 1726–1731.

Hausser, M. and Mel, B. (2003). Dendrites: bug or feature? *Current Opinion in Neurobiology* 13, 372–383.

Hoffman, K.L. and McNaughton, B.L. (2002). Sleep on it: cortical reorganization after the fact. *Trends in Neuroscience* 25(1), 1–2.
Hyvärinen, A., Karhunen, J. and Oja, E. (1999). *Independent Component Analysis*. New York: Wiley.
Insausti, R. and Amaral, D.G. (2004). Hippocampal Formation. In Paxinos G, Mai JK editors. *The Human Nervous System*. Elsevier. 871–914.
Ishizuka, N., Weber, J. and Amaral, D.G. (1990). Organization of intrahippocampal projections originating from CA3 pyramidal cells in the rat. *Journal of Comparative Neurology* 295, 580–623.
Kelley, A. E. (1999). Functional Specificity of Ventral Striatal Compartments in Appetitive Behaviors. Annals of the New York Academy of Sciences 877, 71–90.
Kesner, R.P., Lee, I. and Gilbert, P. (2004). A behavioral assessment of hippocampal function based on a subregional analysis. *Reviews in the Neurosciences* 15, 333–351.
Kesner, R.P. and Hopkins, R.O. (2006). Mnemonic functions of the hippocampus: A comparison between animals and humans. *Biological Psychology* 73, 3–18.
Kensinger, E.A., Ullman, M.T. and Corkin, S. (2001). Bilateral Medial Temporal Lobe Damage Does Not Affect Lexical or Grammatical Processing: Evidence From Amnesic Patient H.M.. *Hippocampus* 11, 347–360.
Lavenex, P. and Amaral, D.G. (2000). Hippocampal-neocortical interaction: a hierarchy of associativity. *Hippocampus* 10, 420–430.
Lavenex, P., Suzuki, W.A. and Amaral, D.G. (2004). Perirhinal and Parahippocampal Cortices of the Macaque Monkey: Intrinsic Projections and Interconnections, *Journal of Comparative Neurology* 472, 371–394.
Leiner, H. C., Leiner, A. L. and Dow, R. S. (1993). Cognitive and language functions of the human cerebellum. Trends in Neuroscience 16, 444–447.
Leutgeb, S., Leutgeb, J.K., Barnes, C.A., Moser, E.I., McNaughton, B.L. and Moser M-B. (2005). Independent codes for spatial and episodic memory in hippocampal neuronal ensembles. *Science* 309, 619–623.
Leutgeb, S., Leutgeb, J.K., Treves, A., Moser, M-B. and Moser, E.I. (2004). Distinct Ensemble Codes in Hippocampal Areas CA3 and CA1. *Science* 305, 1295–1298.
Lisman, J.E. (1999). Relating Hippocampal Circuitry to Function: Recall of Memory Sequences by Reciprocal Dentate-CA3 Interactions. *Neuron* 22, 233–242.
Lisman, J.E. and Otmakhova, N.A. (2001). Storage, Recall, and Novelty Detection of Sequences by the Hippocampus: Elaborating on the SOCRATIC Model to Account for Normal and Aberrant Effects of Dopamine. *Hippocampus* 11, 551–568.
Maguire, E.A., Gadian, D.G., Johnsrude, I.S., Good, C.D., Ashburner, J., Frackowiak, R.S.J. and Frith, C.D. (2000). Navigation-related structural changes in the hippocampi of taxi drivers. *Proceedings of the National Academy of Sciences (USA)* 97(8), 4398–4403.
McClelland, J.L., McNaughton, B.L. and O'Reilly, R.C. (1995). Why there are complementary learning systems in the hippocampus and neocortex: insights from the successes and failures of connectionist models of learning and memory. *Psychological Review* 102(3), 419–457.
McDonald, A. J. (1991). Organization of amygdaloid projections to the prefrontal cortex and associated striatum in the rat. Neuroscience 44(1), 1–14.
Milner, B., Corkin, S. and Teuber, H-L. (1968). Further analysis of the hippocampal amnesic syndrome: 14-year follow-up study of H.M. *Neuropsychologia* 6, 215–234.
Mody, I. (2002). The GAD-given right of dentate gyrus granule cells to become GABAergic. *Epilepsy Currents* 2, 143–145.
Mountcastle, V.H. (2003). Introduction to special issue on cortical columns. *Cerebral Cortex* 13, 2–4.
Muller, W. and Misgeld, U. (1991). Picrotoxin- and 4-aminopyridine-induced activity in hilar neurons in the guinea pig hippocampal slice. *Journal of Neurophysiology* 65, 141–147.
Naber, P.A., Witter, M.P. and Lopes da Silva, F.H. (1999). Perirhinal cortex input to the hippocampus in the rat: evidence for parallel pathways, both direct and indirect. A combined physiological and anatomical study. *European Journal of Neuroscience* 11, 4119–4133.

Nadel, L. and Moscovitch, M. (1997). Memory consolidation, retrograde amnesia and the hippocampal complex. *Current Opinion in Neurobiology* 7, 217–227.

Nauta, W. J. H., Smith, G. P., Faull, R. L. M. and Domesick, V. B. (1978) *Neuroscience* 3, 385–401.

Penttonen, M., Kamondi, K., Sik, A., Acsady, L. and Buzsaki, G. (1997). Feed-forward and feedback activation of the dentate gyrus in vivo during dentate spikes and sharp wave bursts. *Hippocampus* 7, 437–450.

Phelps, E.A. (2006). Emotion and Cognition: Insights from Studies of the Human Amygdala. *Annual Review of Psychology* 57, 27–53.

Ratnayake, U. and Gedeon, T.D. (2003). Extending The Recommendation Architecture Model for Text Mining. *International Journal of Knowledge-Based Intelligent Engineering Systems*, 7, 139–148.

Rekkas, P.V. and Todd Constable, R. (2005). Evidence That Autobiographic Memory Retrieval Does Not Become Independent of the Hippocampus: An fMRI Study Contrasting Very Recent with Remote Events. *Journal of Cognitive Neuroscience* 17, 1950–1961.

Rempel-Clower, N.L., Zola, S.M., Squire, L.S. and Amaral, D.G. (1996). Three Cases of Enduring Memory Impairment after Bilateral Damage Limited to the Hippocampal Formation. *Journal of Neuroscience* 16, 5233–5255.

Rogers, T.T., Hocking, J., Noppeney, U., Mechelli, A., Gorno-Tempini, M.L., Patterson, K. and Price, C.J. (2006). Anterior temporal cortex and semantic memory: Reconciling findings from neuropsychology and functional imaging. Cognitive, *Affective and Behavioral Neuroscience* 6(3), 201–213.

Sagar, J.H., Cohen, N.J., Corkin, S. and Growden, J.H. (1985). Dissociations among processes in remote memory. *Annals of the New York Academy of Science* 444, 533–535.

Schoenbaum, G. and Setlow, B. (2003). Lesions of Nucleus Accumbens Disrupt Learning about Aversive Outcomes. Journal of Neuroscience 23(30), 9833–9841.

Scoville, W.B. and Milner, B. (1957). Loss of recent memory after bilateral hippocampal lesions. *Journal of Neurology, Neurosurgery, and Psychiatry* 20, 11–21.

Sesack, S.R., and Carr, D.B. (2002). Selective prefrontal cortex inputs to dopamine cells: implications for schizophrenia. Physiology and Behavior 77, 513–517.

Shibata, H. (1993). Direct Projections From the Anterior Thalamic Nuclei to the Retrohippocampal Region in the Rat. *Journal of Comparative Neurology* 337, 431–445.

Skaggs, W.E. and McNaughton, B.L. (1996). Replay of Neuronal Firing Sequences in Rat Hippocampus During Sleep Following Spatial Experience. *Science* 271, 1870–1873.

Skaggs, W.E,, McNaughton, B.L., Wilson, M.A. and Barnes, C.A. (1996). Theta phase precession in hippocampal neuronal populations and the compression of temporal sequences. *Hippocampus* 6, 149–172.

Smith, M.L. (1988). Recall of Spatial Location by the Amnesic Patient H.M. *Brain and Cognition* 7, 178–183.

Squire, L.R. and Alvarez, P. (1995). Retrograde amnesia and memory consolidation: a neurobiological perspective. *Current Opinion in Neurobiology* 5, 169–177.

Suzuki, W.A. and Eichenbaum, H. (2000). The Neurophysiology of Memory. *Annals of the New York Academy of Sciences* 911, 175–191.

Suzuki, W. and Amaral, D.G. (1990). Cortical inputs to the CA1 field of the monkey hippocampus originates from the perirhinal and parahippocampal cortex but not from area TE. *Neuroscience Letters* 115, 43–48.

Suzuki, W.A. (1996). Neuroanatomy of the monkey entorhinal, perirhinal and parahippocampal cortices: Organization of cortical inputs and interconnections with amygdala and striatum. *Seminars in Neurosciences* 8, 3–12.

Tamamaki, N., Abe, K. and Nojyo, Y. (1988). Three-dimensional analysis of the whole axonal arbors originating from single CA2 pyramidal neurons in the rat hippocampus with the aid of a computer graphic technique. *Brain Research*, 452, 255–272.

Tanaka, Y., Miyazawa, Y., Akaoka, F. and Yamada, T. (1997). Amnesia Following Damage to the Mammillary Bodies. *Neurology* 48, 160–165.

Tanaka, K. (2003). Columns for complex visual object features in the inferotemporal cortex: clustering of cells with similar but slightly different stimulus selectivities. *Cerebral Cortex* 13, 90–99.

Teyler, T.J. and DiScenna, P. (1986). The hippocampal memory indexing theory. *Behavioral Neuroscience* 100(2), 147–154.

Thompson-Schill, S.L. (2003). Neuroimaging studies of semantic memory: Inferring "how" from "where." *Neuropsychologia*, 41, 280–292.

Tronson, N.C. and Taylor, J.R. (2007). Molecular mechanisms of memory reconsolidation. *Nature Reviews Neuroscience* 8, 262–275.

Tulving, E., Markowitsch, H.J., Craik, F.I.M., Habib, R. and Houle, S. (1996). Novelty and Familiarity Activations in PET Studies of Memory Encoding and Retrieval. *Cerebral Cortex* 6, 71–79.

Veazey, R.B., Amaral, D.G. and Cowan, W.M. (1982). The Morphology and Connections of the Posterior Hypothalamus in the Cynomolgus Monkey (Macaca fascicularis). 11. Efferent Connections. *Journal of Comparative Neurology* 207, 135–156.

Vertes, R.P. and Eastman, K.E. (2000). The case against memory consolidation in REM sleep. *Behavioral and Brain Sciences* 23, 867–876.

Walker, M.C., Ruiz, A. and Kullmann, D.M. (2001). Monosynaptic GABAergic signaling from dentate to CA3 with a pharmacological and physiological profile typical of mossy fiber synapses. *Neuron* 29, 703–715.

Wallenstein, G.V., Eichenbaum, H. and Hasselmo, M.E. (1998). The hippocampus as an associator of discontiguous events. *Trends in Neuroscience* 21, 317–323.

Wyss, J.M., Swanson, L.W. and Cowan, W.M. (1979). A study of subcortical afferents to the hippocampal formation in the rat. *Neuroscience* 4, 463–476.

Zola-Morgan, S., Squire, L.S. and Amaral, D.G. (1986). Human Amnesia and the Medial Temporal Region: Enduring Memory Impairment Following a Bilateral Lesion Limited to Field CA1 of the Hippocampus. *Journal of Neuroscience* 6(10), 2950–2967.

Cognitive Measure on Different Profiles

Marilda Spindola, Giovani Carra, Alexandre Balbinot, and Milton A. Zaro

Abstract Based on neurology and cognitive science many studies are developed to understand the human model mental, getting to know how human cognition works, especially about learning processes that involve complex contents and spatial-logical reasoning. Event Related Potential – ERP - is a basic and non-invasive method of electrophysiological investigation. It can be used to assess aspects of human cognitive processing by changing the rhythm of the frequency bands brain indicate that some type of processing or neuronal behavior. This paper focuses on ERP technique to help understand cognitive pathway in subjects from different areas of knowledge when they are exposed to an external visual stimulus. In the experiment we used 2D and 3D visual stimulus in the same picture. The signals were captured using 10 (ten) Electroencephalogram - EEG - channel system developed for this project and interfaced in a ADC (Analogical Digital System) board with LabVIEW system - National Instruments. That research was performed using project of experiments technique – DOE. The signal processing were done (math and statistical techniques) showing the relationship between cognitive pathway by groups and intergroups.

Keywords ERP-EEG · Learning process · Cognitive measures

M. Spindola (✉), G. Carra, and A. Balbinot
Biomedical Engineering Research Group – NPEngBio, Departamento de Engenharia Elétrica, Universidade de Caxias do Sul – UCS, Alameda João Dal Sasso, 800 – Zip Code: 95700-000 – Bento Gonçalves, RS, Brazil
e-mail: mschiara@ucs.br; gcarra4@ucs.br; abalbinot@gmail.com

A. Balbinot
Universidade Federal do Rio Grande do Sul - UFRGS, Escola de Engenharia, Departamento de Engenharia Elétrica - DELET, Laboratório de Instrumentação Eletro-Eletrônica - IEE, Av Osvaldo Aranha, 103 - Bom Fim - Porto Alegre - RS - Brasil, CEP: 90035-190, Fone: (51) 3308-3326

M.A. Zaro
Universidade Federal do Rio Grande do Sul - UFRGS, Programa de Pós-graduação em Informática na Educação, Av. Paulo Gama, 110 - prédio 12105 - 3° andar sala 332, 90040-060 - Porto Alegre (RS) - Brasil
e-mail: zaro@ufrgs.br

A. Hussain et al. (eds.), *Brain Inspired Cognitive Systems 2008*, Advances in Experimental Medicine and Biology 657, DOI 10.1007/978-0-387-79100-5_19,

1 Introduction

The discovery of organization patterns and how human thought work have enabled the creation of new hypotheses about the learning process that affect the cognitive and motor systems. Considering that the development of cognitive skills such as spatial manipulation of objects, construction of logical and abstract, and full complex knowledge depends on innate pre-requisites as well as on interaction with cultural and geographic environment, family, etc, it is necessary to understand how innate and cultural knowledge is acquired.

With the theoretical principle of scientists Pinker (1997) and Gardner (1999) pointing to the possible cognitive differences between humans and also supported by cognitive science and the empirical observation of the cognitive behavior of students from areas of scientific and technological knowledge, and considering that students should engage in complex cognitive and advanced processes during the course, one can suppose they already have (a priori) or have developed a specific profile of a professional of the area.

That is the premise that the Research Group in Biomedical Engineering from the University of Caxias do Sul (UCS - CARVI - Brazil), along with the neurosciences group at PGIE-UFRGS, has been developing methodology with the objective to identify pathways of brain signals for specific cognitive activities in subjects with different cognitive profiles.

2 Theoretical Considerations

The advances in neurosciences and investigative techniques on cerebral sign patterns associated to cognitive demands have allowed education to discover new methodologies and derive theoretical postulations that can help formulate new models for different pedagogical practices, associated to different cognitive profiles. The latest educational processes have involved results from researchers that point to different kinds of learners and different motivations circumscribing them. To know people's cognitive functioning regarding cognitive aspects related to spatial abilities will enable significant improvement in the inter-relationship between subject and learning object. Considering how important this theme is to understand human learning processes and to prepare better teaching processes, for areas of scientific and technological knowledge, human and social sciences, the research proposed herein is aimed at the understanding of how mental models are formed, especially on those matters regarding spatial abilities (Gardner 1999).

Amongst the several concepts applied to the mental model, the most adequate one for this research refers to the formation and strengthening of the neuronal net through the presence of more synapses associated to stimulus that modify the concepts and the relationship between the subject and his environment (Merrill 2000). The learning process is directly linked to the formation and changes to the mental model of a subject learner according to the stimuli that he is subject to. Moreover, according to the literature, the majority of the subjects can easily alter, remodel and create new mental models when the external stimulus is visual (Desimone and Duncan 1995) (Kastner and Ungerleider 2000). The ease to recognize visual stimuli which is a component of the spatial ability is one of the parameters which cognitive scientists have used to classify or quantify the mental model of a subject learner.

Many researchers have shown it is possible to identify and measure different patterns of cerebral signals, related to external visual stimulus, when subjects are submitted to tasks of recognizing simple geometric images (circles, squares, triangles) presented in virtual format of 2 and 3 dimensions accordingly (Guizhi 2006). The measured signals can be utilized as parameters of identification amongst different mental models of subject learners, relating to spatial ability, for the proposed stimulus.

Relations between cognitive phenomena and human biological mechanisms were more deeply observed and interpreted thanks to state-of-the-art technology in the prospection area for cerebral activity, through non-invasive methods such as quantitative electroencephalograms (EEGqs) (Ribeiro 2005). EEG sources are electrical potentials generated by cortical neurons, which respond to several stimuli on the depth of the brain. Cerebral activity is captured on the surface of the scalp by means of electrodes. The observation of significant alterations in the cerebral signal in healthy subjects, in relation to knowledge of a visual pattern that concern tasks that involve visual stimulus, occur mostly in the frontal and parietal regions, when the signal is measured between 300–600 ms after the stimuli is applied (Bledowski 2004, Luck 2005, Schumacher 2005).

By using specific software for the treatment of signals obtained from the EEG, it has become possible to build maps of cerebral activities at certain moments, making it easier to understand electrical potentials on the surface of the brain. Consistent research show the correlation between certain kinds of waves with alert activities and cerebral processing, such as Beta activity which is characterized by its low amplitude and which appears on the frequency band from 14 to 40 Hz (Roberts and Bell 2000), (Luck 2000), (Guizhi 2006). With quantitative analysis of the cerebral electric activity, which uses technological resources on the evaluation of the EEG, it is possible to overcome the visual exam of the plotting that comprises a significant subjective component (Fonseca 2003). The analysis of EEGs has also indicated that, during the presentation of stimuli, specific changes take place on cerebral signals, represented by a significant increase in the synaptic activity of millions of neurons simultaneously. Changes on electrical potentials of the membrane occur in fractions of a second after stimulus was presented on different regions of the brain. These potentials evoked by a stimulus take place in a synchronized way. The result from the electrical potentials of a neuronal population is known as Event-Related Potentials (ERP) (Luck 2005), which consists on a series of positive and negative waves that can be identified either numerically or according to their latency. The main ERP components are N1, P2, N2 and P3. Each component is identified by the letter that indicates whether the wave is negative or positive and by the number that indicates the time of the occurrence, measured in tenths of seconds (Veiga 2004). The main focus of study on ERP has been the third positive wave, called P3 or

P300 (Luck 2000). Component P300 usually occurs 300 ms after the presentation of stimuli associated to visual and auditory processes, and has been a powerful tool in the study of cognitive processes. As for measure P300, it is possible to quantify two variables: latency, whose measure reflects the time necessary to allocate resources and evoke the memory related to the stimulus, and the amplitude of signal, which allocates resources that focus the cognitive processes of immediate memory. Differences found in P300 measurements among the individuals, when subject to the same processes, and which evoke cognitive demands, make it possible to infer that there are differences between the capacity of processing and the speed rate on the cognition cerebral processes (Veiga 2004).

3 Materials and Methodology

Theoretical assumptions that support this research (Gardner 1999) and (Pinker 1997), and those matters that have already been mentioned concerning a possible relationship between resultants from variables related to the investigation on measures of cognitive abilities, especially and particularly, spatial abilities and the different attention patterns given to virtual visual stimuli lead to carrying out experiments supported by electroencephalography (EEG), developed by Biomedical Engineering Research Group at University of Caxias do Sul (Carra et al. 2007), using especially the ERP (Event-Related-Potential) technique. The goal is to identify different patterns on cerebral signals concerning attention evoked from the visual and spatial stimulus: recognizing 2D and 3D images, when the individual is subject to an experiment with visual stimuli that require attention. The investigation on the attention process during the perception of different visual patterns on volunteers of two different areas of knowledge was carried out as Design of Experiments – DOE (Montgomery 2000).

Design of Experiments is a technique used to plan experiments, it encompasses which data will be used, in which quantity and in which conditions it must be collected for a specific experiment, basically trying to satisfy two main objectives: the best possible results with statistical precision at the smallest expense (Montgomery 2000). This methodology is strongly based on statistical concepts, designed to optimize the planning, execution and analysis of an experiment (Ten Caten 2007). The objective of the Design of Experiments for the scientific investigation in the area of cognitive science includes helping in building a base of trusted knowledge and in this way reducing uncertainties involving which theories, tools and methodologies would be more adequate for certain contexts. Ten Caten (2007) presents some important concepts in the planning of a experiment: output v***ariables that are the output variables (answers)*** of the process which can be measured and that allow for quantification of characteristical behavior of a system; ***process parameter that encompass all the process variables (present in the experiment) which can be changed and which may have an influence on the output variables***; controllable factors that are a subdivision of the process parameters, elected to be studied at several steps of the experiment; levels that correspond to each adjustment of the controllable factors used on the experiment, e.g., the two types of areas of knowledge or the two types of presentation of the visual stimuli; ***constant factors are the process parameters that are not contemplated in the experiment and that are kept constant during its execution***; uncontrollable ***factors (noise) are the variables that cannot be controlled by the research group***, being responsible for the experimental error, that is, an experimental fluctuation, e.g., the power source noise; interactions occur in the study of 2 or more controllable factors, when the effect of one of the factors depends on the adjustment established for another factor.

The *Design of Experiments methodology* facilitates the experimental procedures proposed for neurosciences and allows for the testing with control over the related variables, of quantitative and qualitative character, with the production of information that could point to conclusions, although temporary, regarding the neuroscientific hypothesis. The following described scientific experiment planning involve the methodology and statistical procedures of the *Design of Experiments.*

3.1 Experiment Project

The parameters of the *Design of Experiments are hereby defined*:

- *Subjects*: 16 (sixteen) young volunteers participated in this experiment; they were undergraduate students from two different areas of knowledge: social sciences, along with the students of the College of Law and students of the College of Design. In which area of knowledge, there were eight (8) participants equally divided between males and females. The experiment lasted four days, whereas the number of participants was distributed accordingly per day (with double-blind random assortment for the day's collection). Participants had a normal sight system. The number of participants follows the guidelines of the sampling for the Design of Experiments (Ten Caten 2007). The study was approved by the local ethics committee;
- *Kinds of stimuli*: stimuli presented consisted of two images: a ball shape, in two or three dimensions (2D and 3D), with the same color and same area occupied on the monitor screen, as shown in Figure 1. To obtain more statistical representativeness, the experiment contemplated five repetitions for each visual stimulation: 2D and 3D for each subject in the research. All images were projected by the LabVIEW system (software and hardware belonging to National Instruments), which is also used on the process of digital-analogical conversion the signal captured;
- *Response variables*: response variable to be considered is the cerebral signal measured in the experiment that is related to the alteration in the latent cerebral signal, after applying the visual stimuli. (for the experiment, a circular image was used and presented in 2D e 3D dimensions respectively). The measurement is amplitude parameter of electroencephalographic waves converted in magnitude in the frequency domain (analysis in the frequency domain). Differences related to potential evoked from spatial and visual stimulus are associated to the

Fig. 1 Virtual visual stimuli

Table 1 Controllable factors

Factors	Level numbers	Level kinds
A: volunteers	2	2 (two) Knowledge areas
B: stimuli kinds	2	ball (2D) and ball (3D)
C: scalp points	4	FP1, FP2, P3 and P4

occurrence of different amplitudes and frequencies (magnitude levels between the signals), considered as different forms of allocating resources;

- *Controllable factors*: as for the definition of controllable factors we chose to carry out a complete factorial project with 3-controllable factors at several levels (Ten Caten 2007). The factors identified as controllable and having priority in this research are the subject's areas of knowledge in two levels, the visual stimulus presented in two levels (images 2D and 3D) and the researched scalp areas in 4 levels (we chose to evaluate only the most significant points in relation to the evoked signal for a visual stimuli) (Buschman 2007, Sakai 2008). Table 1 resume factors and number of levels for each one;
- *Experiment project*: subjects were invited to participate and agreed upon all the procedures during the experiment. Acquisition of data in the experiment was carried out in an acclimatized room, without audible noises, without any stimuli that could distract the subjects. Volunteers (subjects of the research) remained on a chair with back and head rest, in the most comfortable sitting position. The chair was positioned at a distance of 90 cm in front of a 15" screen, used to present the visual stimulus. Volunteers were connected to the ten lines of the collecting system (EEG) by a cap with electrodes. Points used on the cap (FP1, FP2, F3, F4, P3, P4, C3, C4, T3 e T4) correspond to the international 10–20 system (Jasper System), used to standardize capturing and identification of the results, according to Figure 2.

The collected data followed the experimental blockage restriction guidelines. The research's subjects followed the pre-established program, considering as blockage factor only the areas of knowledge, since in order to collect signals, it is necessary

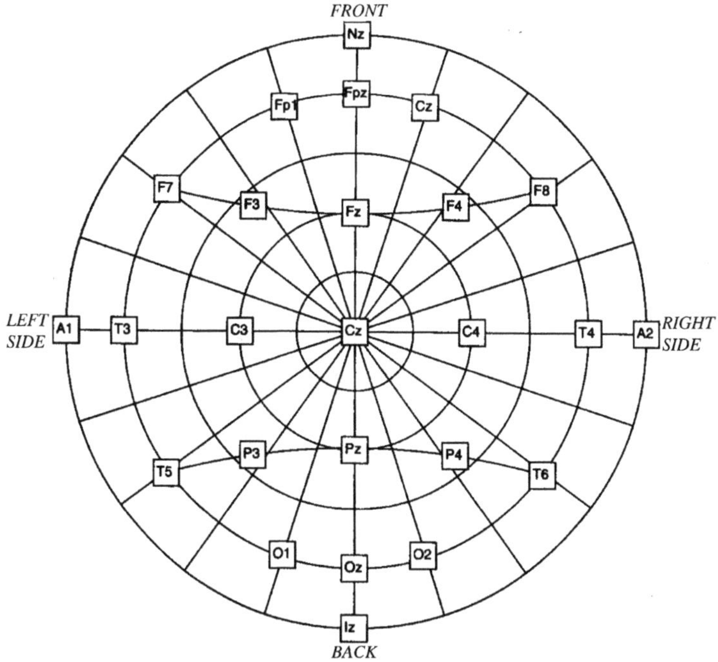

Fig. 2 International System 10–20

to collect them simultaneously from the different areas of the scalp, and being that the procedure follows a protocol that requires the used of a cap with electrodes, the use of a gel, etc, it becomes necessary to collect all the stimuli related signals simultaneously.

During the experiment procedures, some of the external factors that could inadvertently influence the obtained results were kept unchanged, for example: luminosity, temperature, and the sound level of the room. The collection of signals was done by the same operator and electroencephalogram equipment, besides being done at the same time of during the day.

4 Result Analysis

All changes on the EEG tracing were collected, but recordings related to muscular contraction because of winking and facial, neck and forehead muscular movement were excluded from later analysis. To avoid results arisen from anxiety processes, a base recording of the volunteer's neurophysiologic signals was performed during the first five minutes before executing the task proposed by the experiment. During that period the volunteer was asked to remain calm and relaxed and not to move abruptly. Such orientation given to the volunteer causes a change in their electrophysiological signals, making it possible to visualize the changes on the brain frequencies being displayed. This procedure is important because it assures that the system is responding properly to the capture of brain signals and also is necessary to compare the standard signal latency with the evoked signal. The task executed was to visualize two sequences of visual stimuli, each one of them containing the same 5 images (displayed at intervals of 100ms): 2D balls or 3D balls, with time intervals varying between 1800 ms and 2200 ms to avoid the oddball effect which is an alteration in the signal because of the surprise effect or wait conditioning for the same pattern of stimulus. Sequences were drawn before executing the experiment to prevent variable error effects as a result of a subject's fatigue or indisposition at the moment of visualizing the images.

During each test, the experiment volunteers utilized a standard electrode wired helmet, with impedance smaller than 3kΩ, disposed in the international system 10–20. The auricular position was used as reference. In the captured signals, on each measured point on the scalp, a high-pass 0.01Hz filter was used and amplifiers with total gain of 15000 times. After processing the signals in the analogical system, they were also acquired digitally through the LabVIEW system, with an acquisition rate of samples/s. The digitized signals were filtered with a low-pass filter in 55Hz.

Signals recorded each time a shape was repeated were cut into sections at interest time windows within 200ms and 450ms after image exposure. In that period of time, the P300 event occurs and the interest frequency (BETA) is manifested, indicating volunteer's maximum attention to the stimulus proposed. The signals collected were digitized and processed according to the mathematical model of Fast Fourier Transform (FFT). These results indicate the amount of energy demanded for the brain signal at the moment of the experiment.

The response variable analyzed is identified as the magnitude in the (FFT) frequency domain of a time window of the signal that was collected, which corresponds to the change of brain signal due to visual stimulation. Signal FFT confirms that, in the interval when P300 occurs, the Beta rhythm is manifested and it is possible to estimate that, in this rhythm, there might be an intense vigilance stage, as part of a cognitive process.

Variance analysis on the data resulting from the variable "Maximum Magnitude in BETA band: cognitive coefficient" with quadratic sums, respective freedom degrees and quadratic averages of the factors observed can be found in Table 2. Data were statistically analyzed utilizing alpha significance level of 0.05 (5%).

Table 2 Variance Analysis

Source	Sum of square	DF	Mean Square	F value	F tab.	Significant?
SQA	1.23	5	0.25	13.24	2.26	Yes
SQB	0.08	1	0.08	4.36	3.89	Yes
SQC	7.17	3	2.39	128.66	2.65	Yes
SQAB	0.64	5	0.13	6.93	2.26	Yes
SQAC	0.35	15	0.02	1.27	1.72	No
SQBC	0.09	3	0.03	1.54	2.65	No
SQABC	0.18	15	0.01	0.64	1.72	No
Error	3.56	192	0.02	–	–	–
Total	13.30	239	–	–	–	–

The main effects of the factors and the interaction effect of two and three factors were analyzed. Results show that the main effects for factors "A" (volunteers), B (kinds of visual stimuli) and C (scalp points) are significant. That means there is significant difference between the signal patterns of the participants on several acquisitions, regardless of the kind of 2D or 3D stimuli or the scalp point factor. Likewise, there is significant difference to point out between the results concerning experiments 2D and 3D regardless of the area of knowledge and the scalp point. In the analysis of the results obtained at each scalp point, we point out that there are significant differences in the magnitude. That condition was expected due to brain physiology itself, since the points chosen for analysis have distinct functions in the processing of visual stimuli (Luck 2000, Guizhi 2006).

The interaction effects were those from the interaction of two factors. Subjects of the experiment and the kinds of stimuli (A × B), subjects of the experiment and scalp capture points (A × C), kinds of stimuli and scalp capture points (B × C) and the triple interaction among the three control factors (A × B × C). Significant results in those procedures took place when subjects were compared among themselves and the kinds of visual stimuli (A×B), when subjects were compared among themselves and scalp points (A × C) and also when visual stimuli types were compared with scalp points (B × C). Results for these significances point to a difference in the degree of resources allocated with a significant variation of greater magnitude of the BETA frequency band (EEG pattern) for each subject in relation to each kind of stimulus. Next, Figure 3a shows the graphics for two factors of the interactions between participants and the kind of visual stimulus, participants and scalp points, and visual stimulus and scalp points, concerning magnitude.

The graph in Figure 3a shows the difference between the magnitude average in the Beta frequency (through FFTs), of the brain evoked signal of the participants in the experiment that were subject to the visualization of 2D and 3D visual patterns. The graph in Figure 3b shows a significant difference between the kind of stimulus and scalp points. It is important to point out that the subjects of the experiment belong to two large areas of knowledge, here called A1 (scientific-technological area) and A2 (social-human area). By observing the tendencies between the subjects of the experiment relative to the first chart and knowing that the division between the areas of knowledge is split in the following way: participants from 1 to 8 belong to a scientific and technological knowledge area, and participants from 9 to 16 belong to a human and social sciences area, a new statistical analysis was done grouping the participants by area and therefore creating a bi-level factor in each area of knowledge. Table 3 shows variance analysis on the data resulting from the variable "Maximum Magnitude in BETA band: cognitive coefficient" with quadratic sums, respective freedom degrees and quadratic averages of the factors knowledge area in two levels, visual stimulus (2D and 3D) and scalp points (FP1, FP2, P3 and P4). Data were statistically analyzed utilizing a alpha significance level of 0.05 (5%).

The new analysis shows significant results in those procedures that took place when knowledge area were compared between themselves and the kinds of visual stimuli (A × B), when knowledge areas were compared between themselves and scalp points (A × C) and also when the kinds of visual stimuli were compared with

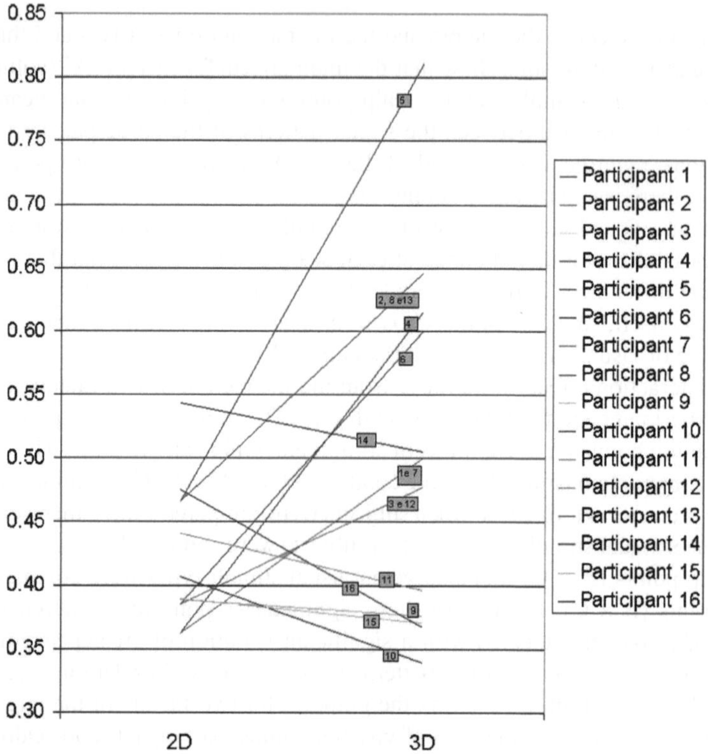

Fig. 3a Factor A (subjects) X Factor B (kind of stimulus)

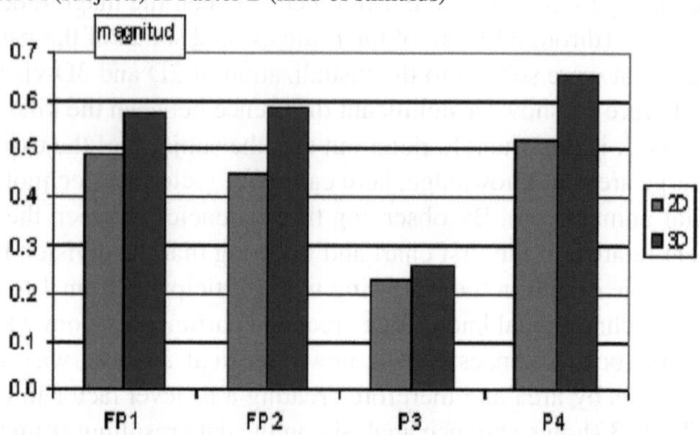

Fig. 3b Factor B (kind of stimulus) X Factor C (scalp points)

scalp points (B × C). Results for these significances point to a difference in the degree of resources allocated with a significant variation of greater magnitude of the BETA frequency band (EEG pattern) for each knowledge area related to each kind of stimulus (Figure 4a) and also for each knowledge area related to points of scalp (Figure 4b).

Table 3 Variance Analysis

Source	Sum of square	DF	Mean Square	F value	F tab.	Significant?
SQA	0.7	1.0	0.7	22.7	3.9	Yes
SQB	1.4	1.0	1.4	45.2	3.9	Yes
SQC	11.4	3.0	3.8	118.7	2.6	Yes
SQAB	1.5	1.0	1.5	47.4	3.9	Yes
SQAC	0.7	3.0	0.2	6.9	2.6	Yes
SQBC	0.3	3.0	0.1	3.2	2.6	Yes
SQABC	0.2	3.0	0.1	2.0	2.6	No
Error	20.0	624.0	0.0			–
Total	36.3	639.0				–

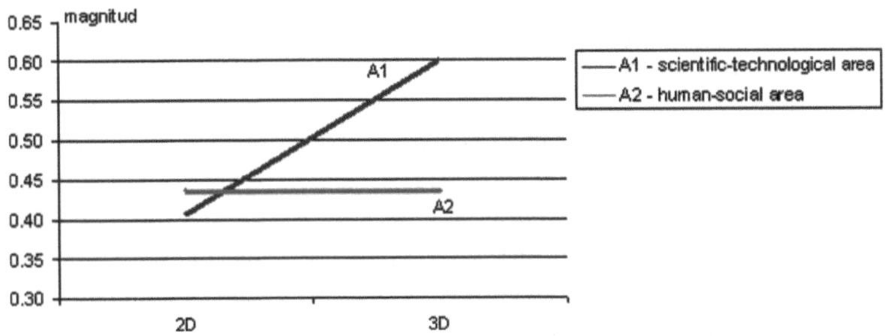

Fig. 4a Factor A (knowledge area) X Factor B (kind of stimulus 2D – 3D)

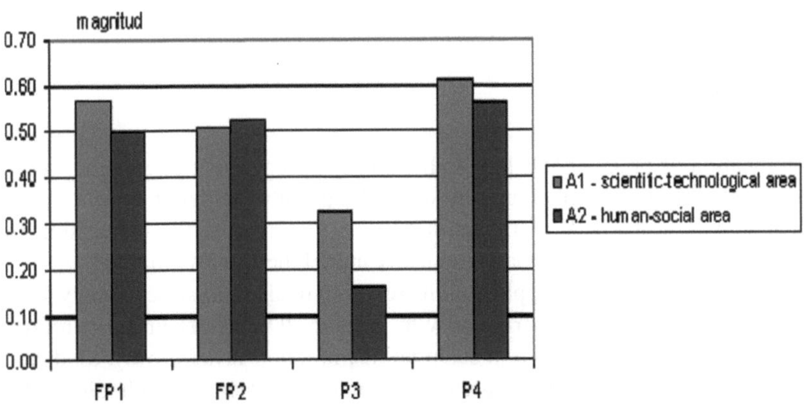

Fig. 4b Factor A (knowledge area) X Factor C (scalp points)

The Figure 4a presents average magnitude values with a crescent linear behavior between 2D and 3D patterns to subjects of knowledge an scientific area (A1), while subjects of area of human and social science (A2) shows average decreasing values for the same stimuli. The FFT average of greater magnitude for the recognition of 2D visual standard (control factor B1), for A1 subject is concentrated

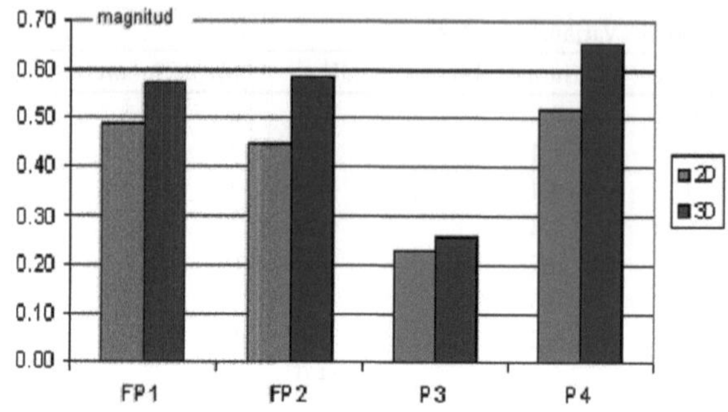

Fig. 4c Factor B (kind of stimulus) X Factor C (scalp points)

around 410 mV, whereas for subject A2, the same reference point is 440 mV. For the 3D stimulus (B2), the FFT averages of greater magnitude in each interval are, respectively, for subjects A1 and A2 at 600 mV and 430 mV. The reference scale for that measure is amplified 15000 times and calibrated in relation to the original signal of the scalp.

Figure 4b relates to each knowledge area the significant difference of magnitude between points of scalp. Figure 4c also relates to the measure to scalp points relative to visual stimulus 2D and 3D.

Those founds can be considered as indicators of cognitive pathways to different subjects of different knowledge areas. Of course those kinds of experiments are to be explored more in depth.

5 Conclusion

Quantification of brain signals is possible through proper instrumentation, such as, for example, electroencephalography used along with digitizing systems of analogical signals. LabVIEW platform enables us to carry out the conversion from analogical into digital and to mathematically model the signals obtained.

Electroencephalography applied in the research made it possible to analyze brain signals evoked from the experiment with graphic 2D and 3D visual stimuli and measure those signals in different scalp points, in a non-invasive way, on different subjects belonging to two large areas of knowledge: scientific technological and human-social.

Concerning the issue of investigating resource demand and allocation in terms of electrical signals for the identification of different stimuli (2D and 3D of the same object) on the same subject and between different subjects, the results point significantly towards a difference in cognitive effort among the subjects, above all those pertaining to different areas of knowledge, when they are subjected to the same visual stimulus. What is observed is that the attention process in the subjects of the experiment, focused on the visualization of objects with different patterns, takes place in a diverse form for that group.

Concordance between frequency band obtained at time windows specific for the evaluation of the ERPs for the subject group and the base determined as Beta band (14 to 40Hz) points to the presence of frequencies that involve cognition during the occurrence of ERPs in this experiment. This finding confirms information cited by other authors (Luck 2000), (Guizhi 2006) about linking P200 and P300 signals with cognition to visual stimulus.

The study carried out intends to aggregate value to scientific and technological areas through scientific methodology applied to educational matters, with multidisciplinary approach, including cognitive sciences, neuroscience and psychometrics, as well as the development of technology for monitoring experimental activities in this area.

References

Bledowski C, Prvulovic D, Hoechstetter K, Scherg M, Wibral M, Goebel R, Linden D E (2004). Localizing P300 Generators in Visual Target and Distractor Processing: A Combined Event-Related Potential and Functional Magnetic Resonance Imaging Study. Department of Psychiatry, Johann Wolfgang Goethe University, Frankfurt, Germany.

Buschman T J, Miller E K (2007). Top-Down Versus Bottom-Up Control of Attention in the prefrontal and Posterior Parietal Cortices.Science 30 March 2007:Vol. 315. No. 5820, pp. 1860–1862.

Carra M, Balbinot A, Chiaramonte M (2007) Desenvolvimento de um protótipo EEG como ferramenta para caracterização de sinais cerebrais em atividades relacionadas a raciocínio lógico. In: II Encontro Nacional de Biomecânica, Évora. Actas do II Encontro Nacional de Biomecânica. Lisboa - Portugal : IST Press, 2007. v. I. p. 387–392.

Cunha J A (2003) Psicodiagnóstico -V. Artmed, 2003. Porto Alegre, Brasil.

Desimone R, Duncan J (1995) Neural Mechanisms of Selective Visual Attention. Annual Review of Neuroscience. Vol. 18: 193–222 (Volume publication date March 1995).

Fonseca L C, Tedrus G M, Martins S M et al "Quantitative electroencephalography in healthy school-age children: analysis of band power". Arq. Neuro-Psiquiatr, Vol. 61, No. 3B, Sept. 2003, pp. 796–801.

Gardner H (1999) Intelligence Reframed: Multiple Intelligences for the 21st Century. BasicBooks.

Guizhi X, Ying Z, Huijuan H and Weili Y (2006) Event-Related Potential Studies of Attention to Shape Under Different Stimuli Tasks.

Kastner S, Ungerleider L (2000) Mechanisms of Visual Attention in the Human Cortex. Annual Review of Neuroscience. Vol. 23: 315–341 (Volume publication date March 2000).

Luck J S, Woodman G F and Vogel E K (2000) "Event-related potential studies of attention", Trends in Cognitive Sciences, Vol. 4, No. 11, November 2000, pp. 432–440.

Luck J S (2005) An introduction to the event-related potencial technique.

Merrill, M.D. (2000), "Knowledge objects and mental-models", in Wiley, D. (Eds), The Instructional Use of Learning Objects, Online Version. available at: www.id2.usu. edu/Papers/ KOMM.PDF (accessed February 22, 2009).

Montgomery D C (2000) Design and Analysis of Experiments. John Wiley and Sons, New York, 5th Edition.

Pinker, S. 1997. How the Mind Works. New York: Norton.

Ribeiro L O M, Timm M I, Becker F and Zaro M A (2005) Monitoramento da atividade cognitiva através de EEG e seu uso potencial na avaliação de ambientes virtuais de aprendizagem e simuladores. GCETE Global Congress on Engineering and Technology Education. March 13–16, 2005, São Paulo, Brazil.

Roberts J E and Bell M A (2000) Sex Differences on a Mental Rotation Task: Variations in Electroencephalogram Hemispheric Activation Between Children and College Students. Department of Psychology, Virginia Polytechnic Institute and State University.

Sakai K (2008) Top-down versus bottom-up control of attention in the prefrontal and posterior parietal cortices. Department of Cognitive Neuroscience, Graduate School of Medicine, The University of Tokyo. Annual Review of Neuroscience.Vol. 31: 219–245.

Schumacher E H, Hendricks M J, D'Esposito M (2005). Sustained involvement of a frontal–parietal network for spatial response selection with practice of a spatial choice–reaction task. School of Psychology, Georgia Institute of Technology, Atlanta, USA.

Ten Caten C (2007) Material de Suporte: Probabilidade e Estatística. Porto Alegre, Brasil.

Veiga H, Deslandes A, Cagy M, McDowell K, Pompeu F, Piedade R, Ribeiro P (2004) Visual Event-Related Potential (P300): A normative study. Arq Neuropsiquiatr 2004, 62(3-A): 575–581.

Index

MIX
Papier aus verantwortungsvollen Quellen
Paper from responsible sources
FSC® C105338

If you have any concerns about our products,
you can contact us on
ProductSafety@springernature.com

In case Publisher is established outside the EU,
the EU authorized representative is:
Springer Nature Customer Service Center GmbH
Europaplatz 3, 69115 Heidelberg, Germany

Printed by Libri Plureos GmbH
in Hamburg, Germany